2D NANOMATERIALS FOR ENERGY APPLICATIONS

2D NANOMATERIALS FOR ENERGY APPLICATIONS

Graphene and Beyond

Edited by

SPYRIDON ZAFEIRATOS

Institut de Chimie et Procédés pour l'Energie, l'Environnement et la Santé (ICPEES), ECPM, UMR 7515 CNRS – Université de Strasbourg, 25 rue Becquerel, 67087 Strasbourg, France

ELSEVIER

Elsevier
Radarweg 29, PO Box 211, 1000 AE Amsterdam, Netherlands
The Boulevard, Langford Lane, Kidlington, Oxford OX5 1GB, United Kingdom
50 Hampshire Street, 5th Floor, Cambridge, MA 02139, United States

British Library Cataloguing-in-Publication Data
A catalogue record for this book is available from the British Library

Library of Congress Cataloging-in-Publication Data
A catalog record for this book is available from the Library of Congress

ISBN: 978-0-12-816723-6

For Information on all Elsevier publications
visit our website at https://www.elsevier.com/books-and-journals

Publisher: Matthew Deans
Acquisition Editor: Simon Holt
Editorial Project Manager: Emma Hayes
Production Project Manager: Sojan P. Pazhayattil
Cover Designer: Greg Harris

Typeset by MPS Limited, Chennai, India

Dedication

To Grace, Peace, Wisdom, and Hope.

Contents

List of contributors

Craig E. Banks Faculty of Science and Engineering, Manchester Fuel Cell Innovation Center, Manchester Metropolitan University, Manchester, United Kingdom

Valery A. Barachevsky Photochemistry Centre RAS, FSRC "Crystallography and Photonics", Moscow, Russia

Konstantinos C. Christoforidis Department of Environmental Engineering, Democritus University of Thrace, Xanthi, Greece

Johnny Deschamps Department of Chemistry and Chemical Engineering (UCP), Ecole Nationale Supérieure de Techniques Avancées (ENSTA IP Paris), Palaiseau, France

Kapil Dhaka Department of Materials Science and Engineering, Technion - Israel Institute of Technology, Haifa, Israel

Michael P. Down Faculty of Science and Engineering, Manchester Fuel Cell Innovation Center, Manchester Metropolitan University, Manchester, United Kingdom

Fabio C. Fonseca Nuclear and Energy Research Institute (IPEN), São Paulo, Brazil

Sujoy Kumar Ghosh Organic Nano-Piezoelectric Device Laboratory, Department of Physics, Jadavpur University, Kolkata, India

Emmanuel Kymakis Department of Electrical and Computer Engineering, Hellenic Mediterranean University (HMU), Heraklion, Greece

Weiwei Lei Institute for Frontier Materials, Deakin University, Waurn Ponds Campus, Geelong, Victoria, Australia

Xin Li State Key Laboratory of Organic–Inorganic Composites, College of Chemical Engineering, Beijing University of Chemical Technology, Beijing, P.R. China

Dan Liu Institute for Frontier Materials, Deakin University, Waurn Ponds Campus, Geelong, Victoria, Australia

Dipankar Mandal Organic Nano-Piezoelectric Device Laboratory, Department of Physics, Jadavpur University, Kolkata, India; Institute of Nano Science and Technology, Mohali, India

Jun Mei School of Chemistry, Physics and Mechanical Engineering, Queensland University of Technology (QUT), Brisbane, QLD, Australia

Leticia P.R. Moraes School of Chemistry, Physics and Mechanical Engineering, Queensland University of Technology (QUT), Brisbane, QLD, Australia; Nuclear and Energy Research Institute (IPEN), São Paulo, Brazil

Santanu Mukherjee Mechanical and Nuclear Engineering Department, Kansas State University, Manhattan, KS, United States

Hiromichi Ohta Graduate School of Information Science and Technology, Hokkaido University, Kita, Sapporo, Japan; Research Institute for Electronic Science, Hokkaido University, Kita, Sapporo, Japan

Apostolos Panagiotopoulos Department of Electrical and Computer Engineering, Hellenic Mediterranean University (HMU), Heraklion, Greece; Department of Materials Science and Technology, University of Crete, Heraklion, Greece

Konstantinos Petridis Department of Electronic Engineering, Hellenic Mediterranean University (HMU), Chania, Greece

Si Qin Institute for Frontier Materials, Deakin University, Waurn Ponds Campus, Geelong, Victoria, Australia

Zhongkan Ren Mechanical and Nuclear Engineering Department, Kansas State University, Manhattan, KS, United States

Gurpreet Singh Mechanical and Nuclear Engineering Department, Kansas State University, Manhattan, KS, United States

Minas M. Stylianakis Department of Electrical and Computer Engineering, Hellenic Mediterranean University (HMU), Heraklion, Greece

Zhenyu Sun State Key Laboratory of Organic–Inorganic Composites, College of Chemical Engineering, Beijing University of Chemical Technology, Beijing, P.R. China

Ziqi Sun School of Chemistry, Physics and Mechanical Engineering, Queensland University of Technology (QUT), Brisbane, QLD, Australia

Maytal Caspary Toroker Department of Materials Science and Engineering, Technion - Israel Institute of Technology, Haifa, Israel

Yu-Qiao Zhang Graduate School of Information Science and Technology, Hokkaido University, Kita, Sapporo, Japan; Research Institute for Electronic Science, Hokkaido University, Kita, Sapporo, Japan

Preface

Two-dimensional (2D) materials are a family of atomically thin nanomaterials with nearly planar morphology. Since the isolation of graphene, in 2004, the development of 2D materials has experienced unprecedented attention from both industry and academia, due to their unusual and remarkable mechanical and electronic properties. A large amount of research has been dedicated to the development of new materials using novel fabrication methods and to the exploration of their properties in various technological applications, ranging from electronics to biomedicine. These studies showed that the physical and chemical properties of 2D materials can be effectively tuned through different strategies such as controlling their dimensions, the crystallographic structure and defects, or by doping with heteroatoms. This flexibility facilitates the design of 2D materials for dedicated applications in the field of energy conversion and storage.

The aim of this book is to provide a comprehensive overview of the current status and future perspectives of 2D materials dedicated to energy conversion and storage technologies. Individual chapters are going across the whole spectrum of the energy-related applications of 2D materials, from nanogenerators and hydrogen storage, up to supercapacitors. Each chapter focuses on a different energy application allowing readers to gain a greater understanding of the most promising 2D materials in the field for future use in energy devices.

The book is written for researchers and graduate students but can be equally used by industry professionals. The readers can be benefited by the fact that all the practical applications of 2D materials in the field of energy conversion and storage are contained in a single book. As the editor of this book, I hope that it can help researchers specialized in fabrication and/or characterization of 2D materials to gain a broader understanding and better visualize the targeted applications. In addition, I expect that professionals will find this book particularly helpful because each chapter is dealing with a single application. This provides them with a concrete reference source of the state-of-the-art materials and technologies in their technological/business field.

I am particularly thankful to the 27 contributors of this book spanning different backgrounds and four different continents. All of them are renowned academics and experts in their chosen fields, which I believe makes the content of the book sufficiently comprehensive as well as of the highest scientific standards.

Spyridon Zafeiratos
Strasbourg, October 2019

1

Piezoelectricity of 2D materials and its applications toward mechanical energy harvesting

Sujoy Kumar Ghosh[1] and Dipankar Mandal[1,2]

[1]Organic Nano-Piezoelectric Device Laboratory, Department of Physics, Jadavpur University, Kolkata, India [2]Institute of Nano Science and Technology, Mohali, India

1.1 Introduction

In the past two decades, with the rapid growth of the Internet of things, enormous small electronics such as sensors, actuators, and wireless transmitters have been integrated into every corner of this world for health monitoring, biochemical detection, environmental protection, remote controls, wireless transmission, and security. Each of these devices requires only small-scale power in microwatt (μW) to milliwatt (mW) level, which also demands the power source with the characteristics of mobility, sustainability, and availability. Traditionally, batteries are commonly used to power these devices. However, monitoring, managing, and recycling the large quantities of batteries with limited lifetime are extremely difficult tasks, and the waste hazard chemicals left in the exhausted batteries has become another crucial threat to the environment. Therefore, new technologies that can harvest energy from the environment as sustainable self-sufficient micro/nanopower sources offer a possible solution [1]. This is a newly emerging field of nanoenergy, which is about the applications of nanomaterials and nanotechnology for harvesting energy to power up micro/nanosystems. It can be used to possibly replace batteries or at least extend the lifetime of a battery. The concept of nanoenergy is quite different from the large gridscale power in megawatt (MW) or gigawatt (GW) as required for a country, a city, or an airplane.

2D Nanomaterials for Energy Applications. DOI: https://doi.org/10.1016/B978-0-12-816723-6.00001-0

Nanogenerator (NG) is a new type of nanotechnology that converts mechanical energy as produced by small-scale physical change into electricity. An NG has two typical approaches:

1. piezoelectric nanogenerator (PENG) and
2. triboelectric nanogenerator (TENG).

The major applications of NG are in three directions [2]:

1. sustainable nano/micro-power source for small devices to achieve self-powering;
2. active sensors for medical, infrastructure, human–machine, environmental monitoring, and security; and
3. basic networks units for harvesting water motion energy at low frequency toward the dream of blue energy.

First proposed idea of self-powering comes in 2006 as a result of discovery of PENG [3], which utilizes piezoelectric effect of ZnO nanowires for converting tiny mechanical energy into electricity. This study inspires the field of nanoenergy. The TENG was first invented in 2012 [4]. The basic mechanism of TENG is the use of the electrostatic charges created on the surfaces of two dissimilar materials when they are brought into physical contact: the contact induced triboelectric charges can generate a potential drop when the two surfaces are separated by a mechanical force, which can drive electrons to flow between the two electrodes built on the top and bottom surfaces of the two materials [5]. However, piezoelectricity of a material relies on the electromechanical coupling behavior indicated by the second rank strain sensor, $S_{ij} = d_{ijk}E_i$, where, d_{ijk} is the piezoelectric charge coefficient and E_i is the electric field [1,2]. Additionally, as piezoelectric materials are usually anisotropic, the piezoelectric constituter equations under a small uniform mechanical strain are given by, $P_i = (\boldsymbol{e})_{ijk}(\boldsymbol{S})_{jk}$, where, P_i is the polarization and $(\boldsymbol{e})_{ijk}$ is the third order piezoelectric tensor. The TENG produces higher output power in comparison to PENG. However, TENG suffers from its own mechanism because its performance depends on air gap and surface potential of the materials which facing many difficulties during implantation in real-life problem. At this point, PENG is highly suitable due to its superior sensitivity and stability in the output performances over prolonged period of time. So far, tremendous efforts have been made to advance PENGs using different established active materials such as zinc oxide (ZnO), lead zirconate titanate (PZT), poly(vinylidene fluoride) (PVDF), and other nanomaterials [1]. Therefore, PENGs have the capability of harvesting mechanical energy and offer a great potential for remote/wireless sensing, charging batteries, and powering electronic devices [6–8].

From the requirement of society for lightweight and miniature electronic devices, atomically thin two-dimensional (2D) materials exhibiting a wide range of unique electrical, optical, mechanical, and thermal properties, which do not exist in their bulk counterparts have grown tremendous research interest [9,10]. Owing to the outstanding advantages of properties such as ultrathin, transparency, flexibility, large surface-to-volume ratio, and stackable layers enlighten the development in lightweight and high-performance multifunctional applications [11]. Particularly, the 2D layered materials with noncentrosymmetric structure have great potential in mechanical energy harvesting applications. In this chapter, we will discuss about 2D layered materials with a noncentrosymmetric structure in the framework of in-plane and out-of-plane piezoelectricity, their energy harvesting capabilities, and practical designs of PENGs.

1.2 Piezoelectricity in 2D materials

The periodic table in Fig. 1.1A shows that there are 40 different transition metal dichalcogenides (TMDs) compound exist [11]. Among them, Group 4–7 TMDs are predominantly layered, whereas some of group 8–10 TMDs are commonly found in nonlayered structures. In layered structures, typically, each layer has a thickness of 0.6–0.7 nm, which consists of a hexagonally packed layer of metal atoms sandwiched between two layers of chalcogen atoms. The intralayer M–X bonds of 2D TMDs—whose generalized formula is MX_2 (where M is a transition metal of groups 4–10 and X is a chalcogen), are predominantly covalent in nature, whereas the sandwich layers are coupled by weak van der Waals forces thus allowing the crystal to readily cleave along the layer surface.

The metal coordination of layered TMDs can be either trigonal prismatic (D_{3h} point group) or octahedral (O_h point group) (typically distorted and sometimes referred to as trigonal-antiprismatic) as shown in Fig. 1.1B and C, respectively. In contrast to graphite, bulk TMDs manifests a wide variety of polymorphs. Most commonly encountered polymorphs are 1T, 2H, and 3R where the letters stand for trigonal, hexagonal, and rhombohedral, respectively, and the digit indicates the number of X–M–X units in the unit cell (i.e., the number of layers in the stacking sequence).

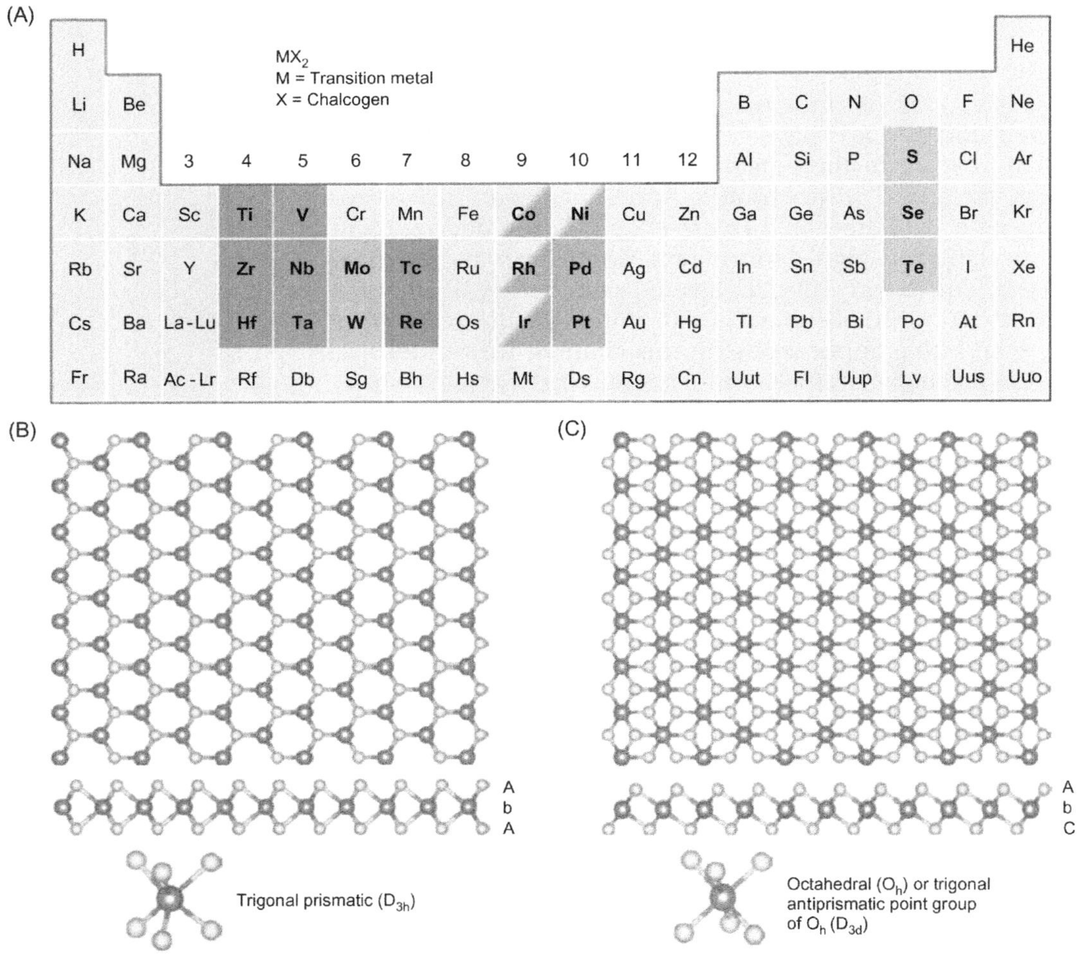

Figure 1.1 (A) Demonstration of 40 different layered TMDs in periodic table, (B and C) *c*-axis and section view of single-layer TMD with trigonal prismatic (B), and octahedral (C) coordinations. Atom color code: purple, metal; yellow, and chalcogen.The labels AbA and AbC represent the stacking sequence where the upper- and lower-case letters represent chalcogen and metal elements, respectively. *Source*: Reproduced with permission from M. Chhowalla, H.S. Shin, G. Eda, L.J. Li, K.P. Loh, H. Zhang, The chemistry of two-dimensional layered transition metal dichalcogenide nanosheets, Nat. Chem. 5 (2013) 263–275.

1.2.1 In-plane piezoelectricity

In 2012 Duerloo et al. first speculated that monolayer TMDs materials are piezoelectric, unlike their bulk crystals, based on DFT calculation [12]. It can be seen from Fig. 1.2 that a single h-BN (hexagonal boron nitride) or 2H-TMDs monolayer does not have an inversion center. Lack of centrosymmetry has the

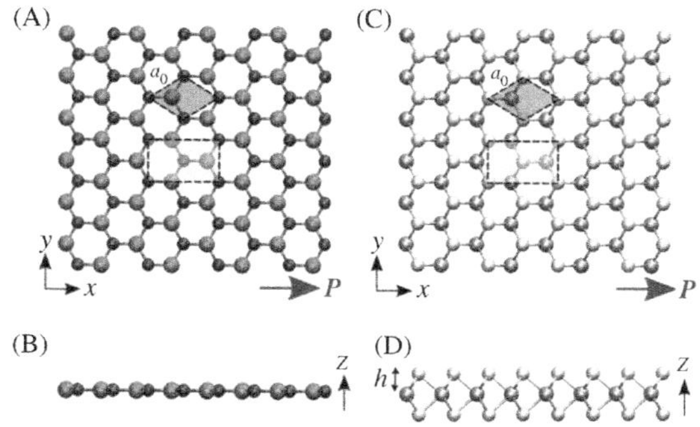

Figure 1.2 Monolayer boron nitride (h-BN) with top (A) and side view (B) geometry and monolayer trigonal prismatic molybdenum disulfide (2H-MoS_2) with top (C) and side view (D) geometry where B atoms are red (bigger atom), N atoms are blue (smaller atom), Mo (transition metal) atoms are silver (middle atoms in figure D), and S (chalcogenide) atoms are yellow (end side atoms in figure D). The polarization direction and axes are labeled, and the hexagonal primitive cell is highlighted in blue. *Source*: Reproduced with permission from K.N. Duerloo, M.T. Ong, E.J. Reed, Intrinsic piezoelectricity in two-dimensional materials, J. Phys. Chem. Lett. 3 (2012) 2871 – 2876.

mathematical consequence that all odd-rank tensor properties, including the third-rank piezoelectric tensor ($\boldsymbol{e}_{ijk}$), may be nonzero.

1.2.1.1 Molybdenum disulfide

In the year of 2014, both Zhu et al. [13] and Wu et al. [14] experimentally observed the piezoelectricity of monolayer 2D molybdenum disulfide (MoS_2) membrane. Although, its application toward mechanical energy harvesting and piezotronic sensing were only first demonstrated by Wu et al.

It is important to note that MoS_2 having crystalline D_{3h} symmetry only possesses in-plane piezoelectricity (e_{11}) and does not involve out-of-plane (e_{33}) piezoelectricity. Since the MoS_2 membrane has one atomic layer of Mo between two identical S layers, packed in a hexagonal lattice. Each rhombic prismatic unit cell is asymmetrically occupied by two S atoms on the left site and one Mo atom on the right (Fig. 1.3A), such that an external electric field pointing from the S site to the Mo site in the hexagonal lattice (armchair direction, E_1) can deform the unit cell by stretching the Mo–S bond and cause internal piezoelectric stress (Fig. 1.3B). The picture will be clearer if we consider the side view in Fig. 1.2D. From the side view each unit cell consists of two sulfur atoms (yellow, *i.e.*, end side atoms) and one molybdenum atom (silver, *i.e.*, middle atom), and

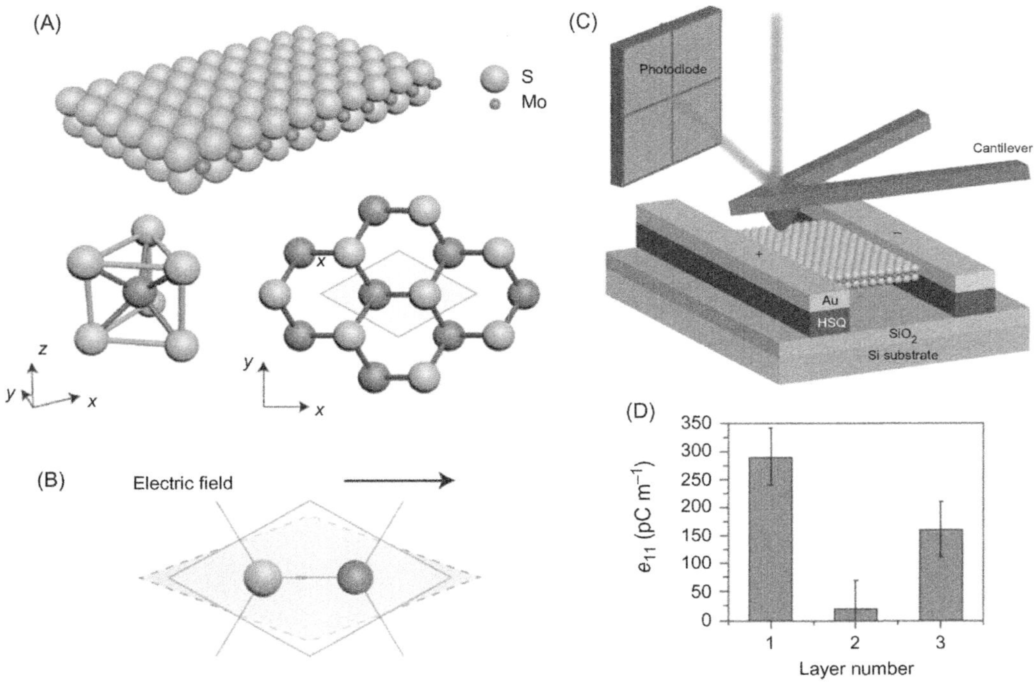

Figure 1.3 (A) Monolayer of MoS_2 consists of S—Mo—S stacking with a total thickness of 0.6 nm, (B) stress generation under electric field, (C) design of PFM: the MoS_2 film was suspended on two hydrogen silsesquioxane (HSQ), posted, and clamped underneath by two Au electrodes. The film was indented with a scanning AFM probe. The induced stress changed the load on the cantilever, which was observed by the deflection of a laser beam, (D) measured piezoelectric coefficient in one-, two-, and three-layer MoS_2 membranes. *Source*: Reproduced with permission from H. Zhu, Y. Wang, J. Xiao, M. Liu, S. Xiong, Z.J. Wong, et al., Observation of piezoelectricity in free-standing monolayer MoS2, Nat. Nanotechnol. 10 (2015) 151—155.

therefore the crystal breaks the inversion symmetry in x–y plane but preserves mirror symmetry in z-direction. The piezoelectricity of monolayer MoS_2 was probed by piezoresponse force microscopy (PFM) which measures picometer deformations with nanometer spatial resolution [15] and allows the quantitative determination of the piezoelectric constant [16]. Unlike in conventional PFM, the MoS_2 membrane is not coupled to the vertical electric field between the tip and substrate due to its mirror symmetry along the z-axis. Therefore, Zhu et al. [13] developed a method that combines a laterally applied electric field and nanoindentation in an atomic force microscope (AFM) to measure the piezoelectrically generated membrane stress (Fig. 1.3C). Another important fact is that piezoelectricity (e_{11}) in this 2D crystal only presents for odd number of layers and almost vanishes for even number of layers (Fig. 1.3D). In addition, e_{11} decreases with increasing thickness. The measured maximum piezoelectric coefficient was of $e_{11} = 2.9 \times 10^{-10}\ \mathrm{C\ m^{-1}}$ for single-layer MoS_2 membrane. Moreover, because of the unique hexagonal structure of monolayer MoS_2, the in-plane piezoelectric field vector owns three equivalent directions that are along the semiconductor's threefold symmetry axis. As verified by the reported SHG signals and piezoelectric coupling strength [13,14] the asymmetry and piezoelectric responses exhibit strong angle dependence and the corresponding maximum occurs at a multiple of 60 degrees (defining the "armchair" direction as 0 degree). Interestingly, bilayer MoS_2 gives the same planar atom projections at the six vertex sites of hexagonal structure and thus keeps the crystal centrosymmetry, resulting in the loss of piezoelectric effect. However, in recent years, Huang et al. showed that piezoelectricity of the MoS_2 films developed by ALD technique critically depends on its grain sizes [17]. For the grain size less than 120 nm, the piezoelectric constant of the MoS_2 film increases with the increase in the grain size, this contradicts to its classical piezoelectricity. They hypothesized that stresses and polar clusters in the grains with different sizes are different, which can affect the piezoelectric response. For example, smaller stress and more polar clusters in the larger grain may provide a greater piezoelectric constant. However, when the grain size increases to a certain value, the piezoelectric constant will decrease, because the small-scale effect will disappear and the classical piezoelectric theory caused by noncentrosymmetric crystal structure will be dominant.

1.2.1.2 Hexagonal boron nitride

In the year of 2009, the wideband insulator ($\simeq$5.9 eV) layered hexagonal boron nitride (h-BN) was theoretically calculated as possessing in-plane piezoelectricity [18]. The point group of 2D h-BN is same as MoS_2 (D_{3h}). The boron and nitride atoms in h-BN crystals also arrange in a hexagonal array. Unlike its counterpart, graphene, monolayer h-BN possesses strong in-plane piezoelectricity due to the alternating arrangement of boron and nitride atoms in the hexagonal vertex site.

Li et al. measured optical second-harmonic generation (SHG) from atomically thin samples of h-BN with one to five layers to experimentally probe the layer and thickness dependence in-plane piezoelectricity [19]. It is evident from Fig. 1.4A that SHG signal only obtained from odd number of layers and

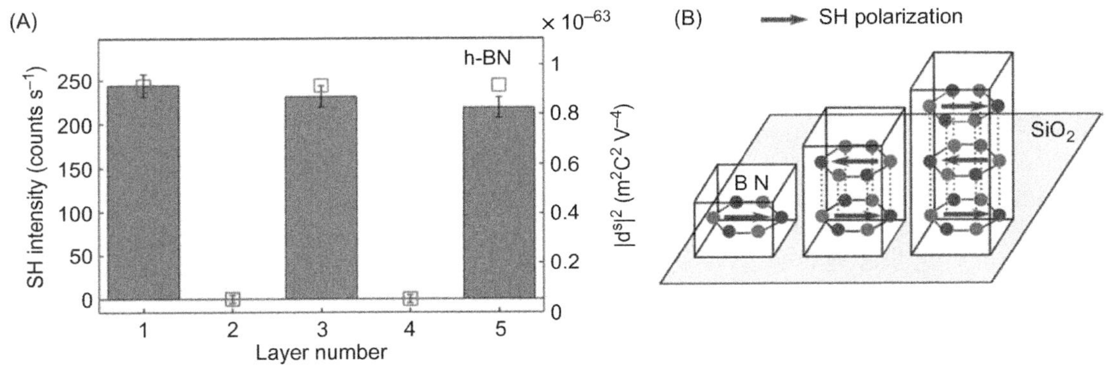

Figure 1.4 (A) The layer dependence of SH intensity and surface nonlinear susceptibility for few-layer h-BN and (B) illustration of the model for the calculation of the SH intensity as a function of layer thickness. *Source*: Reproduced with permission from Y. Li, Y. Rao, K.F. Mak, Y. You, S. Wang, C.R. Dean, et al., Probing symmetry properties of few-layer MoS2 and h-BN by optical second-harmonic generation, Nano Lett. 13 (2013) 3329 – 3333.

no measureable signal has been observed for even number of layers. It means in-plane piezoelectricity for h-BN exists for odd number of layers as shown previously with 2D MoS_2. However, there is no thickness dependency observed in case of h-BN which was found to decrease with increasing thickness for MoS_2. In recent years, it was revealed by first-principles calculations that h-BN with boron/nitrogen atom vacancy shows improved piezoelectricity and even magnetic moment [20].

1.2.1.3 Carbon nitride

On the basis of crystal symmetry considerations, pristine carbon nitride (C_3N_4) in its various forms is nonpiezoelectric. However, Zelisko et al. showed via PFM and quantum mechanical calculations that both atomically thin and layered graphitic carbon nitride (g-C_3N_4) or graphene nitride nanosheets exhibit anomalous piezoelectricity [21].

C_3N_4, is predicted to exist in α, β, cubic, pseudo-cubic, and 2D graphitic (g) forms [22]. The graphitic form of carbon nitride is known as graphene nitride or g-C_3N_4. A single sheet of g-C_3N_4 may be composed of an s-triazine or tri-s-triazine repetition (Fig. 1.1A), with the tri-s-triazine form being the more energetically stable of the two structures [21,22]. The tri-s-triazine sheet of g-C_3N_4 naturally has noncentrosymmetric, triangular holes throughout its structure.

Therefore it is expected that the in-plane, e_{11}, component of the piezoelectric tensor will be nonzero, where the $\hat{e}_1$ direction is aligned with the triangular holes as shown in Fig. 1.5A.

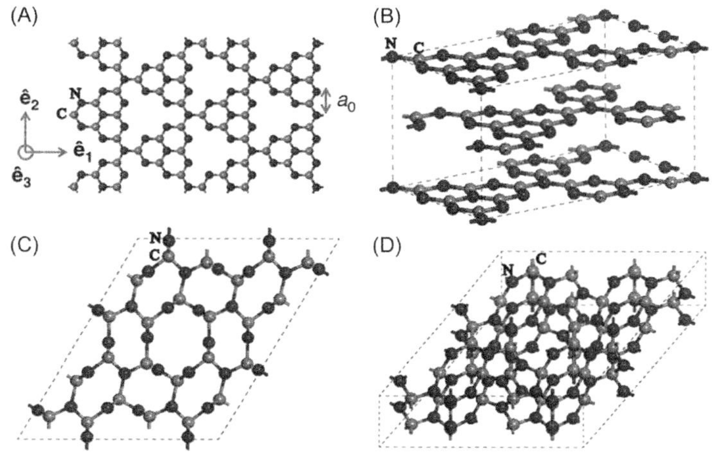

Figure 1.5 C_3N_4 and its different forms. Nitrogen and carbon atoms are represented by blue and gray colors, respectively(as shown in respective figures): (A) single-sheet tri-s-triazine (g-C_3N_4); (B) layered tri-s-triazine sheets and (C and D) β form shown from two perspectives to illustrate the three-dimensionality of the structure. *Source*: Reproduced with permission from M. Zelisko, Y. Hanlumyuang, S. Yang, Y. Liu, C. Lei, J. Li, et al., Anomalous piezoelectricity in two-dimensional graphene nitride nanosheets, Nat. Commun. 5 (2014) 4284–4287.

The preferred stacking method for g-C_3N_4 is by using a repeat unit in the $P\bar{6}m2$ symmetry group (Fig. 1.5B) with 3.29 Å spacing between sheets [23].

The piezoelectricity of the g-C_3N_4 was probed by PFM technique where both out-of-plane and in-plane PFM responses were observed (as shown in Fig. 1.6A and B). The flexoelectricity, together with the presence of noncentrosymmetric holes, is the central reason that explains the emergence of apparent piezoelectricity in g-C_3N_4. However, quantum mechanical calculations only show nonzero in-plane piezoelectric coefficient and zero out-of-plane piezoelectricity. This discrepancy may arise from several experimental facts and during PFM measurement procedures of g-C_3N_4 spherical shape particle ($\simeq$ containing many layers) on conductive substrate which includes topography mapping (Fig. 1.6C), PFM amplitude mapping (Fig. 1.6D), distribution of polarity within the particle (Fig. 1.6E), and resonant frequency mapping (Fig. 1.6F).

1.2.1.4 Other materials

Theoretically, various 2D semiconductors were predicted to have in-plane piezoelectric effect. For example, first-principles

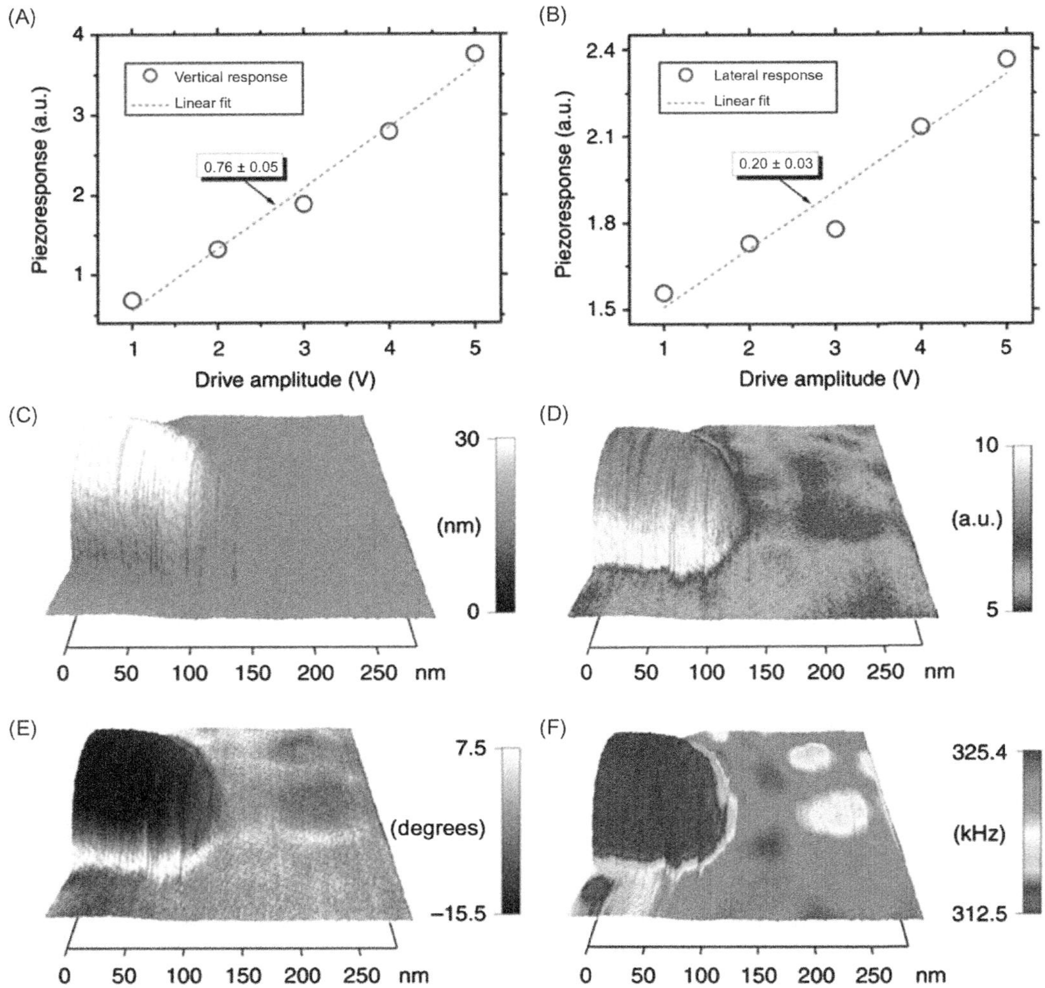

Figure 1.6 (A) Vertical and (B) lateral PFM responses of g-C_3N_4 by displacement versus applied voltage linear graph. PFM response of a g-C_3N_4 particle on conductive substrate includes (C) topography, (D) corrected vertical PFM amplitude, (E) phase, and (F) resonant frequency mapping. *Source*: Reproduced with permission from M. Zelisko, Y. Hanlumyuang, S. Yang, Y. Liu, C. Lei, J. Li, et al., Anomalous piezoelectricity in two-dimensional graphene nitride nanosheets, Nat. Commun. 5 (2014) 4284–4287.

density functional theory (DFT) calculations showed that monolayers of group III monochalcogenides (MX, M = Ga or In, X = S or Se) are intrinsically piezoelectric, such as GaS, GaSe, and InSe possess large in-plane piezoelectric coefficients of 2.06, 2.30, and 1.46 pm V^{-1} respectively [24]. Fig. 1.7 presents top and side views of the atomic structure of a GaS monolayer,

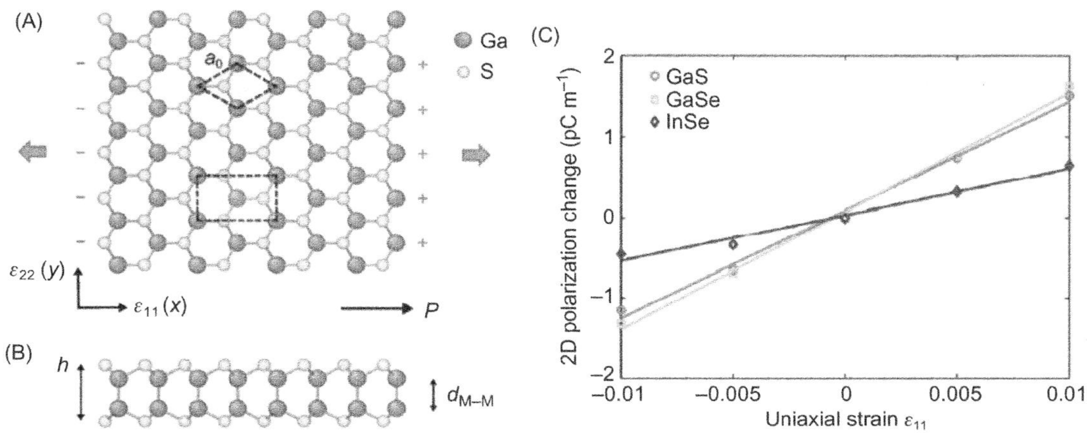

Figure 1.7 Structural model of a GaS monolayer viewed from (A) the top and (B) side. The larger sized gallium atoms are purple, while the smaller sized sulfur atoms are yellow, (C) polarization change in unit cell of MX monolayers per unit of area along the *x* direction under a uniaxial strain ε_{11} along the same direction. *Source:* Reproduced with permission from W. Li, J. Li, Piezoelectricity in two-dimensional group-III monochalcogenides, Nano Res. 8 (2015) 3796 (7 pp).

which is representative of the monolayer structure in all three MX compounds. Two vertically bonded layers of metal atoms are sandwiched between two layers of chalcogen atoms, and when viewed from the top, the M and X atoms each occupy one triangular sublattice of the overall honeycomb lattice. This structure of the MX monolayer therefore bears a strong similarity to monolayers of 2H-type TMDs such as MoS_2. In fact, both MX and MoS_2 monolayers belong to the $D3h$ ($\bar{6}m2$) point group.

Additionally, monolayer group IV monochalcogenides (MX, M = Sn or Ge, X = Se or S), including SnSe, SnS, GeSe, and GeS were shown to be piezoelectric possessing giant piezoelectric coefficients which is about one to two orders of magnitude larger than those of other 2D materials, such as MoS_2 and GaSe, and bulk quartz and AlN which are widely used in industry. As shown in Fig. 1.8, group IV monochalcogenides possesses C_{2v} point group (Fig. 1.8C and D) in contrast to the group III monochalcogenides Fig. 1.8A and B.

The comparison table for in-plane piezoelectric coefficient (d_{11}) between established piezoelectric materials and group IV monochalcogenides is given in Fig. 1.8E, which shows that few group IV monochalcogenides exhibits the d_{11} value beyond 200 pm V^{-1} which is even comparable to PZT [1].

Blonsky et al. predicted the in-plane piezoelectric coefficient d_{11} for 37 materials within the families of 2D metal

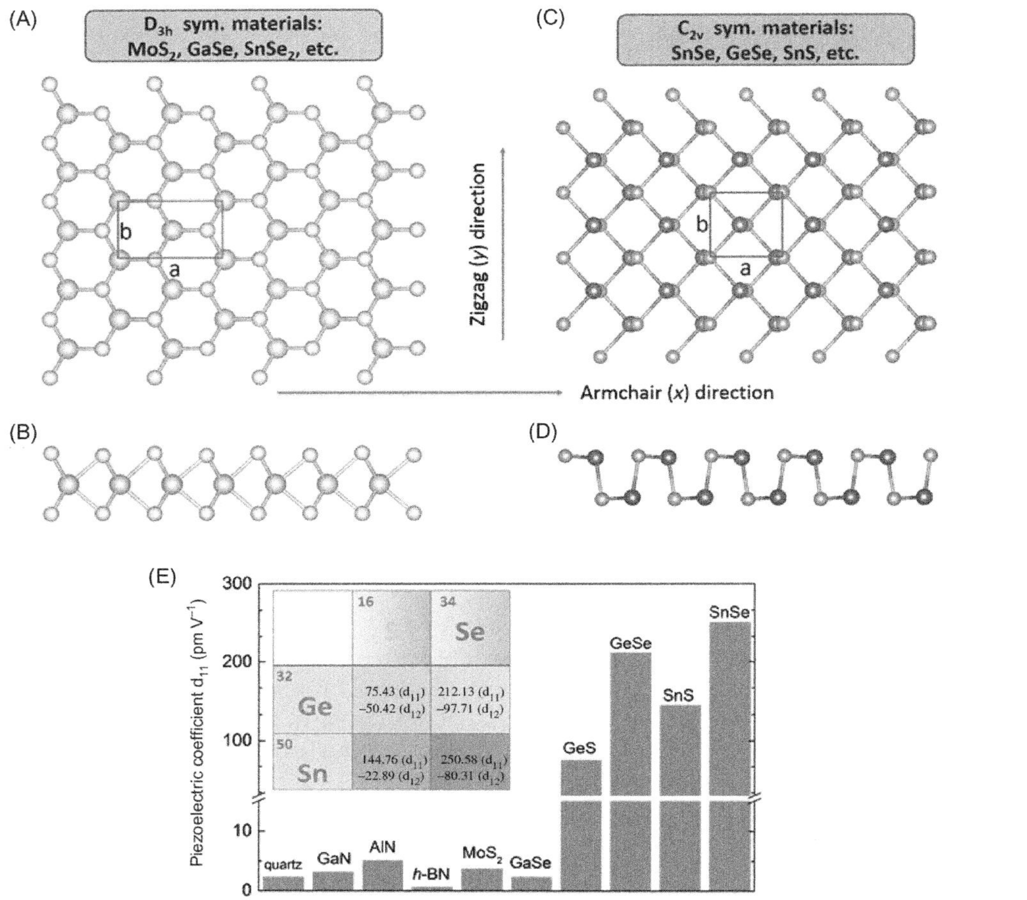

Figure 1.8 (A) The top and (B) side views of the ball-stick atomic structure of D_{3h} hexagonal monolayer, (C) the top and (D) side view of the ball-stick atomic structure of C_{2v} orthorhombic monolayer, and (E) comparison bar diagram of in-plane piezoelectric coefficient (d_{11}) between established piezoelectric materials and group IV monochalcogenides. *Source*: Reproduced with permission from R. Fei, W. Li, J. Li, L. Yang, Giant piezoelectricity of monolayer group IV monochalcogenides: SnSe, SnS, GeSe, and GeS, Appl. Phys. Lett. 107 (2015) 173104–173105 [25].

dichalcogenides (MX_2, M = Cr, Mo, W, Nb, Ta and X = S, Se, Te) (Fig. 1.9A), group IIA and IIB metal oxides (MO, M = Be, Mg, Ca, Zn, Cd, Pb) (Fig. 1.9B), group III——V semiconductors (AX, A = B, Al, Ga, In and X = N, P, As, Sb) (Fig. 1.9C) and out-of-plane piezoelectricity (d_{31}) of group III——V semiconductors (Fig. 1.9D).

All of these materials with nonzero piezoelectric coefficients exhibit hexagonal symmetry and have either a 2H structure (Fig 1.9E), planar hexagonal (Fig. 1.9F) or a buckled

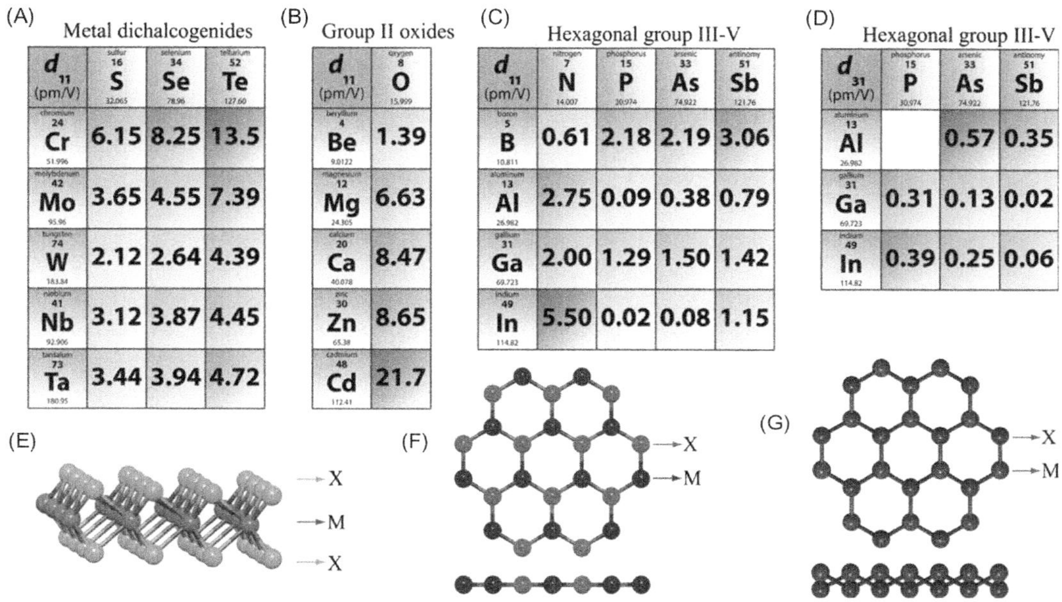

Figure 1.9 Periodic trends for d_{11} in (A) metal dichalcogenides, (B) metal oxides, and (C) group III–V semiconductors as well as for d_{31} in (D) group III–V semiconductors. Structures in each group with large coefficients are shaded red. Material structures are illustrated for (E) 2H, (F) planar hexagonal, and (G) buckled hexagonal structures. *Source*: Reproduced with permission from M.N. Blonsky, H.L. Zhuang, A.K. Singh, R.G. Hennig, Ab Initio prediction of piezoelectricity in two-dimensional materials, ACS Nano 9 (2015) 9885–9891 [26].

hexagonal (Fig. 1.9G). We note that SnX_2 and VX_2 dichalcogenides have a 1T structure, which displays inversion symmetry and therefore no piezoelectricity (Fig. 1.9).

1.2.2 Out-of-plane piezoelectricity

Next, we will turn to the experimentally reported out-of-plane 2D piezoelectric materials.

1.2.2.1 Indium selenide

Zhou et al. reported the first experimental evidence of out-of-plane piezoelectricity and ferroelectricity in van der Waals layered alpha phase indium selenide (α-In_2Se_3) nanoflakes, as a typical III–VI semiconductor [27].

The noncentrosymmetric $R3m$ symmetry of the α-In_2Se_3 samples was evidenced by scanning transmission electron microscopy, SHG, and Raman spectroscopy measurements. The side view of crystal structure is schematically shown in Fig. 1.10A. It can be seen that one quintuple layer consists of

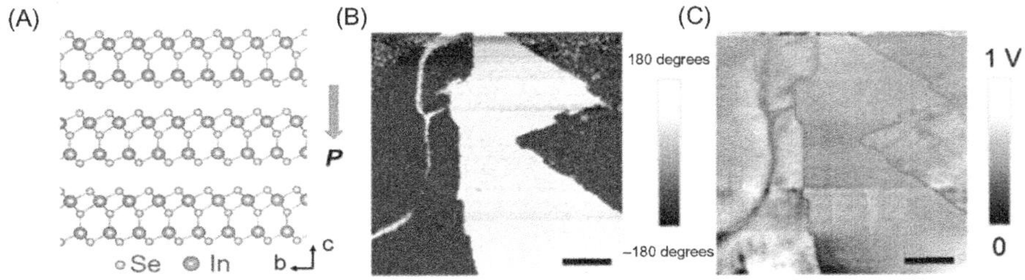

Figure 1.10 (A) Crystal structure of α-In_2Se_3 (side view) with direction of polarization, (B) PFM phase, and (C) amplitude images of a thin α-In_2Se_3 flake ($>$ 100 nm). The scale bars are 1 μm in (B) and (C). *Source*: Reproduced with permission from Y. Zhou, D. Wu, Y. Zhu, Y. Cho, Q. He, X. Yang, et al., Out-of-plane piezoelectricity and ferroelectricity in layered α-In2Se3 nanoflakes, Nano Lett. 17 (2017) 5508 – 5513

five adjacent atom layers (Se-In-Se-In-Se) and the corresponding structure presents the asymmetry property. Under external strain, the produced electric dipole pointing from negative selenide atom to positive indium atom can result in the vertical (out-of-plane) piezoelectric field as the arrow shows. Experimentally, it was shown with exfoliated In_2Se_3 flake via out-of-plane PFM phase and amplitude images, shown in Fig. 1.10B and C. Fig. 1.10B shows two distinct regions with 180-degree phase difference, corresponding to domains with up and down polarization vectors perpendicular to the flake surface, whereas the domain walls appear as darker lines in the PFM amplitude image (Fig. 1.10C).

1.2.2.2 Graphene

Graphene is known as a perfect and centrosymmetric hexagonal material with one type of carbon atoms and does not have piezoelectric effect. However, the supported substrate, such as silica, often introduces impurities such as some chemical substances into graphene, namely doped graphene and its symmetry structure will be unintentionally broken, resulting in the generation of out-of-plane dipole (piezoelectricity) as indicated in Fig. 1.11.

Rodrigues et al. experimentally confirmed the existence of electric polarization induced in graphene deposited on SiO_2 via direct measurements of converse piezoelectric effect by PFM [28]. Piezoelectric activity was mainly observed on the supported graphene regions where van der Waals and/or chemical interaction between the SiO_2 surface and graphene layer can induce an anisotropic strain and detectable PFM signal. The in-plane strain in graphene was evaluated by polarization-dependent

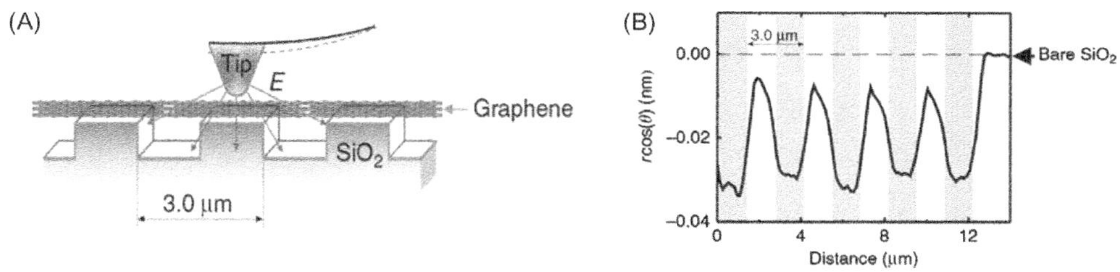

Figure 1.11 (A) Schematic of the PFM measurements on single-layer graphene adsorbed on the TGZ4 grating substrate. (B) Cross section of the piezoresponse along the line on graphene across the grating structure (shaded areas correspond to supported graphene). The dashed red line (passing through 0 ordinate) denotes the baseline corresponding to the signal on the bare SiO_2 substrate. *Source*: Reproduced with permission from G.C. Rodrigues, P. Zelenovskiy, K. Romanyuk, S. Luchkin, Y. Kopelevich, A. Kholkin, Strong piezoelectricity in single-layer graphene deposited on SiO2 grating substrates, Nat. Commun. 7 (2015) 7572 (6 pp).

Raman spectroscopy. Piezoelectric effect was sufficiently high ($d_{33} \simeq 1.4$ nmV^{-1}), that is, more than twice of the best piezoelectric ceramics such as modified lead zirconate titanate (PZT). In another case, ferroelectric-substrate induced extrinsic piezoelectricity in multilayer MoS_2 on $PbTiO_3$ film has been observed [29]. This out-of-plane dipole effect in MoS_2 flakes, trigged by ferroelectric polarization, is more pronounced for the 19 nm specimen.

1.2.2.3 Janus MoSSe monolayer

Recently, Lu et al. synthesized Janus MoSSe monolayer with broken out-of-plane symmetry by the plasma stripping and thermal selenization. In this structure, one layer of S-atom was fully replaced by the layer of Se-atoms as shown in Fig. 1.12A. The confirmation of the Janus structure of MoSSe was made by means of scanning transmission electron microscopy (Fig. 1.12B) and energy-dependent X-ray photoelectron spectroscopy. Additionally, the existence of vertical dipoles was proved by SHG and PFM measurements. As a result of the out-of-plane asymmetry, Janus MoSSe experimentally shows an intrinsic vertical piezoelectric response, the first demonstration in a single-molecular-layer crystal. Because the distortion of Mo–S and Mo–Se bonds under an electric field do not cancel, a net thickness change is induced by applying a vertical voltage. As a result, a clear piezoelectric contrast between the Janus MoSSe monolayer and the substrate has been observed (Fig. 1.12C). The corresponding piezoelectric coefficient d_{33} is $\sim$0.1 pm V^{-1} and can potentially be improved by one order of magnitude by increasing the dipolar contrast of the chemical bonds.

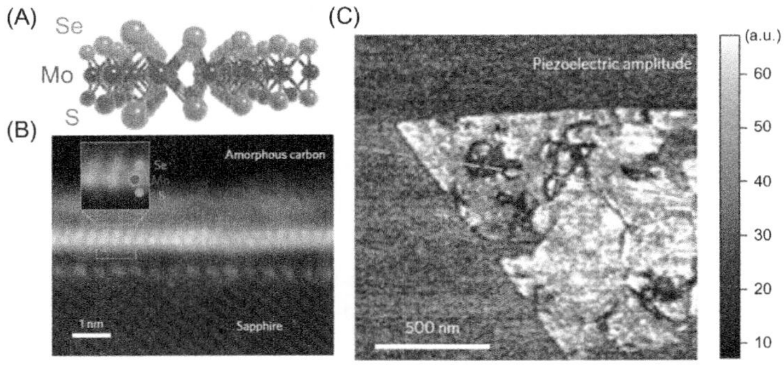

Figure 1.12 (A) Crystal structure of Janus MoSSe monolayer, (B) annular dark-field scanning transmission electron microscopy image of the sample cross section, showing the asymmetric MoSSe monolayer structure with Se (orange) on top and S (yellow) at the bottom of the Mo atoms (blue, which is in the middle of Se and S), and (C) piezoelectric amplitude of an isolated Janus MoSSe monolayer directly grown on highly oriented pyrolytic graphite (HOPG), measured by resonance-enhanced piezoresponse force microscopy. *Source*: Reproduced with permission from A.Y. Lu, H. Zhu, J. Xiao, C.P. Chuu, Y. Han, M.H. Chiu, et al., Janus monolayers of transition metal dichalcogenides, Nat. Nanotechnol. 12, (2017) 744–749 [30].

We have summarized the piezoelectric properties as shown in Table 1.1 for the experimentally confirmed piezoelectric 2D materials. Generally, the feature of noncentrosymmetry is the sufficient requirement for piezoelectric materials (except for the cubic class 432). By symmetry analysis, one can preliminarily conclude whether a 2D material is piezoelectric or not. Although various 2D materials have been theoretically predicted to be piezoelectric, most of them need to be further explored or confirmed by experiments.

1.3 Mechanical energy harvesting applications

1.3.1 MoS_2-based energy harvester

The first application of single-atomic-layer 2D MoS_2 flakes in mechanical energy harvesting and piezotronic sensing was demonstrated by Wu et al. in 2014 [14]. Repeated stretching and releasing of odd-layer MoS_2 flakes produces oscillating electrical outputs, which convert mechanical energy into electricity. The strain-induced polarization charges in single-layer MoS_2 can also modulate charge carrier transport at the MoS_2-metal barrier which enable enhanced strain sensing. In addition, a large piezoresistivity was observed in even-layer MoS_2 with a gauge factor of about 230 for the bilayer material, which indicates a

Table 1.1 Summery of the experimentally confirmed piezoelectricity in 2D materials [9].

Sl. no.	2D materials	Crystal structure	Piezoelectric directions	Piezo-coefficients	Notes
1.	Monolayer MoS_2	Hexagonal	In-plane angel dependence	$d_{11} \simeq$ 2.5–4 pm V^{-1} [12] $e_{11} \simeq$ 250–400 pC m^{-1} [12,13]	Odd–even effect with the thickness increased
2.	Monolayer h-BN	Hexagonal	In-plane angel dependence	$e_{11} \simeq$ 100–400 pC m^{-1} [12]	Odd–even effect with the thickness increased [19]
3.	g-C_3N_4		Out-of-plane and in-plane	$d_{33} \simeq$ 1.4 pm V^{-1} [21] $e_{11} \simeq$ 218 pC m^{-1} [21]	Existence of piezoelectricity regardless of thickness
4.	α-In2Se3	Rhombohedral	Out-of-plane		Confirmed by the ferroelectricity [27]
5.	Doped graphene	Hexagonal	Out-of-plane	$d_{33} \simeq$ 1.4 nm V^{-1} [28]	Extrinsic piezoelectricity
6.	Janus MoSSe	Hexagonal	Out-of-plane and in-plane	$d_{33} \simeq$ 0.1 pm V^{-1} [30]	

possible strain-induced change in band structure. This study demonstrates the potential of 2D nanomaterials in powering nanodevices, adaptive bioprobes and tunable/stretchable electronics/optoelectronics. In this experiment, single-layer MoS_2 flake was transferred to a polyethylene terephthalate (PET) flexible substrate in order to make a flexible device (Fig. 1.13A). The electrical contacts were made by Cr/Pd/Au (1 nm/20 nm/50 nm) deposition, as shown in Fig. 1.13A. When the PET substrate was bent mechanically, uniaxial strain is applied to the MoS_2 with a magnitude proportional to the inverse bending radius (Fig. 1.13B, bending figure). The applied strain was limited to 0.8% to avoid sample slippage. The piezoelectric response was studied by applying strain to a device coupled to an external load resistor (Fig. 1.13B). In this configuration, strain-induced polarization charges at the sample edges can drive the flow of electrons in external circuit. When the substrate is released, electrons flow back in the opposite direction.

Fig. 1.14A shows the piezoelectric current and voltage responses of a single-layer MoS_2. When strain is applied in the x direction ("armchair" direction), positive voltage and

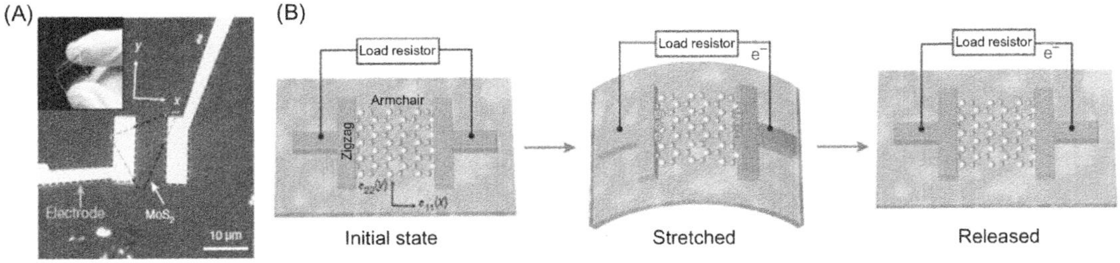

Figure 1.13 (A) The flexible device with single-layer MoS_2 flake and electrodes at its zigzag edges and (B) operation scheme of the single-layer MoS_2 piezoelectric device under stretching and released condition. *Source*: Reproduced with permission from W. Wu, L. Wang, Y. Li, F. Zhang, L. Lin, S. Niu, et al., Piezoelectricity of single-atomic-layer MoS2 for energy conversion and piezotronics, Nature 514 (2014) 470–474.

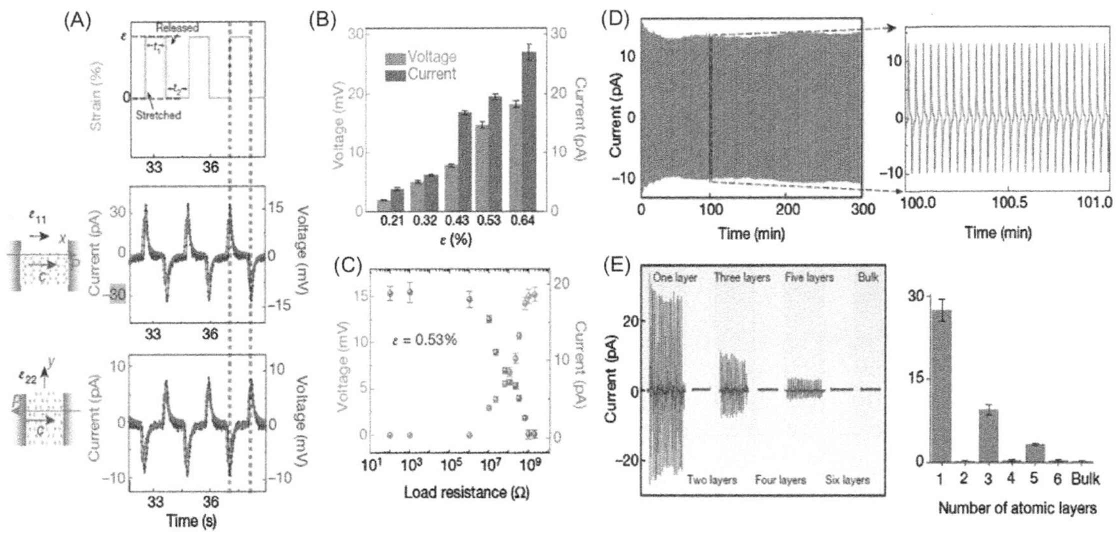

Figure 1.14 (A) Top: applied strain as a function of time. Middle: corresponding piezoelectric outputs (such as voltage response with 1 GΩ external load and short-circuit current response of a single-layer MoS_2 device under periodic strain in two different principal directions) from single-atomic-layer MoS_2 when strain is applied in the *x* direction (armchair direction). Bottom: corresponding piezoelectric outputs from the same device when strain is applied in the *y* direction (zigzag direction). (B) Dependence of piezoelectric outputs from a single-layer MoS_2 device on the magnitude of the applied strain, (C) dependence of voltage and current outputs from a single-layer MoS_2 device under 0.53% strain as a function of load resistance, (D) cyclic test showing the stability of single-layer MoS_2 device for prolonged period, and (E) evolution of the piezoelectric outputs with increasing number of atomic layers (n) in MoS_2 flakes. *Source*: Reproduced with permission from W. Wu, L. Wang, Y. Li, F. Zhang, L. Lin, S. Niu, et al., Piezoelectricity of single-atomic-layer MoS2 for energy conversion and piezotronics, Nature 514 (2014) 470–474.

current output were observed with increasing strain, and negative output was observed with decreasing strain. Both responses increased with the magnitude of the applied strain.

In particular, for single-layer MoS_2 the peak open-circuit voltage reached 18 mV and the peak short-circuit current reached 27 pA (Fig. 1.14B). The output power was estimated by studying the voltage and current outputs as a function of load resistance (Fig. 1.14C). The maximum instantaneous power delivered to the load at 0.53% strain was achieved for a load resistance of $\simeq$220 MΩ and reached 55.3 fW (5.53×10^{-14} W), with a corresponding power density of 2 mW m^{-2}. The conversion efficiency of the single-layer MoS_2 NG, which is the ratio of the electric power delivered to the load to the total mechanical deformation energy stored in the single-layer MoS_2 after being strained, can therefore be estimated as $\simeq$5.08%. This energy conversion was stable over time, as shown for cyclic loading up to 0.43% strain at 0.5 Hz for 300 min (Fig. 1.14D). In addition, the piezoelectric signal was found to decrease with an increasing number of atomic layers (n) and almost no detectable output can be seen for bulk flake and even-layer samples (Fig. 1.14E).

Furthermore, the changes in direct-current electrical transport properties of the devices with strain was characterized in a two-terminal configuration with the polarity of the applied voltage defined with respect to the drain electrode. The metal–semiconductor–metal device was fabricated, consisted of two back-to-back Schottky barriers, and transport across the reverse-biased Pd–MoS_2 Schottky barrier limited the current flow. In this configuration, changes in transport behavior arose largely from two effects: the piezotronic effect, in which strain-induced charge asymmetrically modulated the Schottky barriers, and a piezoresistive effect, in which strain-induced bandgap change modulated the entire resistance of the device. For the single-layer device, the current (I)–voltage (V) curve shifted leftward (toward negative drain bias) under tensile strain, and rightward with compressive strain. The opposite trend was observed under negative drain bias, showing the characteristic of piezotronic effect.

In bilayer and bulk MoS_2 devices, the response is purely piezoresistive: the current increases symmetrically with applied strain, and the gap region in the I–V curve shrinks, consistent with a lowering of both source and drain Schottky barriers as a result of a decrease in the bandgap and/or a change in carrier density. In this case, piezoelectric polarization charges at the zigzag edges were able to affect the metal–MoS_2 contacts directly. Controllable modulation of M–S contacts or p–n junctions in 2D nanomaterials by strain-induced polarization may offer a novel approach unavailable to conventional technologies

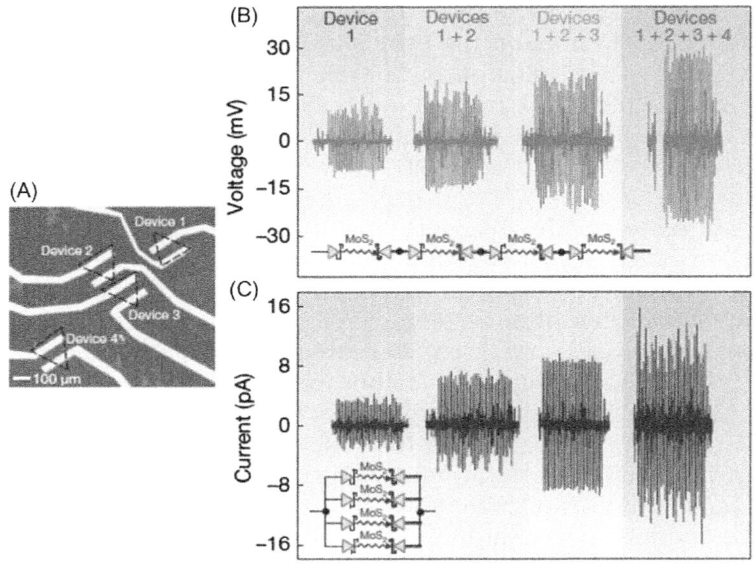

Figure 1.15 (A) Optical image of an array integration of four CVD growth single-layer MoS_2 flakes, (B) constructive voltage outputs by serial connection, and (C) constructive current outputs by parallel connection of the individual flakes in the circuit. *Source*: Reproduced with permission from W. Wu, L. Wang, Y. Li, F. Zhang, L. Lin, S. Niu, et al., Piezoelectricity of single-atomic-layer MoS2 for energy conversion and piezotronics, Nature 514 (2014) 470–474.

using electrical control signals, without modifying the interface structure or chemistry, for implementing tunable electronics/optoelectronics, enhanced photovoltaics, hybrid spintronics, and catalysis.

Finally, the array integration of single-layer MoS_2 flakes to boost the piezoelectric output for energy conversion was demonstrated (Fig. 1.15A). Using constructive connectivity of the four flakes either in series (Fig. 1.15B) or in parallel (Fig. 1.15C), consistent enhancements in output voltages or currents were observed, respectively. This may open up possibilities of achieving practical technology at an even larger scale with 2D piezoelectric nanomaterials for powering nanodevices, tactile imaging, and wearable electronics.

After this, the output performance of MoS_2 NGs including up to seven atomic layers of MoS_2 was theoretically examined [31]. Fig. 1.16A shows a typical structure of a MoS_2 PENG. In this model, a MoS_2 flake consisting of three atomic layers is sandwiched between the left-hand and right-hand electrodes. MoS_2 PENGs have three components: R_{In} is the PENG internal resistance, C_{NG} is the PENG capacitance, and V is the voltage source, which utilizes the piezo-potential to drive the electron flow through the circuit. The equivalent circuit of the MoS_2 PENG with R_{Ex} is shown in Fig. 1.16B.

Under periodic tensile strain of 0.5% with square-wave frequency of 0.5 GHz (Fig. 1.16C), the MoS_2 PENGs of different

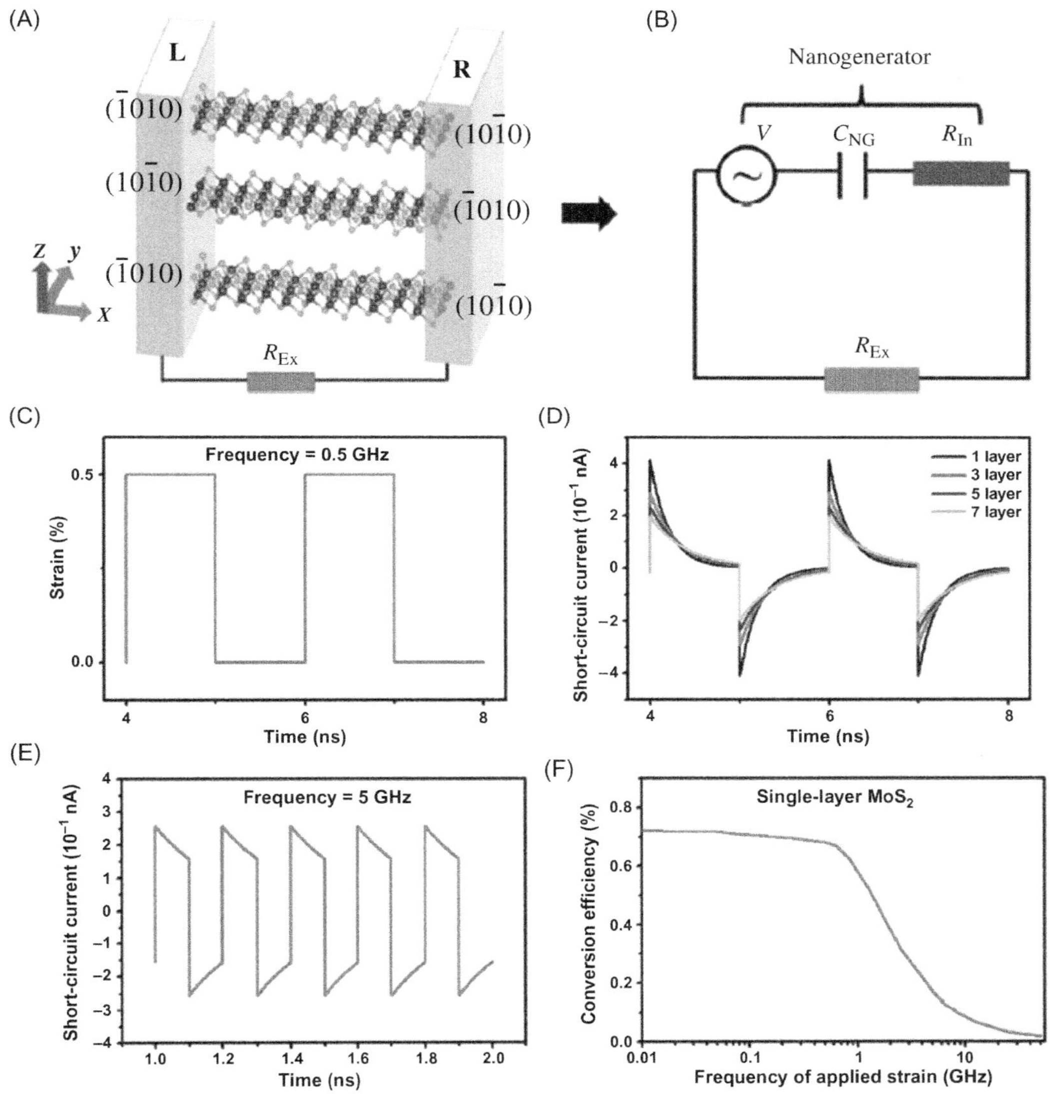

Figure 1.16 (A) Schematic illustration of a three-layer MoS_2 PENG in connection with an external load resistor R_{Ex}, (B) equivalent circuit of the three-layer MoS_2 nanogenerator, (C) square-wave external strain applied on and released from the NG. (D) Corresponding short-circuit currents of odd-layer MoS_2 NGs under applied strain, (E) short-circuit output current of single-layer MoS_2 PENG under 5 GHz square-wave applied strain, and (F) energy conversion efficiency of single-layer MoS_2 PENG versus frequency of the square-wave applied strain. *Source*: Reproduced with permission from Y. Zhou, W. Liu, X. Huang, A. Zhang, Y. Zhang, Z.L. Wang, Theoretical study on two-dimensional MoS2 piezoelectric nanogenerators, Nano Res. 9 (2016) 800–807.

layers showed significant short-circuit output current which decreases with increasing number of MoS_2 layers (Fig. 1.16D). Additionally, the MoS_2 PENGs was capable to harvest very high frequency 5 GHz square-wave applied strain (Fig. 1.16E). Compared with the case of the 0.5 GHz applied strain, shown in Fig. 1.16C and D, the peak current under a 5 GHz strain hardly relaxed in a half periodicity (0.1 ns) before the next peak current appeared, which caused a negative influence on the energy conversion efficiency (Fig. 1.16E). In Fig. 1.16F, the energy conversion efficiency of the single-layer MoS_2 PENG under different frequencies of the square-wave applied strain is shown.

In addition, Kim et al. fabricated chemical vapor deposition (CVD) grown monolayer MoS_2-based flexible PENG and showed that output performances with the armchair direction of MoS_2 is about two times higher than that from the PENG with the zigzag direction of MoS_2 under the same strain of 0.48% and the strain velocity of 70 mm s^{-1} [32]. The lateral electrode configuration of monolayer MoS_2-based flexible PENG with armchair and zigzag atomic orientation axis illustrated in Fig. 1.18A and B, respectively. The fabricated flexible PENG on PET substrate was shown in Fig. 1.17C.

Fig. 1.17E and F show optical images showing mechanical energy harvesting active regions with the armchair and zigzag atomic orientations of the PENGs, respectively. The piezoelectric output performances were investigated by applying mechanical strain (Fig. 1.17G–J). The voltage and current output shown in Fig. 1.17G–J were obtained with bending strain of 0.48% at a frequency of 0.5 Hz. The measured output voltage approached up to 20 mV (Fig. 1.17G) and the output current was over 30 pA (Fig. 1.17I) from the PENG with the armchair direction of MoS_2. However, the output voltage (Fig. 1.17H) and current (Fig. 1.17J) are less than 10 mV and 20 pA, respectively, from the PENG with the zigzag direction of MoS_2. This result suggests that manipulation of the MoS_2 atomic orientation along an armchair direction in a large scale is very critical to dramatically enhance piezoelectric power output performance from single-crystalline MoS_2 monolayer-based PENGs.

1.3.2 WSe_2-based energy harvester

Lee et al. showed that WSe_2 bilayers fabricated via turbostratic stacking have reliable piezoelectric properties that cannot be obtained from a mechanically exfoliated WSe_2 bilayer with Bernal stacking [33]. The flexible piezoelectric WSe_2 bilayers exhibit a prominent mechanical durability of up to 0.95% of

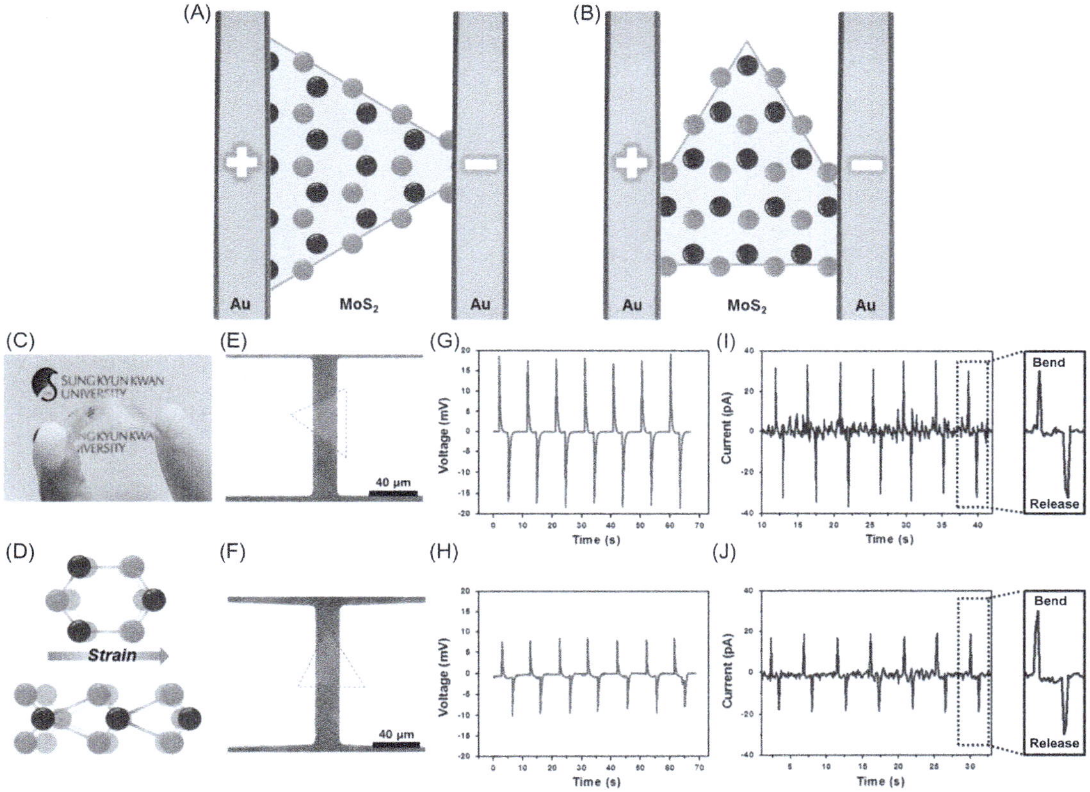

Figure 1.17 (A) Armchair and (B) zigzag directions of MoS_2 in lateral electrode configuration, (C) digital photograph of the CVD grown monolayer MoS_2-based PENG, (D) top (upper portion) and side view (lower portion) of the atomic structure of monolayer MoS_2, (E–F) optical images showing energy harvesting active regions with the armchair and zigzag atomic orientations of the PENGs, respectively, (G) output voltage and (I) output current under armchair orientation and (H) output voltage and (J) output current under zigzag orientation of MoS_2-based PENG. *Source*: Reproduced with permission from S.K. Kim, R. Bhatia, T.H. Kim, D. Seol, J.H. Kim, H. Kim, et al., Directional dependent piezoelectric effect in CVD grown monolayer MoS2 for flexible piezoelectric nanogenerators, Nano Energy 22 (2016) 483–489.

strain as well as reliable energy harvesting performance, which is adequate to drive a small liquid crystal display without external energy sources, in contrast to monolayer WSe_2 for which the device performance becomes degraded above a strain of 0.63%. Fig. 1.18A shows a schematic illustration of the m-WSe_2-based NG and optical image of Cr/Au electrode on a PET substrate used for fabrication of the device is shown in Fig. 1.18B.

The piezoelectric coefficient d_{11} of the m-WSe_2 was estimated to be 3.26 ± 0.3 pm V^{-1}, which is a reasonable value

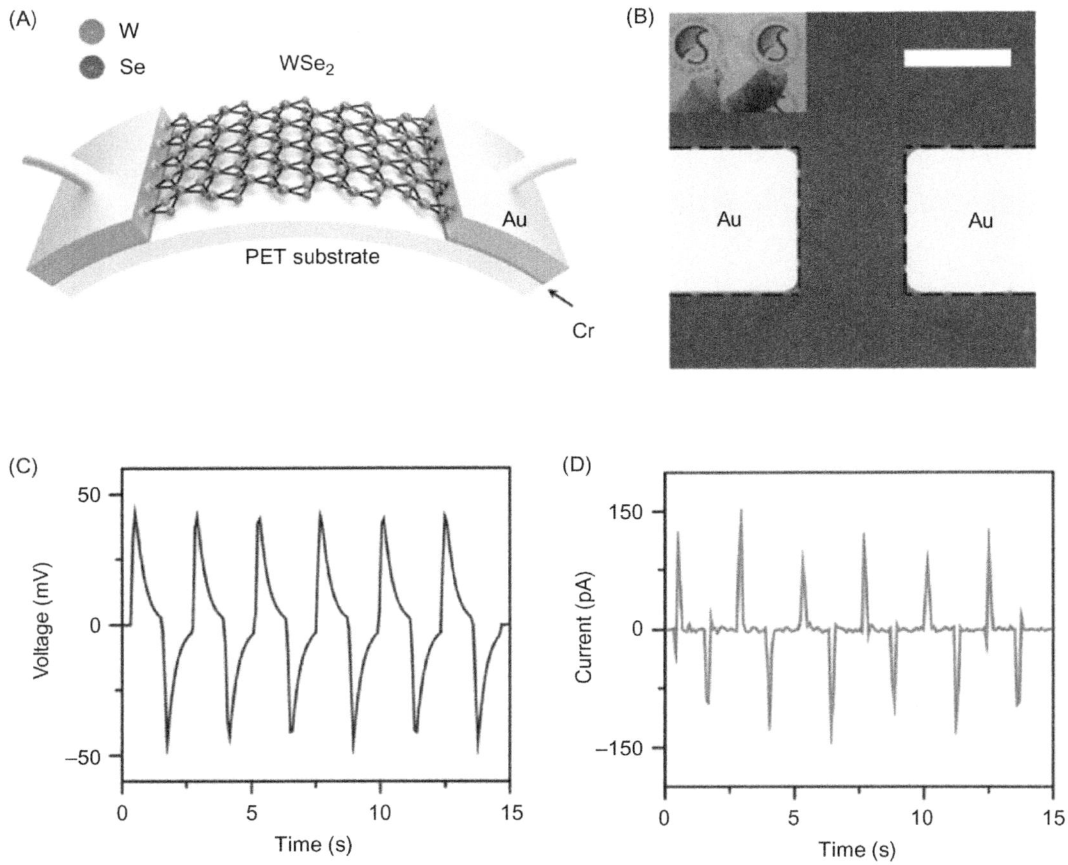

Figure 1.18 (A) Schematic of the m-WSe2-based PENG, (B) optical image of the electrode configuration with the original PENG in the inset, and (C) output voltage and (D) short-circuit current from the PENG under periodic tensile strain. *Source*: Reproduced with permission from J.H. Lee, J.Y. Park, E.B. Cho, T.Y. Kim, S.A. Han, T.H. Kim, et al., Reliable piezoelectricity in bilayer WSe_2 for piezoelectric nanogenerators, Adv. Mater. 29 (2017) 1606667 (7 pp).

compared to previously reported simulation result (2.79 pm V^{-1}) [12]. The m-WSe_2 was also characterized in terms of their piezoelectric output voltage and current response. The peak voltage generated with 1 GΩ of load resistance reached 45 mV (Fig. 1.18C) and the peak short-circuit current reached 100 pA (Fig. 1.18D) for 0.39% strain and 40 mm s^{-1} strain rate. The maximum instantaneous power reached 2.54 pW at a load resistance of 500 MΩ and the conversion efficiency of the flexible m-WSe_2-based PENG reached 2.41% and sustained for over 1000 cycles.

In addition, direct piezoelectric response of the bilayer WSe_2 (b-WSe_2) was investigated by preparing two types of b-WSe_2. The first type consists of b-WSe_2 directly grown on a sapphire substrate via CVD (db-WSe_2), and the second consists of m-WSe_2 made via CVD and subsequently transferred onto another m-WSe_2, resulting in bilayer WSe_2 (tb-WSe_2). It was found that mechanical durability of the tb-WSe_2 was higher than that of m-WSe_2.

The m-WSe_2 can sustain up to 0.64% of strain and generated maximum output voltage of 90 mV (Fig. 1.19A) but tb-WSe2 found to sustain up to 0.95% of strain and resulted 85 mV peak voltage (Fig. 1.19B). The maximum instantaneous power output

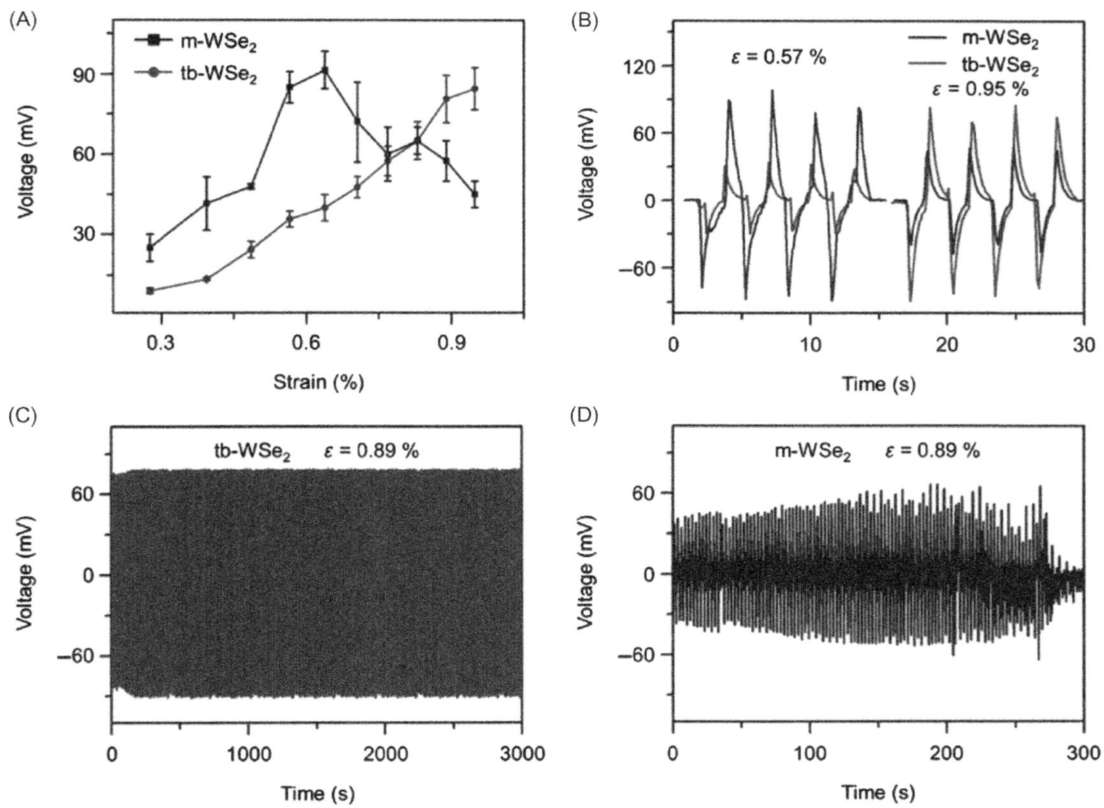

Figure 1.19 (A) Piezoelectric peak output voltages of m-WSe_2 and tb-WSe_2 as a function of strain, (B) output voltage of m-WSe2 (dark black line) and tb-WSe2 (faded black line) with a low strain (0.57%) and high strain (0.95%), and (C and D) mechanical stability of the tb-WSe2 and m-WSe_2-based PENGs with 0.89% of strain for more than 1000 cycles. *Source*: Reproduced with permission from J.H. Lee, J.Y. Park, E.B. Cho, T.Y. Kim, S.A. Han, T.H. Kim, et al., Reliable piezoelectricity in bilayer WSe_2 for piezoelectric nanogenerators, Adv. Mater. 29 (2017) 1606667 (7 pp).

was 4.05 pW with tb-WSe2 at 0.89% of strain, and it was stable for over 1000 cycles (Fig. 1.19C). However, a maximum power of only 2.05 pW was achieved with m-WSe_2 at a strain of 0.89%, and the power decreased significantly with repeated strain (Fig. 1.19D). Therefore, tb-WSe_2 was found to be suitable for mechanical energy harvesting in spite of low piezoelectric coefficient than that of m-WSe_2.

In order to apply for real-life applications with improved output power, multielectrode patterning design was applied on large-area tb-WSe_2 and NGs were integrated in a package. Five integrated NGs were fabricated in a single substrate (Fig. 1.20A).

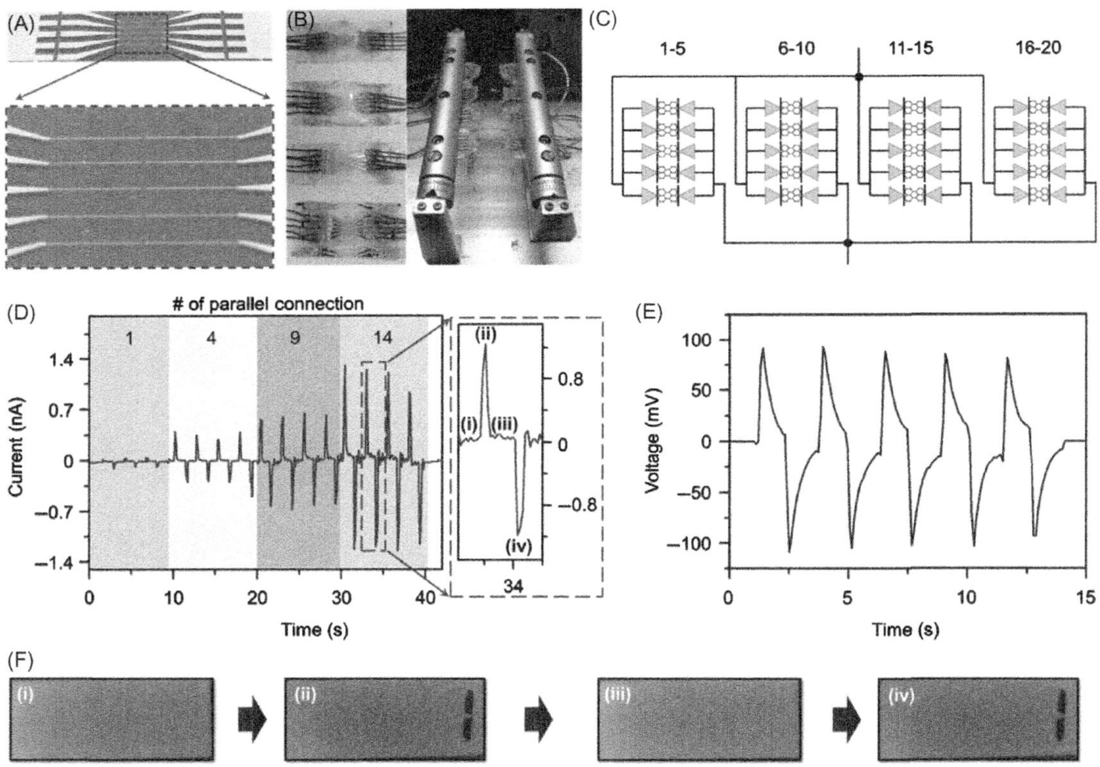

Figure 1.20 (A) Image of an array consisting of five tb-WSe2 PENG, (B) Cu wire connected 20 tb-WSe_2-based PENGs (left) and PENGs on the bending machine (right), (C) schematic of the parallel connection of the 20 PENGs to improve the output current, (D) measured output currents for the integrated tb—WSe_2 PENGs as a function of the number of parallel connections, (E) measured output voltages of the integrated tb-WSe2 PENGs with 1 GΩ of load resistance, and (F) four snapshots taken from a full cycle driving of an LCD by the 14 integrated monolayer WSe_2 PENGs. *Source*: Reproduced with permission from J.H. Lee, J.Y. Park, E.B. Cho, T.Y. Kim, S.A. Han, T.H. Kim, et al., Reliable piezoelectricity in bilayer WSe2 for piezoelectric nanogenerators, Adv. Mater. 29 (2017) 1606667 (7 pp).

The 14 working PENGs were successfully integrated among the 20 PENGs (Fig. 1.20B and C), and the measured output current linearly increased up to 1.4 nA as the number of parallel connections increased (Fig. 1.20D). The output voltages of the parallel-connected PENGs show an almost similar output voltage for the single PEHs (Fig. 1.20E). A very small, commercially available LCD was operated by the PENGs (Fig. 1.20F). The number "1" was selected as the output for the LCD screen, and the LCD was directly connected to the PENGs. Fig. 1.20F shows a series of snapshots taken from the LCD when the PENGs were bent and released with 0.95% of strain and 40 mm s^{-1} of strain rate, showing a blinking number "1" that corresponded to the piezoelectric signal. This is the first demonstration of a 2D WSe_2-based PENG for self-powered electronics, indicating the need for further research for sensors, actuators, and mechanical energy harvesting applications.

1.3.3 Lead (II) iodide (PbI_2)-based energy harvester

Song et al. have demonstrated a flexible 2D piezoelectric device based on lead (II) iodide (PbI_2) nanosheets, fabricated with a recrystallization method [34]. Fig. 1.21A shows the atomic structure and mechanism of piezoelectric charge generation under strain. The microscope image of the device is depicted in Fig. 1.21B. It is evident that the PbI_2 nanosheets have been used to construct flexible 2D piezoelectric devices by depositing gold (Au) electrodes on both sides of the hexagonal nanosheets.

As a result, the PbI_2 nanosheet will form Schottky contact with Au electrodes, directly reflecting by the I–V scanning curves in Fig. 1.21C. Mechanical compression and subsequent relaxation process change the dipole moments within PbI_2 nanosheets on PET substrate, to generate piezoelectric potentials (Fig. 1.21D). As a result output current is back and forth through the external circuit. The preliminary tests show peak open-circuit voltage (Fig. 1.21E) and short-circuit current (Fig. 1.21F) reaching 29.4 mV and 20.9 pA, respectively.

Importantly, the piezoelectricity is not affected by the numbers of layer as observed from the voltage and current output (Fig. 1.21G). The loading power and energy conversion efficiency of the device was 0.12 pW (Fig. 1.21H) and about 3.2%, respectively.

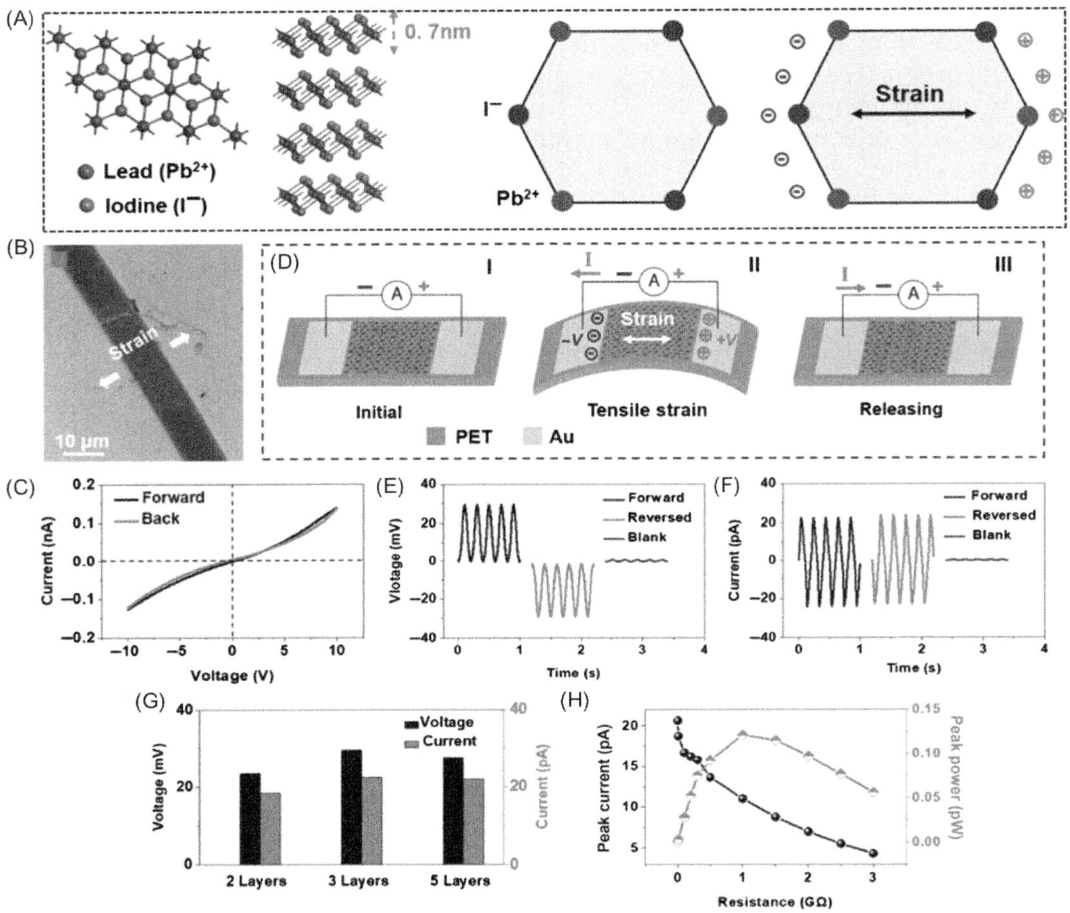

Figure 1.21 (A) Atomic structure of PbI_2 2D nanosheet and its behavior under strain, (B) microscopic image of the device under investigation, (C) I—V characteristic of the devices indicating Schottky contact between PbI_2 nanosheet and Au electrode, (D) piezoelectric charge generation mechanism under repeated tensile strain, (E) open-circuit voltage and (F) short-circuit current from the device, (G) dependence of the voltage and current output on number of layers of PbI_2 nanosheet, and (H) peak current and output power as a function of load resistance. *Source*: Reproduced with permission from H.B. Songa, I. Karakurt, M. Wei, N. Liu, Y. Chu, J. Zhong, et al., Lead iodide nanosheets for piezoelectric energy conversion and strain sensing, Nano Energy 49 (2018) 7—13.

Moreover, the electrical energy generated by the PENG under constant stimulating tensile strain of 0.339% and frequency of 5 Hz was accumulated within a capacitor (≃ 560 pF) via bridge rectifier (Fig. 1.22A–C) and can be enhanced by integration (Fig. 1.22D).

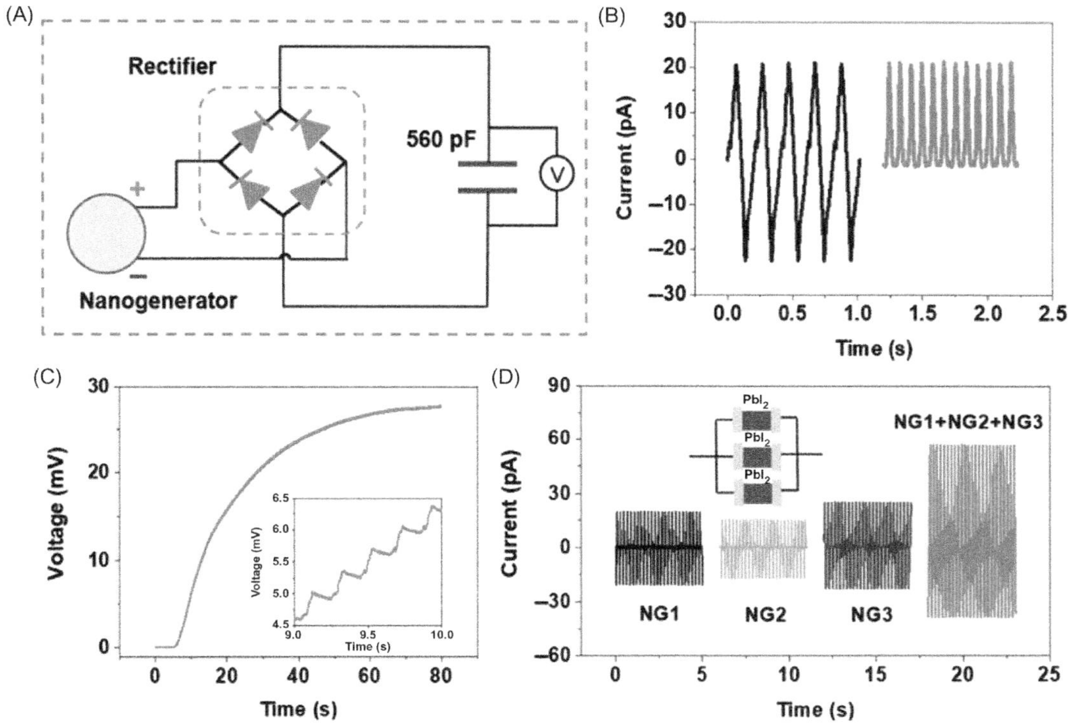

Figure 1.22 (A) The circuit diagram used for charging up the capacitor using the 2D PENG, (B) unrectified (black, *i.e.,* first one) and rectified (red, *i.e.,* second one) output current, (C) charging voltage response across the capacitor, and (D) linear superposition test by three PbI_2 nanosheets-based 2D PENGs. *Source*: Reproduced with permission from H.B. Songa, I. Karakurt, M. Wei, N. Liu, Y. Chu, J. Zhong, et al., Lead iodide nanosheets for piezoelectric energy conversion and strain sensing, Nano Energy 49 (2018) 7–13.

1.3.4 α-In_2Se_3-based energy harvester

Xue et al. experimentally reported the coexistence of out-of-plane and in-plane piezoelectricity in monolayer to bulk α-In_2Se_3, attributed to their noncentrosymmetry originating from the hexagonal stacking [35]. Specifically, the corresponding d_{33} piezoelectric coefficient of α-In_2Se_3 increases from 0.34 pm V^{-1} (monolayer) to 5.6 pm V^{-1} (bulk) without any odd–even effect. Furthermore, α-In_2Se_3-based flexible PENG as an energy harvesting cell and electronic skin was demonstrated. To fabricate the flexible α-In_2Se_3 PENG, multilayer samples with thicknesses of 100 − 200 nm are transferred to the PET substrates and silver paste was used as the two metal electrodes (Fig. 1.23A). Due to the fluidity of silver paste, two terminals of

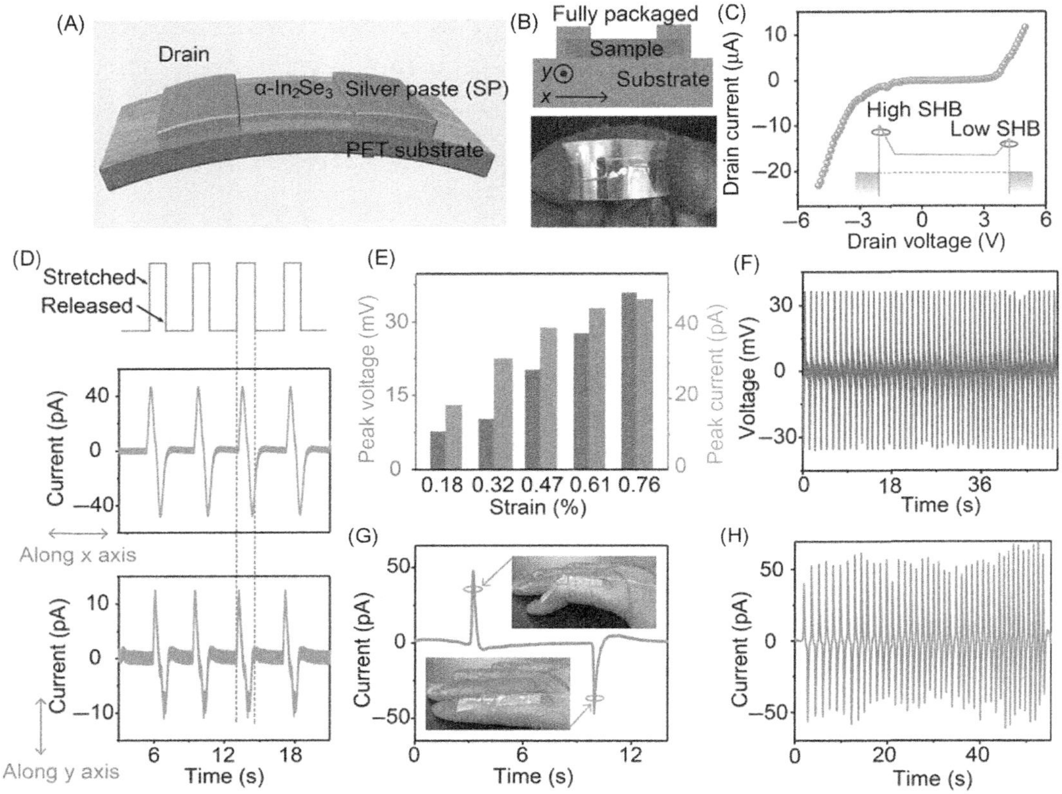

Figure 1.23 (A) Schematic of the α-In_2Se_3-based flexible PENG where α-In_2Se_3 semiconductors are fully encapsulated by silver paste, and (B) side view of the sketched devices in (A). Here, the *x* and *y* axes are defined as the directions parallel and normal with the long side of substrate, respectively. The bottom inset of (B) shows the optical image of the nanogenerator. (C) I–V curve showing metal–semiconductor Schottky contacts of nanogenerator. (D) Short-circuit current response of the nanogenerator under periodic strain (0.76%) in two different directions. (E) Strain dependence of the piezoelectric output. (F) Typical open-circuit voltage outputs upon a strain value of 0.76%. (G) Integration of the piezoelectric nanogenerator on an index finger to scavenge the mechanical energy induced by finger motions. (H) Current outputs when the index finger repeatedly curves and extends. *Source*: Reproduced with permission from F. Xue, J. Zhang, W. Hu, W.T. Hsu, A. Han, S.F. Leung, et al., Multidirection piezoelectricity in mono- and multilayered hexagonal α-In2Se3, ACS Nano 12 (2018) 4976 – 4983.

the multilayer α-In_2Se_3 were often fully encapsulated, as shown at the top of Fig. 1.23B, which was beneficial to exploit the vertical and lateral polarization charges for the piezoelectric outputs.

The *x*-axis was chosen as the length direction of the substrate and the *y*-axis as the width direction. Optical image of the flexible and transparent PENG is shown at the bottom panel of

Fig. 1.23B. I–V measurement in Fig. 1.23C shows that both metal – semiconductor junctions of the PENG are Schottky contacts. The piezoelectric current responses of the as-fabricated PENG are shown in Fig. 1.23D. The dependence of piezoelectric outputs (peak voltage and peak current) with the magnitude of applied strain is shown in Fig. 1.23E. A stable voltage signal under a fixed (0.76%) and short strain interval of 1 s is displayed in Fig. 1.23F, demonstrating excellent output reproducibility. Under maximum strain of 0.76%, the maximum voltage and current output of 35.7 mV and 47.3 pA, respectively, were observed. Furthermore, the PENG was demonstrated as e-skin by integrating the PENG onto human skin, such as the back of an index finger. To do that, the PET substrate was changed to highly flexible and transparent common "scotch" tape. After attaching the index finger to the as-fabricated PENG, the device can immediately give a peak current response (Fig. 1.23G) under repeated curving or extending motion as shown in Fig. 1.23H. This demonstrates that this type of device has great potential for harvesting biomechanical energy or acting as electronic skin.

1.3.5 Other 2D materials-based energy harvester

In recent years, 2D materials as filler are found to be attractive for piezoelectric energy harvesting. For example, 2D SnO_2 nanosheet [36], MoS_2 [37], rGO [38], and GO [39] found to be used as filler materials for improving the piezoelectric energy harvesting performance of the piezoelectric polymer materials such as PVDF and its copolymer. The high surface-to-volume ratio of the 2D materials efficiently increases the capacitance of the micro-capacitor inside the material. Moreover, interfacial interactions between 2D materials and the polymer chain effectively arrange the chain, resulting in high piezoelectric composite material.

The development of composites with aforementioned materials can be potential approach to obtain the high-quality flexible piezoelectric material. Kar et al. demonstrated that without any electrical poling 2D SnO_2 nanosheet/PVDF composite-based PENG can generate superior output voltages (42 V), adequate current density (6.25 $\mu A\ cm^{-2}$), and high power density (4900 $W\ m^{-3}$) which can directly light up 85 commercial LEDs and charging up the capacitors [36]. Maity et al. successfully incorporated 2D MoS_2 into PVDF fiber (Fig. 1.24) in order to fabricate highly sensitive PENG [37].

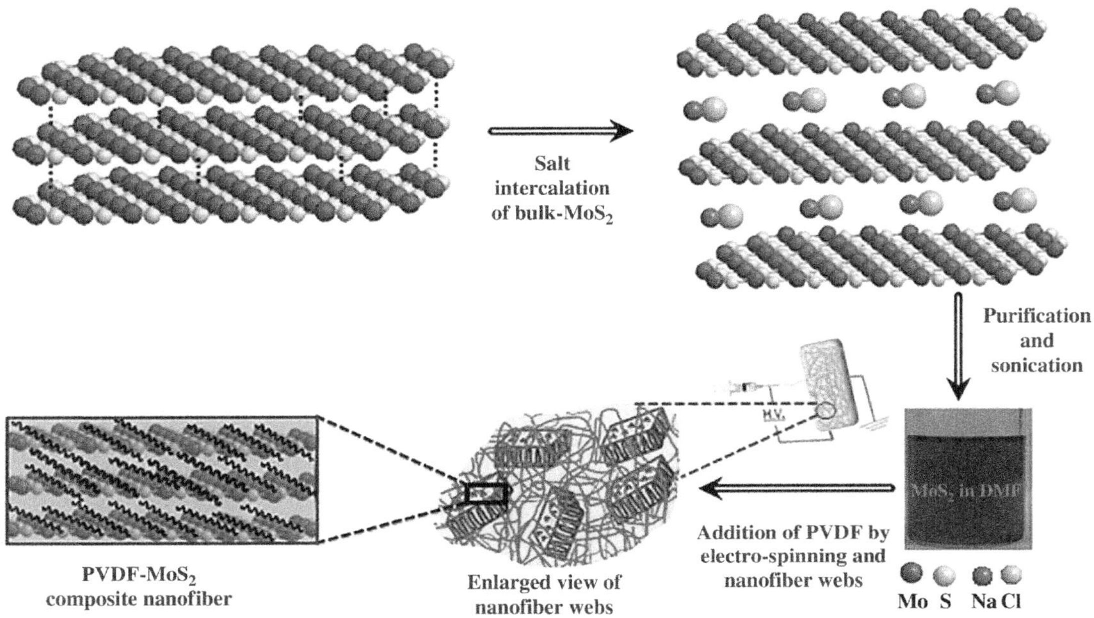

Figure 1.24 Schematic representation of MoS_2 nanosheets incorporation into PVDF to prepare the PVDF-MoS_2 composite nanofiber. *Source*: Reproduced with permission from K. Maity, B. Mahanty, T.K. Sinha, S. Garain, A. Biswas, S.K. Ghosh, et al., Two-dimensional piezoelectric MoS_2-modulated nanogenerator and nanosensor made of poly(vinlydine fluoride) nanofiber webs for self-powered electronics and robotics, Energy Technol. 5 (2017) 234–243.

The as-fabricated NG was so much sensitive that it can even transduce very lightweight leaf (0.68 g) (Fig. 1.25A) and matchstick (0.09 mg) (Fig. 1.25B) falling from different heights. This sensitivity was further used for speech signal recognition/acoustic energy harvesting. As a result, the PENG successfully detects several English letters (e.g., A, B, C, D, and E) and output response was observed (Fig. 1.25C). Furthermore, the fast Fourier transform (FFT)-processed frequency profile spectrum was further used to identify the frequency range of the acoustic signals (Fig. 1.25C and D). Additionally, the PENG was further used to detect and discriminate the acoustic signals from several musical instruments, such as solo piano, violin, and flute.

In another study, the role of rGO in improving the mechanical energy harvesting potentiality of P(VDF-TrFE) matrix without applying the conventional polling process was explored [38]. Five different piezoelectric sheets were prepared by varying the rGO contents in P(VDF-TrFE) matrix to realize its optimum concentration in the matrix. The effect of rGO was realized by enhanced electroactive polar beta (β) phase content of

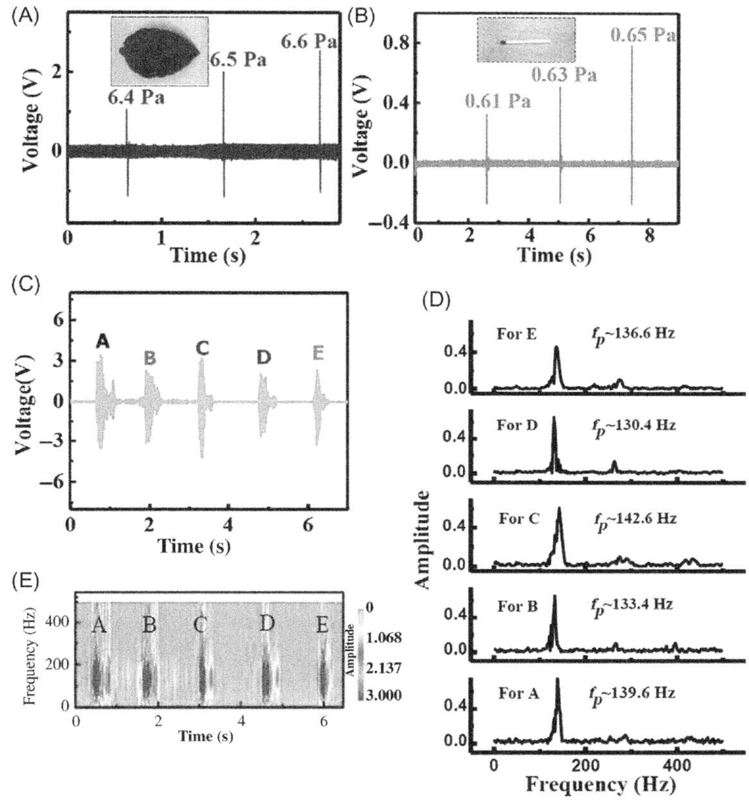

Figure 1.25 (A) Responses of a falling leaf (illustrated in the inset) from different heights (corresponds to different external pressure impacts). (B) Responses of a matchstick (illustrated in the inset) falling from different heights (which corresponds to different external pressure impacts). (C) Output responses from the PENG as different letters (A, B, C, D, and E) are pronounced. (D) The corresponding FFT-processed frequency spectrum and (E) FFT-processed time-dependent spectrogram (the left side color scale bar denotes the amplitude). *Source*: Reproduced with permission from K. Maity, B. Mahanty, T.K. Sinha, S. Garain, A. Biswas, S.K. Ghosh, et al., Two-dimensional piezoelectric MoS2-modulated nanogenerator and nanosensor made of poly(vinlydine fluoride) nanofiber webs for self-powered electronics and robotics, Energy Technol. 5 (2017) 234–243.

P(VDF-TrFE). Among the several fabricated PENGs, the PENG with 0.1% rGO content reveals the maximum open-circuit voltage of 2.4 V and highest peak of short-circuit current around 0.8 μA at an applied force of 2 N. Moreover, it exhibits the highest output power of 3.2 μW at 1.8 MΩ load resistance.

Roy et al. [39] found GO as a highly potential 2D material to improve the sensitivity of PVDF nanofibers-based PENG. The device can generate a maximum output power density of ~6.2 mW m^{-2} under mechanical pressure and power up an array of LEDs. In addition, a self-powered wearable sensor was

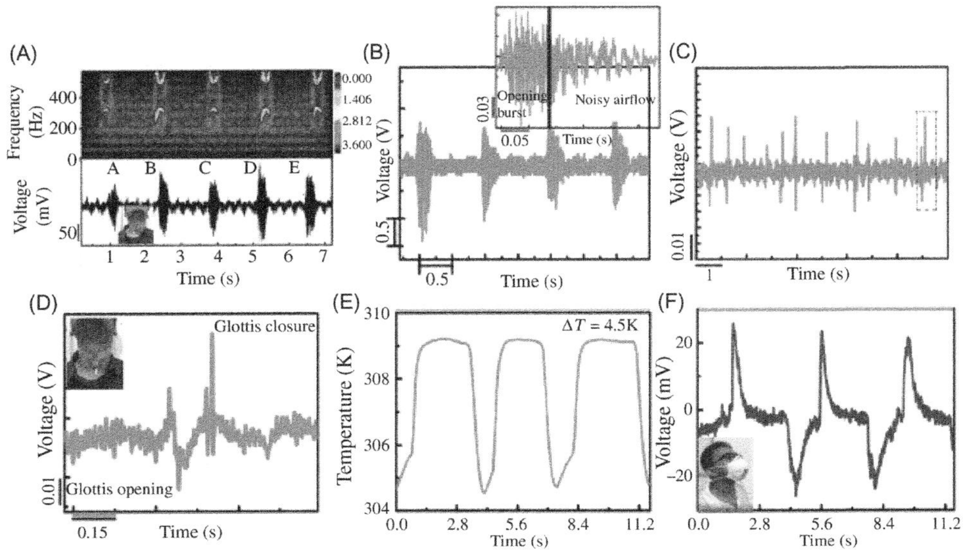

Figure 1.26 (A) Photograph of the GPPNG attached to the throat of a male participant during speaking of "A," "B," "C," "D," and "E." Recorded output voltage response curves during pronunciation of these letters with FFT signal in the upper part are also shown. (B) Relative changes in the output voltage signal as a function of time when monitoring a coughing action (enlarged view is depicted in the inset). (C) Responsive curves of a swallowing exercise. (D) Magnified view of one swallowing signal marked in part (C). Inset shows the application of GPPNG on the human throat. (E) The temperature fluctuation and (F) recorded open-circuit voltage of the GPPNG (photo of the GPPNG attached to a N95 breathing mask is shown in the inset) driven by human respiratory at room temperature. *Source*: Reproduced with permission from K. Roy, S.K. Ghosh, A. Sultana, S. Garain, M. Xie, C.R.Bowen, et al., A self-powered wearable pressure sensor and pyroelectric breathing sensor based on GO interfaced PVDF nanofibers, ACS Appl. Nano Mater. 2 (2019) 2013–2025.

fabricated that can detect and discriminate both pressure and temperature. For example, the pressure sensor accurately and rapidly detected pressures as low as 10 Pa with a high sensitivity (4.3 V kPa^{-1}). Furthermore, the sensor was further used as human physiological signal monitoring, such as, joint movements, coughing signals, swallowing and monitoring throat muscle movement in real-time during repeated speaking of different alphabetic characters (Fig. 1.26A–D), and pyroelectric breathing sensor (Fig. 1.26E and F) by attaching it to different portion of human body. It implies that the fabricated sensor has the potential utility from wearable portable electronics to biomedical e-skin sensors.

1.4 Conclusion and perspectives

There are plenty of materials that are theoretically predicted to be piezoelectric in their 2D forms owing to the 2D confinement and spontaneous breaking of inversion symmetry. The integration of piezoelectricity and other intriguing properties in 2D materials provide a new platform for the exploration of novel physics at the atomic scale as well as potential device applications. The 2D materials have transparency, flexibility, and a high surface-to-volume ratio. Owing to the very low thickness of the atomic unit, a stacking structure using 2D materials can be also made to form a very thin device, which is applicable for insertion into the body or wearable electronic devices. Therefore, it is believed that the 2D piezoelectricity and their application toward mechanical energy harvesting will be one of the important research branches in nanoscience and nanotechnology. Although the fundamental principle of piezoelectric effects has been well established, the understandings on the coupling behavior between piezoelectricity and semiconductor properties are still elusive. For example, diverse 2D materials have been predicted to possess intrinsic piezoelectricity theoretically, such as SnS and GaSe are not yet investigated experimentally due to the huge difficulties in materials preparations using exfoliation or CVD methods. In addition, so far, most of the previous numerical studies on piezoelectric effects are based on the finite element method. Now the observation of piezoelectricity in the fundamental thickness limit allows researchers to investigate these effects from the first-principles calculations, which may provide a more accurate interpretation of the underlying mechanism. For future fundamental research or industrial applications, the following efforts may be made for 2D piezoelectric materials, such as

1. The piezoelectric charges can affect the related quantum yield, therefore, it is important to investigate the influence of strain on electronic structure of piezoelectric 2D materials.
2. Another important fact for the improvement of piezoelectric coefficient is the controlled synthesis of 2D piezoelectric materials with ideal and uniform carrier concentrations.
3. Seeking more stable 2D piezoelectric materials that can be chemically stable in the ambient environment.

2D ferroelectrics enjoy their inherent advantages, such as bandgap tenability and flexibility, promising its applications as electronic sensors and data memories in comparison to the traditional ferroelectric ceramics. Furthermore, the research of 2D ferroelectric is still in its infancy stage. To promote the research, further works on fabrication of high-quality 2D ferroelectrics in a large scale with low costs must be sought.

However, varieties of proof-of-concept devices have been demonstrated so far by using piezoelectricity in the 2D materials for potential applications in mechanical energy harvesting, ultrathin actuators, adaptive electronics, e-skin, and optoelectronics. There is no doubt that outstanding opportunities for innovation exist in this emerging field but also with the great challenges on the road ahead.

References

[1] F.R. Fan, W. Tang, Z.L. Wang, Flexible nanogenerators for energy harvesting and self-powered electronics, Adv. Mater. 28 (2016) 4283–4305.
[2] Z.L. Wang, On Maxwell's displacement current for energy and sensors: the origin of nanogenerators, Mater. Today 20 (2017) 74–82.
[3] Z.L. Wang, J. Song, Piezoelectric nanogenerators based on zinc oxide nanowire arrays, Science 312 (2006) 242–246.
[4] F.R. Fan, Z.Q. Tian, Z.L. Wang, Flexible triboelectric generator, Nano Energy 1 (2012) 328–334.
[5] Z.L. Wang, Triboelectric nanogenerators as new energy technology for self-powered systems and as active mechanical and chemical sensors, ACS Nano 7 (2013) 9533–9557.
[6] J.A. Paradiso, T. Starner, Energy scavenging for mobile and wireless electronics, IEEE Pervasive Comput. 4 (2005) 18–27.
[7] G.K. Ottman, H.F. Hofmann, A.C. Bhatt, G.A. Lesieutre, Adaptive piezoelectric energy harvesting circuit for wireless remote power supply, IEEE Trans. Power Electr. 17 (2002) 669–676.
[8] G. Jonathan, J. Feenstra, H.A. Sodano, K. Farinholt, Energy harvesting from a backpack instrumented with piezoelectric shoulder straps, Smart Mater. Struct. 16 (2007) 1810–1820.
[9] C. Cui, F. Xue, W.J. Hu, L.J. Li, Two-dimensional materials with piezoelectric and ferroelectric functionalities, npj 2D Mater. Appl. 2 (2018) 18 (14 pp.).
[10] F. Yi, H. Ren, J. Shan, X. Sun, D. Wei, Z. Liu, Wearable energy sources based on 2D materials, Chem. Soc. Rev. 47 (2018) 3152–3188.
[11] M. Chhowalla, H.S. Shin, G. Eda, L.J. Li, K.P. Loh, H. Zhang, The chemistry of two-dimensional layered transition metal dichalcogenide nanosheets, Nat. Chem. 5 (2013) 263–275.
[12] K.N. Duerloo, M.T. Ong, E.J. Reed, Intrinsic piezoelectricity in two-dimensional materials, J. Phys. Chem. Lett. 3 (2012) 2871–2876.
[13] H. Zhu, Y. Wang, J. Xiao, M. Liu, S. Xiong, Z.J. Wong, et al., Observation of piezoelectricity in free-standing monolayer MoS_2, Nat. Nanotechnol. 10 (2015) 151–155.
[14] W. Wu, L. Wang, Y. Li, F. Zhang, L. Lin, S. Niu, et al., Piezoelectricity of single-atomic-layer MoS_2 for energy conversion and piezotronics, Nature 514 (2014) 470–474.
[15] A. Gruverman, O. Auciello, H. Tokumoto, Nanoscale investigation of fatigue effects in $Pb(Zr,Ti)O_3$ films, Appl. Phys. Lett. 69 (1996) 3191–3193.
[16] J.A. Christman, R.R. Woolcott, A.I. Kingon, R.J. Nemanich, Piezoelectric measurements with atomic force microscopy, Appl. Phys. Lett. 73 (1998) 3851–3853.
[17] Y. Huang, L. Liu, J. Sha, Y. Chen, Size-dependent piezoelectricity of molybdenum disulfide (MoS_2) films obtained by atomic layer deposition (ALD), Appl. Phys. Lett. 111 (2017) 063902–063904.

[18] K.H. Michel, B. Verberck, Theory of elastic and piezoelectric effects in two dimensional hexagonal boron nitride, Phys. Rev. B Condens. Matter 80 (2009) 308–310.
[19] Y. Li, Y. Rao, K.F. Mak, Y. You, S. Wang, C.R. Dean, et al., Probing symmetry properties of few-layer MoS_2 and h-BN by optical second-harmonic generation, Nano Lett. 13 (2013) 3329–3333.
[20] L.S. Zhao, C.P. Chen, L.L. Liu, H.X. Yu, Y. Chen, X.C. Wang, Magnetism and piezoelectricity of hexagonal boron nitride with triangular vacancy, Chin. Phys. B 27 (2018). 016301–6.
[21] M. Zelisko, Y. Hanlumyuang, S. Yang, Y. Liu, C. Lei, J. Li, et al., Anomalous piezoelectricity in two-dimensional graphene nitride nanosheets, Nat. Commun. 5 (2014) 4284–4287.
[22] Y. Zheng, J. Liu, J. Liang, M. Jaroniec, S. Qiao, Graphitic carbon nitride materials: controllable synthesis and applications in fuel cells and photocatalysis, Energy Environ. Sci. 5 (2012) 6717–6731.
[23] J. Lowther, Relative stability of some possible phases of graphitic carbon nitride, Phys. Rev. B 59 (1999) 11683–11686.
[24] W. Li, J. Li, Piezoelectricity in two-dimensional group-III monochalcogenides, Nano Res. 8 (2015) 3796 (7 pp.).
[25] R. Fei, W. Li, J. Li, L. Yang, Giant piezoelectricity of monolayer group IV monochalcogenides: SnSe, SnS, GeSe, and GeS, Appl. Phys. Lett. 107 (2015) 173104–173105.
[26] M.N. Blonsky, H.L. Zhuang, A.K. Singh, R.G. Hennig, Ab initio prediction of piezoelectricity in two-dimensional materials, ACS Nano 9 (2015) 9885–9891.
[27] Y. Zhou, D. Wu, Y. Zhu, Y. Cho, Q. He, X. Yang, et al., Out-of-plane piezoelectricity and ferroelectricity in layered α-In_2Se_3 nanoflakes, Nano Lett. 17 (2017) 5508–5513.
[28] G.C. Rodrigues, P. Zelenovskiy, K. Romanyuk, S. Luchkin, Y. Kopelevich, A. Kholkin, Strong piezoelectricity in single-layer graphene deposited on SiO_2 grating substrates, Nat. Commun. 7 (2015) 7572 (6 pp.).
[29] H.J. Jin, W.Y. Yoon, W. Jo, Virtual out-of-plane piezoelectric response in MoS_2 layers controlled by ferroelectric polarization, ACS Appl. Mater. Inter. 10 (2017) 1334–1339.
[30] A.Y. Lu, H. Zhu, J. Xiao, C.P. Chuu, Y. Han, M.H. Chiu, et al., Janus monolayers of transition metal dichalcogenides, Nat. Nanotechnol. 12 (2017) 744–749.
[31] Y. Zhou, W. Liu, X. Huang, A. Zhang, Y. Zhang, Z.L. Wang, Theoretical study on two-dimensional MoS_2 piezoelectric nanogenerators, Nano Res. 9 (2016) 800–807.
[32] S.K. Kim, R. Bhatia, T.H. Kim, D. Seol, J.H. Kim, H. Kim, et al., Directional dependent piezoelectric effect in CVD grown monolayer MoS_2 for flexible piezoelectric nanogenerators, Nano Energy 22 (2016) 483–489.
[33] J.H. Lee, J.Y. Park, E.B. Cho, T.Y. Kim, S.A. Han, T.H. Kim, et al., Reliable piezoelectricity in bilayer WSe_2 for piezoelectric nanogenerators, Adv. Mater. 29 (2017) 1606667 (7 pp.).
[34] H.B. Songa, I. Karakurt, M. Wei, N. Liu, Y. Chu, J. Zhong, et al., Lead iodide nanosheets for piezoelectric energy conversion and strain sensing, Nano Energy 49 (2018) 7–13.
[35] F. Xue, J. Zhang, W. Hu, W.T. Hsu, A. Han, S.F. Leung, et al., Multidirection piezoelectricity in mono- and multilayered hexagonal α-In_2Se_3, ACS Nano 12 (2018) 4976–4983.

[36] E. Kar, N. Boseb, B. Dutta, S. Banerjee, N. Mukherjee, S. Mukherjee, 2D SnO_2 nanosheet/PVDF composite based flexible, self-cleaning piezoelectric energy harvester, Energy Convers. Manag. 184 (2019) 600–608.
[37] K. Maity, B. Mahanty, T.K. Sinha, S. Garain, A. Biswas, S.K. Ghosh, et al., Two-dimensional piezoelectric MoS_2-modulated nanogenerator and nanosensor made of poly(vinlydine fluoride) nanofiber webs for self-powered electronics and robotics, Energy Technol. 5 (2017) 234–243.
[38] R.M. Habibur, U. Yaqoob, S. Muhammad, A.S.M.I. Uddin, H.C. Kim, The effect of rGO on dielectric and energy harvesting properties of P(VDF-TrFE) matrix by optimizing electroactive β phase without traditional polling process, Mater. Chem. Phys. 215 (2018) 46–55.
[39] K. Roy, S.K. Ghosh, A. Sultana, S. Garain, M. Xie, C.R. Bowen, et al., A self-powered wearable pressure sensor and pyroelectric breathing sensor based on GO interfaced PVDF nanofibers, ACS Appl. Nano Mater. 2 (2019) 2013–2025.

2

Two-dimensional metal oxide nanomaterials for sustainable energy applications

Leticia P.R. Moraes[1,2,*], Jun Mei[1,*], Fabio C. Fonseca[2] and Ziqi Sun[1]

[1]School of Chemistry Physics and Mechanical Engineering, Queensland University of Technology (QUT), Brisbane, Australia [2]Nuclear and Energy Research Institute (IPEN), São Paulo, Brazil

2.1 Introduction

The production and consumption of energy are determinant for improving our life quality. The intense economic development of the last decades has led to a growing energy demand, which is mostly being supplied through nonrenewable sources such as oil, coal, and natural gas [1,2]. The enormous dependence on these traditional energy sources has caused some major concerns related with the uneven resource distribution and discontinuous supply of fossil fuels, and even the resultant environmental problems such as global warming and atmospheric pollution [3,4]. In this regard, innovative technologies capable of converting and conserving energy on a large scale are needed to promote the sustainable development of our society and economy [5–7]. The generation of environmentally clean and renewable energies have been developed and improved in the last years, leading to an extensive and interdisciplinary research on materials science. Functional nanomaterials with the potential to be employed in energy conversion and storage devices, such as fuel cells, solar cells, supercapacitors, and rechargeable batteries, are widely studied [8–19]. Electrical power generation and catalysis are the areas that are taking more advantage of the potential offered by the use of nanomaterials not only with the well-established nanoparticles-based

*These authors contributed to this chapter equally.

2D Nanomaterials for Energy Applications. DOI: https://doi.org/10.1016/B978-0-12-816723-6.00002-2

catalyst but also with new solar films, green coatings, high-efficiency fuel, and sensing [20].

In terms of nanomaterials, it is well known that the size and shape of the particle play an import influence on the materials' properties. The same chemical element or compound can exhibit different properties in different shapes and dimensions. Generally, nanomaterials are usually defined as materials with length of 1–100 nm in at least one dimension and can be categorized depending on their compositions (inorganic and organic), dimensions (0D, 1D, 2D, and 3D), and origins (natural or synthetic) [21]. It has proven that decreasing the size of the particles to nanoscale can affect the structural characteristics, in the lattice symmetry and cell parameters [22]. As the particle size decreases, the lattice constants changes, and these changes can modify many physical, chemical, and/or mechanical properties. The size effect may be large enough to cause a structural transition by modifying the transition energies [23], and consequently, some materials that are not stable in the bulk and do not exist in ambient atmosphere, can be stable at the nanoscale levels [24,25]. In an oxide nanoparticle, the relative number of atoms at the surfaces and interfaces increases with decreasing size, strongly influencing chemical reactivity in comparison with the bulk material. Solid–gas or solid–liquid chemical reactions can be mostly confined to the surface and/or subsurface regions of the solid [26,27]. Another important effect of size is related to the electronic properties of the oxides. The electronic configurations of nanomaterials are significantly different from that of their bulk counterpart. The density of the energy levels changes with the decreasing of particles size, which lead to important modifications in the optical and electrical properties of the material [28]. Therefore, considering that the nanoscale structure can effectively improve electrochemical reactions efficiency, nanomaterials became the main focus in research, aiming to tailoring the surface morphology of materials, adapting and combining functional properties to produce highly potent multifunctional active materials for the next generation of energy devices with improved energy and power densities to increase their performances.

Among several nanomaterials used for energy applications, metal oxides has been focused due to their unique properties such as low molecular weight, favorable electrochemical and solid-state properties, and low toxicity [19,26,29–31], making them a class of inorganic compounds very important in many areas of materials science with a wide range of applications, such as sensor, catalysis, ceramics, absorbents, and

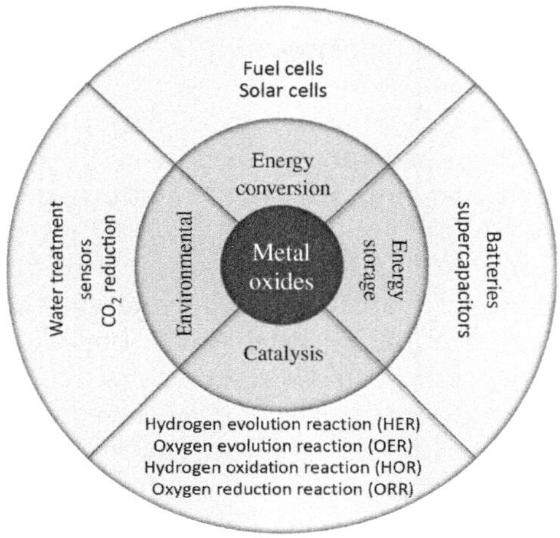

Figure 2.1 Application fields of metal oxides.

superconductors (Fig. 2.1) [15,32–36]. Metals elements are able to form a wide diversity of metal oxides compounds, and they can adopt different structural geometries that can exhibit metallic, semiconductor, or insulator character [37]. The metal oxides are attracting special attention due to their easy mode of formation and multifunctional behavior and therefore have been studied for the development of new energy conversion and storage systems, especially because of the possibility of a large-scale fabrication of a higher electrode surface area, leading to higher charge/discharge rates, and their rich redox reactions involving different ions [27] (Fig. 2.1).

Many strategies have been developed for designing nanomaterials for energy applications, with different dimensions, morphologies, and properties [15–17,32,36,38–42]. After the successful exfoliation of graphene from graphite in 2004 [43], 2D nanomaterials have attracted tremendous attention of researchers, due to its exceptional properties, such as high thermal conductivity, specific surface area, carrier mobility, strength, chemical activity, and tunable electronic properties [36,44]. 2D nanomaterials present sheet-like structures with the lateral size larger than 100 nm and atomic or molecular thickness (typically less than 5 nm) [45] and combine excellent mechanical properties, light transmittance, and electronic properties that make 2D nanomaterials extremely attractive in the fabrication of next-generation electronic/optoelectronic devices [46]. Compared to other dimensional nanomaterials, such as nanowires or nanoparticles, the two most important features of 2D nanomaterials for electrocatalysis are their uniformly exposed lattice plane and unique electronic state [47]. It is therefore that a magic combination of metal oxides and two-dimensionality generates incredible synergetic effects beyond the material itself and thus boosts the performances in energy applications.

In this chapter, we intend to discuss the state-of-the-art progress of 2D metal oxides (MOs) in energy-related applications. First, we summarize the most important reactions involved in energy storage and conversion systems, with a brief introduction about some energy devices. Second, we discuss about 2D MOs and their applications on the mentioned systems.

2.2 Sustainable energy applications

Energy storage plays an important role in the renewable energy scenario, since the most types of renewable energy sources, like wind and solar, are intermittently and unpredictable, making them not reliable for constant power supply. To solve this issue, these new technologies need to store the energy surplus when the sources are available to deliver when the energy production is low. The improvement of these methods is crucial to the use of renewables energies in large-scale, with high performance, efficiency, and reliability. Different types of energy storage systems are being studied in the last years, and usually these systems are classified based on the form of storage, which can be chemical, electrochemical, mechanical, electrical, or thermal modes [48,49]. Batteries and supercapacitors are the most common devices for portable applications, such as electric vehicles and electronic devices.

An alternative energy source to fossil fuels is to use hydrogen as a fuel. Hydrogen (H_2) has been considered as the most promising fuel for the future due to its abundance and chemical energy density (142 $MJ.kg^{-1}$), which is the highest of all chemical fuels [5,50,51]. H_2 is a clean and reproducible energy source, being the only carbon-free chemical energy carrier as it only emits water when burnt with oxygen. However, to be considered truly as a type of clean fuels, hydrogen must be produced from water-splitting process, using renewable energy, through photocatalytic and electrocatalytic reactions, instead of the usual production through steam-reforming reactions [50–53].

In order to improve the next-generation energy devices, it is necessary to get a deeper understanding of the fundamental principles of electrocatalysis, which is responsible for accelerate electrochemical reactions on the electrode surface [54]. The production of H_2 by electrochemical water splitting has been extensively studied since its first report in 1789 [55], when Troostwijk and Deiman using two gold wires immersed in water and connected to an electrostatic generator observed gas production (hydrogen and oxygen). The electrochemical water splitting is typically processed in an electrolyzer, which is composed by three component parts: an electrolyte (ionic conductor), a cathode, and an anode, through the application of an external voltage. However, the electrolysis of water is thermodynamically disfavored, requiring an input of energy to accelerate the reaction. This extra energy (overpotential) is due to the high activation energy required to split water and the low conductivity of water and can be minimized by using efficient electrocatalysts and improving the conductivity of water (adding salts, acids, or bases). The electrochemical decomposition of water in an electrolyzer is composed of two half-reactions: hydrogen evolution reaction (HER) on the cathode and oxygen evolution reaction (OER) on the anode. Inversely, in a hydrogen-air fuel cell, the hydrogen oxidation reaction (HOR) occurs at the anode and the oxygen reduction reaction (ORR) at the cathode (see Table 2.1). The hydrogen obtained through water electrolyzers has a high purity (99.999 vol.%), thus it can be used directly in fuel cells (Fig. 2.2), being an ideal technological loop for the water cycle, where the hydrogen is produced by water splitting through the HER and OER for fuel generation, and then used by power generation through the ORR and HOR in fuel cells [47,56].

The most frequently used commercial water electrolyzers are based on the alkaline media with a diaphragm or under acidic conditions with a proton exchange membrane. The water-splitting reaction can be expressed chemically in different ways, depending on whether the reactions take place in acidic or neutral/basic electrolytes. In an acid electrolyzer, such as a proton exchange membrane electrolysis cell (PEMEC), water molecules are oxidized in the anodic compartment, to produce oxygen molecules and release protons and electrons. Protons travel through the electrolyte to the cathode where they are reduced by the electrons coming from the external circuit to form hydrogen. In a proton exchange membrane fuel cell (PEMFC) reverse reaction of water electrolysis takes place to generate electricity. Hydrogen molecules are oxidized at the anode (HOR),

Table 2.1 Typical characteristics of the main electrolyzers and fuel cell technologies.

	Electrolyzers			Fuel cells		
	Alkaline	**Acid**	**SOE**	**Alkaline**	**Acid**	**SOFC**
Charge carrier	OH^-	H^+	O^{2-}	OH^-	H^+	O^{2-}
Temperature	40–90	20–200	500–1000	50–200	50–90	500–1000
Electrolyte	KOH	Nafion	YSZ or GDC	KOH	Nafion	YSZ or GDC
Anode material	Ni oxides Ni-/Co-based spinels	Pt	LSM-YSZ	Carbon/Platinum	Pt	Ni-YSZ
Cathode material	Ni alloys Stainless steel	Pt, Ru, Ir	Ni-YSZ	Carbon/Platinum	Pt, Ru, Ir	LSM-YSZ
Anode reaction	$4OH^- \rightarrow 2H_2O + O_2 + 4e^-$	$2H_2O \rightarrow O_2 + 4$ $H^+ + 4e^-$	$2O^{2-} \rightarrow O_2 + 4e^-$	$2H_2 + 4OH^- \rightarrow 4$ $H_2O + 4e^-$	$2H_2 \rightarrow 4\,H^+ + 4e^-$	$2H_2 + 2O^{2-} \rightarrow 2H_2O + 4e^-$
Reaction	OER	OER	OER	HOR	HOR	HOR
Cathode reaction	4 $H_2O + 4e^- \rightarrow 2H_2 + 4OH^-$	$4H^+ + 4e^- \rightarrow 2H_2$	$2H_2O + 4e + ^- \rightarrow 2H_2 + 2O^{2-}$	$O_2 + 4e^- + 2H_2O \rightarrow 4OH^-$	$O_2 + 4H^+ + 4e^- \rightarrow 2H_2O$	$O_2 + 4e^- \rightarrow 2O^{2-}$
Reaction	HER	HER	HER	ORR	ORR	ORR
Overall reaction	$2H_2O \leftrightarrow 2H_2 + O_2$	$2H_2O \leftrightarrow 2H_2 + O_2$	$2H_2O \leftrightarrow 2H_2 + O_2$	$2H_2 + O_2 \rightarrow 2H_2O$	$2H_2 + O_2 \leftrightarrow 2H_2O$	$2H_2 + O_2 \leftrightarrow 2H_2O$
Efficiency	59%–80%	65%–82%	Up to 100%	50–55%	40–50%	60–80%

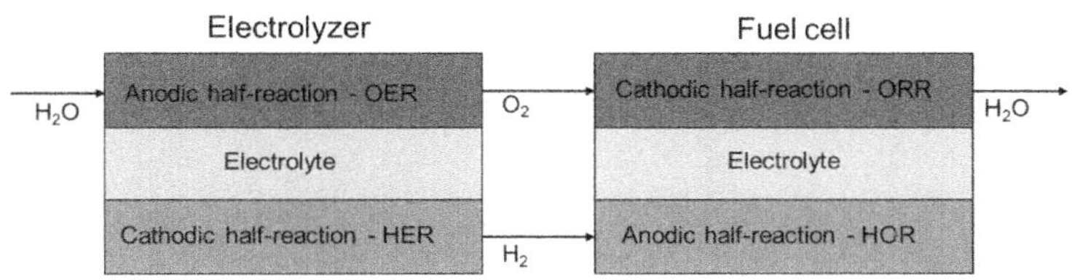

Figure 2.2 Water splitting and fuel cell scheme.

producing protons and electrons that will be used to produce water at the cathode (ORR) [57–61]. In an alkaline electrolyzer water is reduced in the cathode compartment producing protons and hydroxide anions, which recombine in the anode to release oxygen and water.

The water splitting is not thermodynamically favored at standard temperature and pressure conditions, thus an input of energy is needed. This energy input required energy for the water electrolysis reaction comes in the form of a potential difference between the electrodes and can be calculated using the Gibbs free energy ($\Delta G^0 = 237.2$ kJ.mol^{-1}). For an ideal catalyst, the application of 1.23 V is required; however, an additional voltage is necessary to overcome the operational issues, such as resistance of the electrolyte and electrodes and charge and mass transfer [62]. This extra potential is known as overpotential (η) and referred to a value that has to be applied to achieve a specified current density (normally 10 mA.cm^{-2}). It can be calculated as the difference between the applied potential (E) and potential under equilibrium conditions (E_{eq}) as expressed by the equation $\eta = E - E_{eq}$ [63]. Minimizing these overpotentials through the development of efficient electrocatalysts for each half-reaction is a key step in developing highly efficient water splitting devices.

The HER mechanism usually proceeds via the Volmer–Tafel–Heyrovsky steps, where the different pH value leads to different reactants and products in each step [64]. Although the mechanism is similar for both acidic and alkaline media, the catalytic activity for HER in alkaline media is usually much lower than that of the same catalyst in acid media, resulted from the more complicated forming process of adsorbed hydrogen atom in alkaline media [65]. While the reaction is determined by the hydrogen recombination in acidic media, in alkaline solution is determined by a balance between the water dissociation and the interaction of these products with a surface [66].

Another method to produce H_2 from water splitting is using solar energy. Since the first report of HER through water splitting with TiO_2 in 1972 [67], the conversion of solar energy to chemical energy or solar fuels via photocatalytic processes has been considered as one of the most potential solutions to solve global energy crisis and many studies involving TiO_2 have been made [68–78]. To develop efficient photoelectrochemical cells semiconductor materials with long-term stability and able to support rapid charge transfer at a semiconductor/aqueous interface are needed [79]. Usually, a photoelectrolysis cell is composed by an n-type semiconductor photoanode associated with a p-type metal oxide cathode and aqueous electrolyte. The process of water splitting can involve two half-reactions: water oxidation to dioxygen and proton reduction to hydrogen [79]. A general photocatalytic process consists of light absorption, charge separation and migration, and reduction reaction on the catalyst surface [80]. In a semiconductor system, when the energy of the photons is equal to or higher than the bandgap energy, electrons are excited from the valence band (VB) to the conduction band (CB).

A wide range of semiconductors such as titanium dioxide (TiO_2) [81–84], tungsten trioxide (WO_3) [85–87], and zinc oxide (ZnO) [88–91] have been studied for photoelectrochemical water oxidation due to their natural abundance, easy synthesis, superior chemical stability, and relatively low cost. However, there are still some challenges concerning these materials such as the absorption of the ultraviolet light due to their wide bandgap [92], improper band position and only exhibit either water reduction or oxidation activity [93], charge recombination, and lower exposed active site in the bulk photocatalyst [94]. Therefore, the control of the photocatalyst features such as their size and morphology (shape), crystallinity, defects, and dopants is extremely important to overcome these issues. 2D MOs are attracting much attention as new photoactive materials for presenting unique physical–chemical properties, optical absorption, carrier transport, reactive site, among others [95]. The bandgap and the light absorption can be adjusted by controlling the number of layers, the ultrathin nature of these materials reduces the recombination of electrons and holes, and the specific surface can be improved as they present more exposed active sites comparing to their 3D counterparts [96].

In acidic solution, it is known that platinum group metals and their alloys are the most efficient electrocatalysts for HER, while Ru and Ir are best for OER [97]. However to become an economically viable industry, cost reduction is essential, and precious-metal catalysts must be avoided due to their high cost, insufficient activity and durability, low abundance, and toxicity [98,99]. In alkaline media on the contrary, the usage of noble metal catalysts can be avoided because of their lower equilibrium potentials on each electrode at high pH, allowing the use of a large variety of metals as electrode material due to their high corrosion stability in this media [100]. However, as discussed above, the alkaline electrolyzer presents generally lower voltage efficiencies at high current densities compared to acid electrolyzer, due to its relatively low ion conductivity and because it is susceptible to the formation of carbonate as a contaminant by the reaction with carbon dioxide [101]. Traditionally, alkaline water electrolyzers use stainless steel and Ni alloys as catalysts, which are cheap and have a high resistance to corrosion but are not active enough [65]. In the last years, nickel, cobalt and iron-based electrodes have been extensively studied for alkaline electrolysis, since they proved to be relatively stable and active in this media [100].

The common issue among these electrochemical reactions is the requirement of a high-performing electrocatalyst, with high activity, high surface area, good electrochemical conductivity, and stability thus making the overall water splitting more practically viable [102–106]. In recent years, there has been a significant effort in decreasing the Pt loadings and increasing the efficiency of the Pt catalyst, as well as in developing new materials, as alternative for the precious-metal-based electrocatalysts. For a catalyst with a core-shell geometry, the electronic and structural properties can be tailored by changing the thickness, core chemical composition, diameter, and shape, improving the activity and durability of the catalyst [107].

Accordingly, reducing the size of the Pt nanoparticle allows reducing the amount of Pt. For instance, Paunovic et al. [108] decreased the size of Pt nanoparticles in almost three times, increasing the surface area and active sites, thus improving the efficiency, despite the fact that the amount of used platinum was decreased up to five times. Luo et al. [109] reported an electrocatalysts with trace loads of Pt (1.26 wt%) as decoration to MoS_2 nanosheets, with a comparable HER activity to the commercial Pt/C catalysts (20 wt% Pt on Vulcan carbon black). The combination of Pt nanoparticles and two-dimensional MoS_2 nanosheet thus boosted the HER activity and reduced the amount of Pt. This is one typical example of how 2D nanomaterials boost the performance of materials in energy applications.

2.3 2D metal oxides for sustainable energy applications

In the past decades, many types of 2D materials have been synthesized for energy applications. The unique structural advantages of 2D materials give direct benefit to push this class of materials to the edge of research for both materials and energy-related applications, including sustainable fuel generation and energy harvesting, conversion, and storage. To date, many 2D nanomaterials have been discovered, including typical IV group monolayers, transition metal oxides, transition metal dichalcogenides, layered double hydroxides, and so on, which have presented superior properties to the materials in other dimensionalities [36,38]. For example, as typical dichalcogenides, WS_2 and MoS_2 have presented superior activity for HER [110–113], in which the HER activity originates from the sulfur edges of MoS_2 plates and makes the nanosized MoS_2 more active for HER than bulk materials [110,114].

Similar to well-studied transition metal dichalcogenides, transition metal oxides have been widely studied as electrode materials due to their multiple oxidation states that can enhance electrochemical redox reactions [115]. To date, a wide range of 2D MO materials, including binary metal oxides (e.g., TiO_2, Co_3O_4, NiO, MnO_2, V_2O_5, Cr_2O_3, etc.) and ternary metal oxides (e.g., $Li_4Ti_5O_{12}$, $NiCo_2O_4$, $ZnCo_2O_4$, etc.), have been successfully fabricated via either direct exfoliation from their corresponding bulk counterparts or self-assembly from salt precursor-containing solutions (e.g., metal hydroxide, sulfide, nitrate, and carbonate) [47]. Metal oxides present high energy density, excellent electrochemical performance, and thermal and chemical stability, being considered as a promising option as electrode materials [116–120]. For instance, ruthenium oxide, RuO_2, has been extensively studied due to its fast, reversible electron transfer, and electro-adsorption of protons on its surface [121]. Nickel oxide (NiO) is one of the most promising MOs for OER electrocatalysts in alkaline media due to its environmental friendliness, low cost, easy preparation, high theoretical specific capacitance (3750 $F.g^{-1}$) and excellent chemical and thermal stability [122–124]. Iron–nickel oxide porous nanosheets were synthesized via a controlled transformation from LDH precursors, which exhibit advanced OER performance with a low overpotential of 213 mV at 10 mA cm^{-2}, small Tafel slope (32 mV dec^{-1}), and long-term stability, due to their high specific surface area, abundant active sites, small charge transfer resistance, and suitable adsorption energy for intermediates [125].

With the synthesis of a novel family of 2D MOs with ultrathin thickness, this family of materials has been successfully applied in high-performance photocatalytic, photovoltaic, and electrochemical devices, by taking advantage of their unique characteristics of large surface-to-volume ratio, distinctive

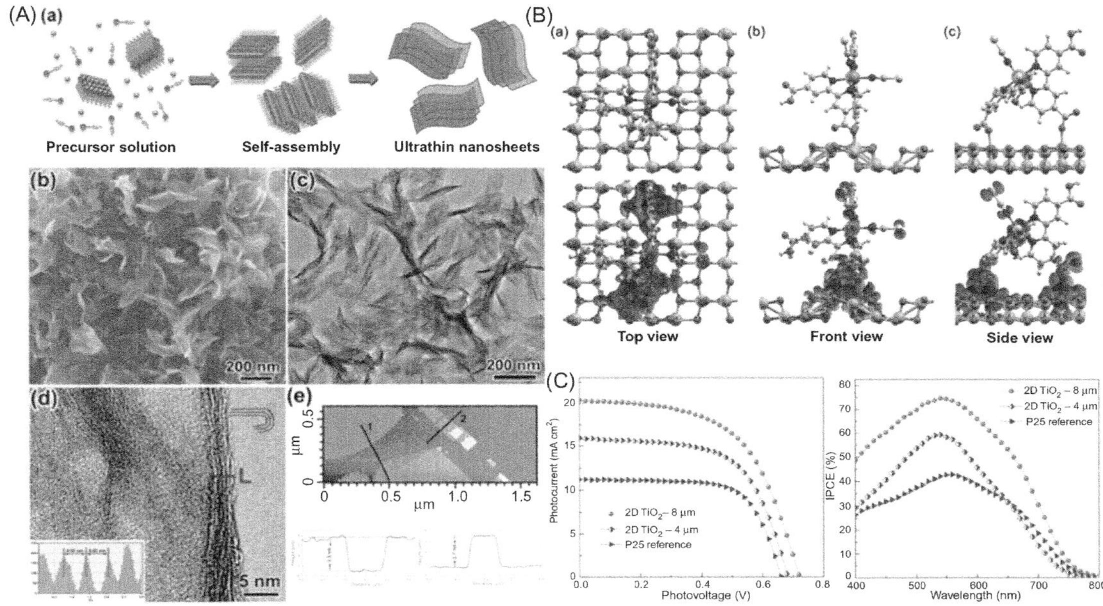

Figure 2.3 Ultrathin 2D TiO_2 nanosheets for solar energy conversion [126]. (A) Synthesis and characterization of ultrathin 2D TiO_2 nanosheets, (B) electronic coupling between N719 dye-molecule and 2D TiO_2 nanosheet, and (C) C–V and IPCE curves of dye-sensitized solar cells with 2D TiO_2 nanosheets and P25 nanoparticles.

electronic properties, and intriguing chemical reactivity. Sun et al. have fabricated single-crystalline ultrathin anatase/rutile 2D TiO_2 nanosheets through a confined growth of the nanosheets in the surfactant templates [16]. The obtained ultrathin 2D TiO_2 nanosheets feature highly chemically active and nonequilibrium exposed surfaces. When used as the photoanode for dye-sensitized solar cells (DSSCs), a much-enhanced conversion efficiency of 8.28% has been achieved, which was an almost two times improvement compared to the reference solar cells with commercial P25 TiO_2 nanoparticle photoanode (5.12%) and other reported TiO_2 nanostructures [126,127]. The DFT calculations concluded that the strong interfacial interaction between N719 and the 2D nanosheets facilitates an efficient charge transfer at the interfaces and leads to superior performance in photovoltaic devices [126] (Fig. 2.3).

The 2D nature of this emerging family of materials gives also additional flexibility of nanostructuring and manipulating the structures, which is otherwise challenging in the 3D bulk form. Liao et al. have performed a systematic study to design multifunctional 2D nanomaterials with optimized electronic

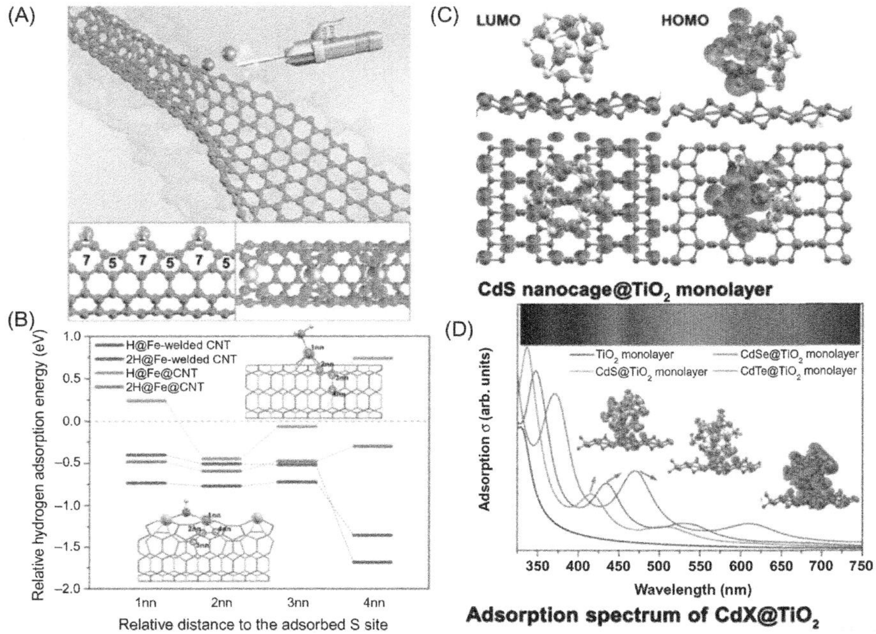

Figure 2.4 Electronic structure manipulation of 2D nanomaterials for energy applications. (A) Metal-atoms welding of graphene into transition metal modified carbon nanotube [132]; (B) the H_2S catalytic reactions of metal atom welded carbon nanotubes [133]; (C) electronic structure of CdS loaded 2D TiO_2 nanosheet [134]; and (D) the calculated light absorption spectra of CdX@TiO_2 nanosheets [134].

structures to modulate the associated optical, catalytic, and surface properties by engineering quantum-size, orientation, strain, and surface modifications on the nanoscale to meet the requirements of diverse sustainable energy applications [128–135]. Fig. 2.4 presents some examples on theoretically manipulating the nanostructural and the electronic structures of 2D nanomaterials. Graphene in its pristine form has been found unable to satisfy some diverse specific demands arising from particular applications. As shown in Fig. 2.4A, Liao et al. theoretically demonstrated that carbon nanoribbons can be welded from graphene by a variety of metal-atoms, such as alkali metals, III-IV group metals, and transition-metals, to form functionalized metal-welded carbon nanotubes (MW-CNTs), which represent a new family of carbon-based nanostructures and can also be extended to other foldable two-dimensional systems, such as BN, MoS_2, and TiO_2 2D nanosheets, and so on [132] They further demonstrated the modified CNTs welded from graphene has a promoting effect of under-coordinated carbon site on the overall H_2S splitting process and can be effectively used for H_2S sensing and dissociation (Fig. 2.4B) [133]. The strategy of decoration of heterostructures on 2D nanomaterials to improve their physical and chemical properties has also been successfully applied to 2D MOs. Liao et al. explored a nanocontact system formed by a 2D TiO_2 monolayer and II–VI semiconductor $(CdX)_{13}$ (X = S, Se, and Te) nanocages for engineering the visible-light absorption (Fig. 2.4C) [134]. It is applausive that the electronic coupling between the CdX nanocages and 2D TiO_2 nanosheets can be strengthened via the formation of Ti-X, which helps couple to more electrons, therefore, leads to an enhancement of the absorption peaks in the visible frequency range. Moreover, on changing the element X in $(CdX)_{13}$ from S to Se then Te, a red-shift of the visible-light absorption peaks to cover the whole visible light range can be achieved (Fig. 2.4D). It is concluded that this low-dimensional $(CdX)_{13}$ nanocage@TiO_2 monolayer nanocontact system significantly promotes charge separation and optical absorption in the visible range and opens a general visible-light absorption strategy for 2D MOs.

Experimentally, 2D nanomaterials for energy storage or conversion can be easily tailored via surface functionalization, nanostructure engineering, and defects and doping engineering to achieve a wide range of electronic functionalities and catalytic properties [136,137]. For example, Seger et al. [138] demonstrated an n-type to p-type switchable photoelectrode with a multilayered thin film of exfoliated $Ti_{0.91}O_2$ nanosheets and polyaniline layers. The conductivity of polyaniline is dependent on the redox state and the pH value, acting as an n-type semiconductor under basic conditions and as a p-type semiconductor in acidic conditions. The growth of multiple layers allows not only a controllable strategy for the type but also the intensity of the generated photocurrent. Another exciting research field for 2D materials is the possibility to stack them layer-by-layer in heterojunctions bonded by van der Waals forces [139]. Haigh et al. [140] confirmed the concept of complex heterostructures by stacking graphene and boron nitride monolayers with atomically sharp and clean interfaces.

When used as catalysts for fuel generation or active electrodes of energy devices, 2D nanomaterials exhibited high specific surface area and exposed surface atoms, and thus are promising candidates for high-performance catalysts. With a large number of exposed surface atoms, it is easy to produce vacancy-type defects in 2D materials, thus to decrease the coordination number [141], leading to more dangling bonds and higher catalytic activity [142]. The control of structural defects allows the tuning of the catalytic activities by modifying the electronic structures [143]. Anion defects such as oxygen vacancies have been studied for different approaches, such as doping and thermal annealing, and is recognized as the most effective way to improve electrical conductivity of metal oxides [144,145]. The most usually observed defects in metal oxides are various types of vacancies or deformation of a bond that might happen when two metals or nonmetals are consecutively bonded together [146]. For example, Wang et al. [147] showed an improvement of electronic properties of TiO_2, WO_3, and α-Fe_2O_3 by increasing oxygen vacancies through thermal treatment, assisting the charge transfer at the interface between metal oxide and substrate and electrolyte. Kim et al. showed that the introduction of oxygen vacancies into the α-MoO_3 lattice leads to a larger interlayer spacing, promoting faster charger storage kinetics and retains structural integrity during the reversible insertion of Li ions [148]. The utilization of ultrathin 2D ternary materials can extensively enlarge the oxygen vacancy and further the charge redistribution and structural distortion, thus improve their energy storage performance [149].

Dou et al. examined Co_3O_4 nanostructures for catalytic performance in different morphologies, that is, solid nanocubes, free-standing mesoporous ultrathin nanosheets, and 3D bundled 2D flowers (Fig. 2.5) [150]. These Co_3O_4 nanostructures presented different surface chemical states, as determined by the XPS. The positions of the core level orbitals of the cobalt element (Co $2p_{1/2}$ and $2p_{3/2}$) for mesoporous 2D nanosheets and 3D bundled 2D flowers show 1.0 and 0.6 eV shifts, respectively, to lower binding energy compared with those of solid nanocubes, as a result of the enhanced electron densities concentrated around the cobalt atoms aroused by the dramatic distortion of the crystal structure of the 2D nanosheets. Owing to the unsaturated electronic states of the metal-atoms and the highly accessible surface areas contributed by the atomic thickness and mesoporosity, the ultrathin 2D Co_3O_4 nanosheets presented decreased energy barriers for mass conversion and exhibit excellent OER catalytic performance with low onset potential, high current density, and long-term cycling stability. This result demonstrated that 2D MO nanomaterials have significant structural advantages in catalytic applications compared to those in other dimensionalities (Fig. 2.5).

2D Co_3O_4 nanosheets were further studied as the anode for high-rate sodium-ion batteries (SIBs) [151]. Different to LIBs,

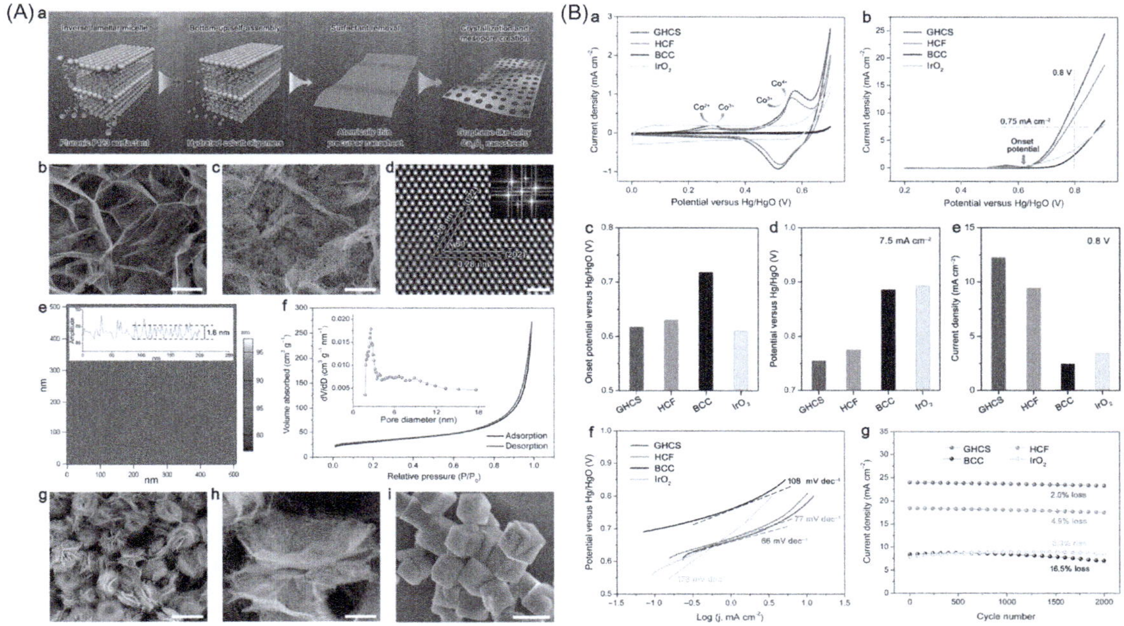

Figure 2.5 Electrocatalytic application of ultrathin 2D Co_3O_4 nanosheets [150]. (A) Synthesis and characterization of ultrathin 2D Co_3O_4 nanosheets together with 3D bundled flowers and solid nanocubes with different chemical stated, and (B) electrocatalytic performance of ultrathin 2D Co_3O_4 nanosheets, 3D bundled flowers, and solid nanocubes.

SIBs have very poor electrochemical properties ascribed to the slow Na^+ diffusion, the sluggish Na^+ insertion/deinsertion kinetics, and the high volume expansion of the host materials caused by the large radius (1.02 Å) of Na^+. As a result, developing effective host materials for fast and reversible Na^+ storage is a challenging but highly desirable task at this moment. In this work, ultrathin 2D Co_3O_4 nanosheets have been developed to provide a capacitive charge storage mechanism, which mainly occurs at the surface or interface of the active materials through electrostatic adsorption, redox reaction, or intercalation, avoids the slow ion diffusion within the crystal lattice, and thus leads to facile reaction kinetics during charging/discharging. By grown 2D ultrathin Co_3O_4 nanosheets on stainless steel mesh as an anode material for SIBs, this novel anode delivered a discharge capacity as high as 509.2 mAh g^{-1} at 50 mA g^{-1}, and reached 427.0 mAh g^{-1} at a high rate of 500 mA g^{-1}, together with high cycling stability, which are significantly superior to the conventional Co_3O_4 nanostructures and again proved the advantage of 2D MOs for high-rate SIB applications.

Very recently, Mei et al. synthesized ultrathin 2D Bi_2O_3 nanosheets that form a-phase/b-phase heterostructures within single nanosheets fabricated by a solution-based molecular self-assembly approach for the applications of both LIBs and SIBs [152]. Bi_2O_3 is a semiconductor with a bandgap of 2.5–2.8 eV and has received considerable attention in energy-related fields, owing to its suitable working window and exceptional theoretical specific capacitance/capacity. This type of dual-phase Bi_2O_3 nanosheets feature mixed crystals of both α- and β-phases and rich surface/edge-active sites, and thus can significantly enhance the surface redox and/or interfacial capacitance for both lithium and sodium ions storage. As expected, the dual-phase Bi_2O_3 nanosheets utilized as anode materials for rechargeable batteries delivered a high initial discharge capacity of 647.6 mAh g^{-1} and a reversible capacity of over 200 mAh g^{-1} after 260 cycles for LIBs, and a stable cycling capacity at ~50 mAh g^{-1} after 500 cycles for SIBs, which exhibited great potential as an exceptional anode material for practical rechargeable batteries and provided another proof of 2D materials for energy applications.

Based on the above discussion, it is clear that 2D MOs are interesting materials for future energy applications, such as electrolysis [150], LIBs [153], SIBs [151], etc., due to their electrochemical stability within their electrochemical operation windows and their relatively fast intercalation/deintercalation kinetics. The charge carries in a 2D nanomaterial are confined along the thickness and can move along the plane. Reducing the scattering effect of free charged carries make the 2D MOs candidates for produce more efficient electronic devices, adjusting electronic properties by controlling the thickness, which is not possible in 0D, 1D, and 3D nanomaterials [154,155]. The 2D MOs, however, also possess some disadvantages as other materials and other dimensionalities, such as the intrinsic low conductivity of some typical metal oxides, easy restacking of the 2D structures, difficulty in transplane diffusion of liquid electrolytes in some applications, etc., which have also impeded the practical application of this family of promising materials.

2.4 Strategies to further improve the performance of 2D MO materials

To further promote the performance of 2D MOs, as illustrated in Fig. 2.6, some strategies have been proposed, such as

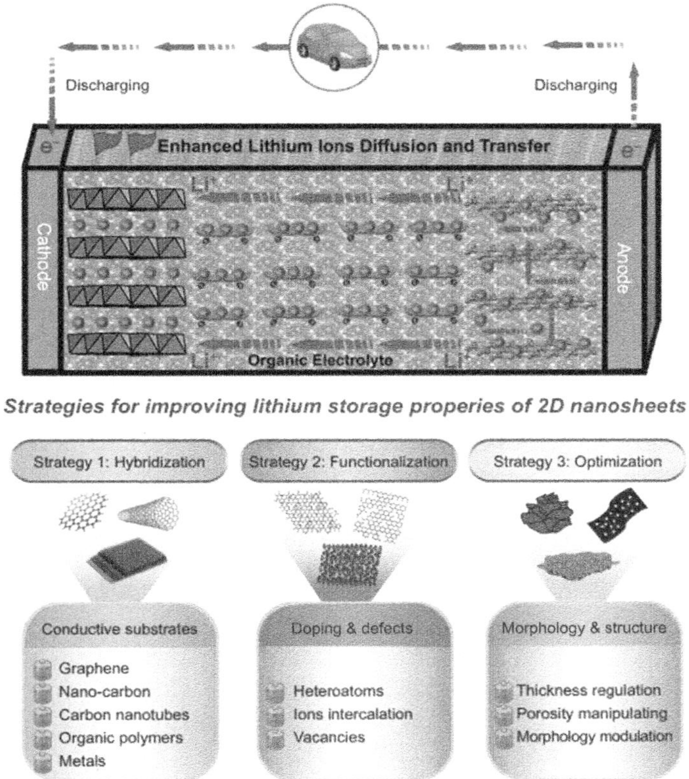

Figure 2.6 Strategies to enhance the performance of 2D MOs in energy storage applications [156].

hybridizing with property-complementary materials, functionalization via doping or creating defects, and morphology optimizations [156]. Among these strategies, hybridizing of 2D transition MOs or dichalcogenides with property-complementary nanomaterials into multifunctional nanocomposites exhibits good synergistic effects in combination with the merits of each constituent and has been regarded as one of the most widely used strategies for improving the electrochemical properties of 2D nanomaterials. The hybridizing materials usually are conductive matters, including graphene, nanocarbon, CNTs, conductive polymers, and metallic nanoparticles, which can significantly improve the overall conductivity, maintain the good dispersion and prevent the agglomeration of 2D materials, accommodate the volume expansion of 2D nanomaterials during the charging/discharging cycles, and provide extra surfaces or interface for extra solid ions storage.

Co_3O_4 is a widely studied material for energy storages. This promising material, however, yet suffers from its intrinsic low electric conductivity and large volume change during cycles. Dou et al. reported the synthesis of an atomic layer-by-layer Co_3O_4/graphene hybrid nanomaterial as anode materials for LIBs via a surfactant-assisted self-assembly method [153]. The as-prepared Co_3O_4 nanosheets with numerous mesopores on the surface possessed a thickness of ~2.0 nm, a lateral size of about 2.5 μm, and a surface area of ~157 $m^2\ g^{-1}$. When utilized as anode for LIBs, this hybrid nanomaterial delivered high discharge capacities of 2014.7 and 1134.4 mA h g^{-1} at 0.11 and 2.25C (1C = 890 mA g^{-1}), respectively. More attractive, 92.1% of the original capacity remained after 2000 cycles at a rate of 2.25C, suggesting exceptional cycling stability compared to other Co_3O_4/C composites. This outstanding lithium-ion storage performance of the 2D Co_3O_4/graphene hybrid nanomaterial mainly is attributed by the unique structure, in which the atom-level thickness and the mesoporous structure of the 2D Co_3O_4 nanosheets allows adequate contact and fast transport of electrode/electrolyte and shortened Li^+ diffusion length, while the graphene substrate provides high electrical conductivity and excellent flexibility as well as enhanced structural stability. It is also found that the formation of new C − O − Co bonds between Co_3O_4 and graphene also contributes to interlayer charge transport.

Carbon nanotubes (CNTs) are typical one-dimensional carbon nanostructures and feature excellent electric conductivity, outstanding mechanical properties, and superior chemical stability, which make CNTs are excellent hybridizing materials to strengthen the structure and increase the electrochemical properties of 2D material-based electrodes. In the study of 2D Bi_2O_3 nanosheets for energy storage, as shown in Fig. 2.7, CNTs have been used as a hybridizing material to form a 2D Bi_2O_3-CNTs-2D graphene (2D-1D-2D) hybrid structures, owing to that the 2D Bi_2O_3 nanomaterial presented rapid fading of the specific capacity aroused by their intrinsic low electronic conductivity and inadequate active sites [152]. The innovative 2D-1D-2D configuration successfully avoids the self-aggregation of the individual large-plane-size 2D graphene, the large-aspect-ratio 1D CNTs, and the 2D Bi_2O_3 nanosheets, and thus effectively increases the reversible capacity, rate capability, and cycling stability for both LIBs and SIBs. For Li^+ ions storage, the initial discharge capacity of the 2D-1D-2D anode enhanced to 823.5 mAh g^{-1}, corresponding to a ~27% enhancement compared to bare 2D Bi_2O_3 anode. Moreover, the Coulombic efficiencies of the

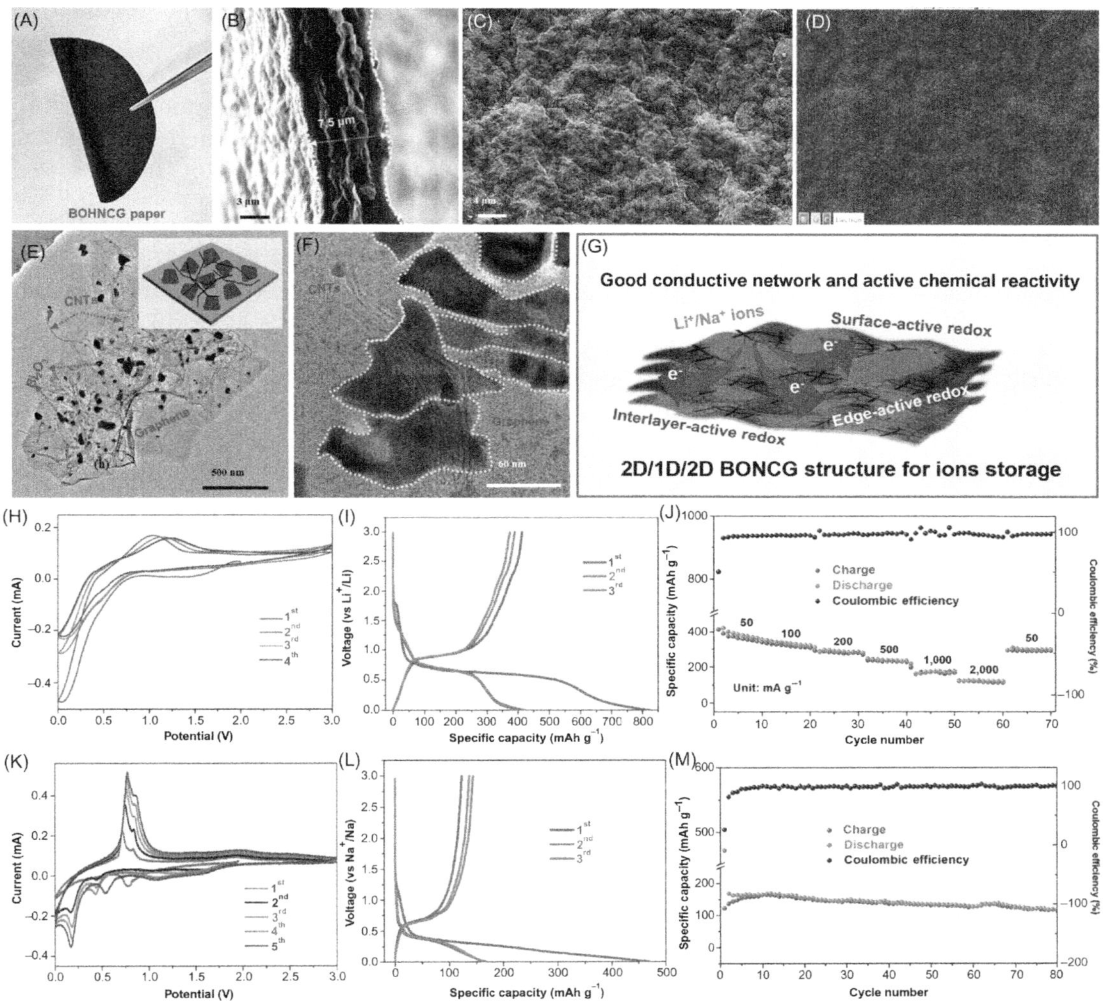

Figure 2.7 2D-1D-2D hybrid heterostructures consisting of 2D Bi_2O_3 nanosheets, 1D carbon nanotubes (CNTs), and 2D graphene for both lithium and sodium ion storages [152]. (A) Optical photograph of 2D-1D-2D paper showing its flexibility. (B) Sectional SEM image of 2D-1D-2D paper. (C) SEM image of 2D-1D-2D paper surface and (D) the corresponding element mapping distribution. (E and F) TEM image of 2D-1D-2D paper showing the presence of 2D graphene, 1D CNTs, and 2D Bi_2O_3. Inset in (E) shows the schematic illustration of 2D/1D/2D heterogeneous structure in the paper. (G) Schematic illustration of the structural advantages of the 2D-1D-2D hybrid paper for ions storage. (H) CV profiles at a scan rate of 0.5 mV s^{-1} in the first four cycles: (I) charge and discharge profiles in the first three cycles and (J) rate capability of free-standing 2D-1D-2D paper anode for LIBs. (K) CV profiles at a scan rate of 0.5 mV s^{-1} in the first five cycles, (L) charge and discharge profiles in the first three cycles, and (M) cycle stability at a rate of 50 mA g^{-1} of free-standing 2D-1D-2D paper anode for SIBs.

2D-1D-2D anode were close to 100% during cycles. For SIB application, the Coulombic efficiencies reached ~100% from the second cycle and the capacity remained ~110 mAh g^{-1} after 80 cycles. The performances of the 2D nanomaterials thus have presented dramatic enhancements by hybridizing with other 2D and 1D nanostructures, demonstrating the success of this strategy for improving the performance of 2D MO nanomaterials.

Metal oxides have not only been used as the 2D nanostructure to form 2D hybrid or heterostructures but also been employed as bridge or spencer materials to form 2D-MO-2D hybrid nanostructures. Mei et al. developed a novel 2D-TiO_2-2D van der Waals (vdW) heterostructured (BPNs@TiO_2@G) hydrogel for lithium-ion batteries, where the 2D black phosphorus nanosheets (BPNs) and porous graphene were separated by TiO_2 nanoparticles [157]. In this structure, TiO_2 nanoparticles play a crucial role as effective spacers and bridges between two heterogeneous graphene and BPN. This unique 2D-TiO_2-2D vdW heterostructure effectively prevents close restack of the 2D nanosheets, provides rapid interlayer transfer paths and enhanced interfacial storage, and receives the inherited advantages of both BPNs and graphene. As a result, this unique 2D-MO-2D heterostructure illustrated shortened diffusion pathway, improved conductivity, suppressed volume changes and lithium dendrite growth over cycling, and widened potential window, which delivered an initial discharge capacity as high as 1336.1 mAh g^{-1} at 0.2 A g^{-1}, superior rate capability (271.1 mAh g^{-1} at 5.0 A g^{-1}), good cycling life (502 mAh g^{-1} for 180 cycles), and a wide potential window of 0.01–3.0 V.

Very recently, the concept of bio-inspiration has brought further enhancement on the performance of 2D MO nanostructures in energy applications. With millions of years of evolution, natural species have evolved with some unique structures and specialized functions to suit with the living environment, which provide researchers enormous inspirations to design novel structures and materials by learning the best from nature. Zhang et al. recently summarized the progress of bio-inspired 2D nanomaterials and their applications in sustainable energy and environmental technologies (Fig. 2.8A), where well presented some examples on how to achieve extraordinary properties which cannot be obtained from the materials in conventional forms [158].

To survive in the wild world, many natural species have evolved with fascinating optical properties, such as glitzy structural colors for attracting prey or mates, which are known as

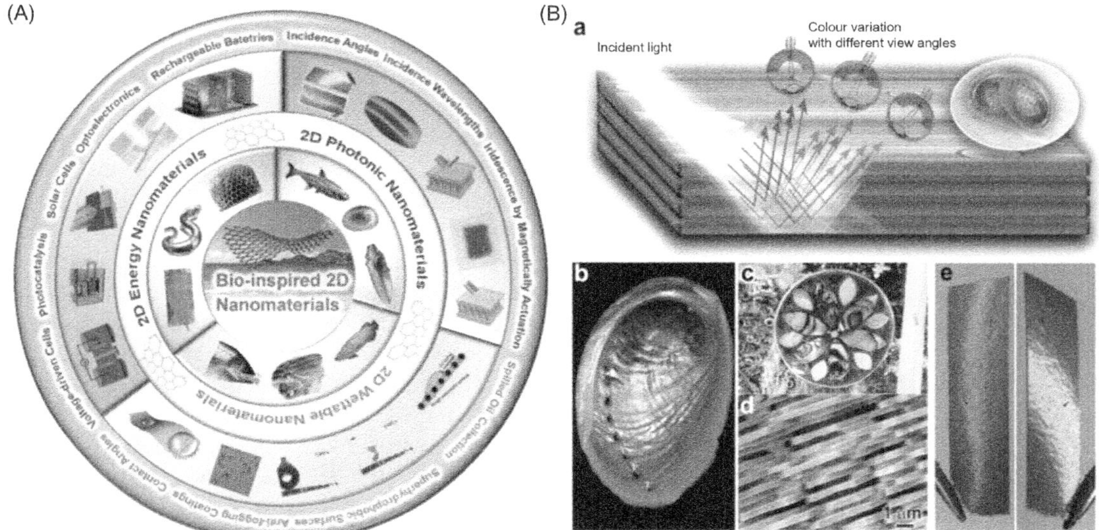

Figure 2.8 Bio-inspired 2D nanomaterials for sustainable energy applications. (A) A summary of bio-inspired 2D nanomaterials for typical sustainable applications [158], and (B) seashell inspired 2D TiO_2-2D graphene nanostructures with extraordinary optical properties for solar energy harvesting and conversion [159].

photonic crystal structures. Among them, one class consisting of periodically stacked 2D multilayers, or known as Bragg Stacks, can generate iridescent colors. By learning the Bragg Stacks, Sun et al. fabricated a seashell inspired 2D photonic nanostructure via a facile vacuum-filtering technique [159]. In this work, as shown in Fig. 2.8B, atomically thin 2D TiO_2 nanosheets and graphene were deposited in a layer-by-layer manner to mimic well-arranged, layered brick-and-mortar microstructures of natural sea-shells. The fabricated bio-inspired photonic structure exhibited significant and beautiful green-red strip-like colors under both dispersion-dominated and reflection-dominated conditions, resulted from its well-designed thickness and well-aligned 2D interfaces. Furthermore, this unique bio-inspired photonic nanostructure presented significantly enhanced photocurrent and much-improved responsibility and sensitivity when used in photoelectronic conversion devices, attributed by the significantly enhanced interlayer charge transfer of the 2D-2D heterostructures and homostructures of the bio-inspired nanostructures.

For energy storage devices, nanostructures have been proven an efficient approach for improving electrochemical properties for energy storage. The well-designed fine nanostructures, unfortunately, are usually destroyed during long-term cycles and finally lose their structural advantages. By learning from the natural honeycombs or beehives which feature extraordinary structural stability, robust mechanical properties, and salient ventilation capacity, Mei et al. developed bio-inspired heterogeneous bimetallic Co-Mo oxide ($CoMoO_x$) nanoarchitectures with 2D nanosheets as walls to take the advantages of both 2D nanostructures and the structural stability and permeability of the honeycombs for improving the rate capability of electrochemical lithium storage devices [160]. Attributed to the unique structural advantages inherited from honeycombs, the $CoMoO_x$ anode presents superior electrochemical performance, such as high discharge capacity (1388.6 mAh g^{-1} @ 0.2 A g^{-1}) and excellent rate capability (597.1 mAh g^{-1} at 5.0 A g^{-1}), which are much higher than those reported Co_3O_4 anodes (i.e., 1271.6 and 213.4 mAh g^{-1} at 0.2 and 5.0A g^{-1}), and no obvious decay after 2000 cycles. This work provides another example on the design of novel nanomaterials for high-performance energy devices by learning from nature.

2.5 Conclusion

2D MOs have attracted great interest in recent years because of their unique electronic, electrical, mechanical, and surface properties that are not observed in bulk materials and opens new possibilities in energy applications. This class of materials has presented an important role in energy conversion and storage devices. The 2D MO nanomaterials, however, also suffer from some intrinsic defects that impeded their practical applications in energy devices, such as low electric conductivity, large volume changes during solid–state ion storage, and restack and loss of the 2D nanostructure in service. These challenges, as proposed in this chapter, can be solved by hybridizing with property-complementary materials, forming of multilayer and multidimensional heterostructures, and developing more sophisticated structures by learning from nature.

Acknowledgments

This work was supported by an ANT-FAPESP collaborative project (2016/50339-2), an ARC Future Fellowship project (FT180100381), and an ARC Discovery project (DP160102627).

References

[1] O. Ellabban, H. Abu-Rub, F. Blaabjerg, Renewable energy resources: current status, future prospects and their enabling technology, Renew. Sustain. Energy Rev. 39 (2014) 748–764. Available from: https://doi.org/10.1016/j.rser.2014.07.113.

[2] Varun, R. Prakash, I.K. Bhat, , Energy, economics and environmental impacts of renewable energy systems, Renew. Sustain. Energy Rev. 13 (2009) 2716–2721. Available from: https://doi.org/10.1016/j.rser.2009.05.007.

[3] N.L. Panwar, S.C. Kaushik, S. Kothari, Role of renewable energy sources in environmental protection: a review, Renew. Sustain. Energy Rev. 15 (2011) 1513–1524. Available from: https://doi.org/10.1016/j.rser.2010.11.037.

[4] I. Dincer, Renewable energy and sustainable development: a crucial review, Renew. Sustain. Energy Rev. 4 (2000) 157–175. Available from: https://doi.org/10.1016/S1364-0321(99)00011-8.

[5] C. Liu, F. Li, L.-P. Ma, H.-M. Cheng, Advanced materials for energy storage, Adv. Mater. 22 (2010) E28–E62. Available from: https://doi.org/10.1002/adma.200903328.

[6] M. Ball, M. Wietschel, The future of hydrogen – opportunities and challenges, Int. J. Hydrog. Energy 34 (2009) 615–627. Available from: https://doi.org/10.1016/j.ijhydene.2008.11.014.

[7] S. Chu, A. Majumdar, Opportunities and challenges for a sustainable energy future, Nature. 488 (2012) 294–303. Available from: https://doi.org/10.1038/nature11475.

[8] P. Simon, Y. Gogotsi, Materials for electrochemical capacitors, Nat. Mater. 7 (2008) 845–854. Available from: https://doi.org/10.1038/nmat2297.

[9] J.N. Coleman, M. Lotya, A. O'Neill, S.D. Bergin, P.J. King, U. Khan, et al., Two-dimensional nanosheets produced by liquid exfoliation of layered materials, Science. 331 (2011) 568–571. Available from: https://doi.org/10.1126/science.1194975.

[10] H. Tong, S. Ouyang, Y. Bi, N. Umezawa, M. Oshikiri, J. Ye, Nano-photocatalytic materials: possibilities and challenges, Adv. Mater. 24 (2012) 229–251. Available from: https://doi.org/10.1002/adma.201102752.

[11] M. Pelaez, N.T. Nolan, S.C. Pillai, M.K. Seery, P. Falaras, A.G. Kontos, et al., A review on the visible light active titanium dioxide photocatalysts for environmental applications, Appl. Catal. B Environ. 125 (2012) 331–349. Available from: https://doi.org/10.1016/j.apcatb.2012.05.036.

[12] B.D. Alexander, P.J. Kulesza, I. Rutkowska, R. Solarska, J. Augustynski, Metal oxide photoanodes for solar hydrogen production, J. Mater. Chem. 18 (2008) 2298–2303. Available from: https://doi.org/10.1039/B718644D.

[13] A.P. Dral, J.E. ten Elshof, 2D metal oxide nanoflakes for sensing applications: review and perspective, Sens. Actuators B Chem. 272 (2018) 369–392. Available from: https://doi.org/10.1016/j.snb.2018.05.157.

[14] L. Dai, D.W. Chang, J.-B. Baek, W. Lu, Carbon nanomaterials for advanced energy conversion and storage, Small. 8 (2012) 1130–1166. Available from: https://doi.org/10.1002/smll.201101594.

[15] A.P. Sharma, P. Dhakal, D.K. Pradhan, M.K. Behera, B. Xiao, M. Bahoura, Fabrication and characterization of SnO_2 nanorods for room temperature gas sensors, AIP Adv. 8 (2018) 095219. Available from: https://doi.org/10.1063/1.5050991.

[16] Z. Sun, T. Liao, Y. Dou, S.M. Hwang, M.-S. Park, L. Jiang, et al., Generalized self-assembly of scalable two-dimensional transition metal oxide

nanosheets, Nat. Commun. 5 (2014) 3813. Available from: https://doi.org/10.1038/ncomms4813.
[17] N.G. Sahoo, Y. Pan, L. Li, S.H. Chan, Graphene-based materials for energy conversion, Adv. Mater. 24 (2012) 4203–4210. Available from: https://doi.org/10.1002/adma.201104971.
[18] X. Chen, L. Liu, P.Y. Yu, S.S. Mao, Increasing solar absorption for photocatalysis with black hydrogenated titanium dioxide nanocrystals, Science. 331 (2011) 746–750. Available from: https://doi.org/10.1126/science.1200448.
[19] J. Desilvestro, O. Haas, Metal oxide cathode materials for electrochemical energy storage: a review, J. Electrochem. Soc. 137 (1990) 5C–22C. Available from: https://doi.org/10.1149/1.2086438.
[20] F. Bensebaa, Chapter 5: Clean energy, in: F. Bensebaa (Ed.), Interface Sci. Technol., Elsevier, 19, 2013, pp. 279–383. Available from: https://doi.org/10.1016/B978-0-12-369550-5.00005-7.
[21] Z. Sun, T. Liao, L. Kou, Strategies for designing metal oxide nanostructures, Sci. China Mater. 60 (2017) 1–24. Available from: https://doi.org/10.1007/s40843-016-5117-0.
[22] P. Ayyub, V.R. Palkar, S. Chattopadhyay, M. Multani, Effect of crystal size reduction on lattice symmetry and cooperative properties, Phys. Rev. B. 51 (1995) 6135–6138. Available from: https://doi.org/10.1103/PhysRevB.51.6135.
[23] J.-C. Nièpce, L. Pizzagalli, Structure and phase transitions in nanocrystals, in: C. Bréchignac, P. Houdy, M. Lahmani (Eds.), *Nanomaterials and Nanochemistry*, Springer, Berlin, Heidelberg, 2007, pp. 35–54. Available from: https://doi.org/10.1007/978-3-540-72993-8_2.
[24] M. Yin, C.-K. Wu, Y. Lou, C. Burda, J.T. Koberstein, Y. Zhu, et al., Copper oxide nanocrystals, J. Am. Chem. Soc. 127 (2005) 9506–9511. Available from: https://doi.org/10.1021/ja050006u.
[25] F.X. Redl, C.T. Black, G.C. Papaefthymiou, R.L. Sandstrom, M. Yin, H. Zeng, et al., Magnetic, electronic, and structural characterization of nonstoichiometric iron oxides at the nanoscale, J. Am. Chem. Soc. 126 (2004) 14583–14599. Available from: https://doi.org/10.1021/ja046808r.
[26] M. Fernández-García, A. Martínez-Arias, J.C. Hanson, J.A. Rodriguez, Nanostructured oxides in chemistry: characterization and properties, Chem. Rev. 104 (2004) 4063–4104. Available from: https://doi.org/10.1021/cr030032f.
[27] E. Lucas, S. Decker, A. Khaleel, A. Seitz, S. Fultz, A. Ponce, et al., Nanocrystalline metal oxides as unique chemical reagents/sorbents, *Chem. Eur. J.* 7 (2001) 2505–2510. Available from: https://doi.org/10.1002/1521-3765(20010618)7:12 < 2505::AID-CHEM25050 > 3.0.CO;2-R.
[28] G. Cao, Y. Wang, Nanostructures and Nanomaterials: Synthesis, Properties, and Applications, 2nd edition, World Scientific, Hackensack, NJ, 2011. Available from: https://doi.org/10.1142/7885.
[29] J. Socratous, K.K. Banger, Y. Vaynzof, A. Sadhanala, A.D. Brown, A. Sepe, et al., Electronic structure of low-temperature solution-processed amorphous metal oxide semiconductors for thin-film transistor applications, Adv. Funct. Mater. 25 (2015) 1873–1885. Available from: https://doi.org/10.1002/adfm.201404375.
[30] X. Yu, T.J. Marks, A. Facchetti, Metal oxides for optoelectronic applications, Nat. Mater. 15 (2016) 383–396. Available from: https://doi.org/10.1038/nmat4599.

[31] M.P. Browne, Z. Sofer, M. Pumera, Layered and two dimensional metal oxides for electrochemical energy conversion, Energy Environ. Sci. 12 (2019) 41–58. Available from: https://doi.org/10.1039/C8EE02495B.
[32] G. Shen, P.-C. Chen, K. Ryu, C. Zhou, Devices and chemical sensing applications of metal oxide nanowires, J. Mater. Chem. 19 (2009) 828–839. Available from: https://doi.org/10.1039/B816543B.
[33] X. Wei, Y. Li, H. Peng, D. Gao, Y. Ou, Y. Yang, et al., A novel functional material of Co_3O_4/Fe_2O_3 nanocubes derived from a MOF precursor for high-performance electrochemical energy storage and conversion application, Chem. Eng. J. 355 (2019) 336–340. Available from: https://doi.org/10.1016/j.cej.2018.08.009.
[34] A. Seliverstov, J. Forster, J. Tucher, K. Kastner, C. Streb, Molecular metal oxides for energy conversion and energy storage, Discovering the Future of Molecular Sciences, Wiley, 2014, pp. 217–242. Available from: https://doi.org/10.1002/9783527673223.ch9.
[35] Y. Wu, Metal Oxides in Energy Technologies, Elsevier, 2018. Available from: https://doi.org/10.1016/C2015-0-06082-8.
[36] J. Mei, T. Liao, L. Kou, Z. Sun, Two-dimensional metal oxide nanomaterials for next-generation rechargeable batteries, Adv. Mater. 29 (2017) 1700176. Available from: https://doi.org/10.1002/adma.201700176.
[37] J. Zemann, Crystal Structures. 2nd edition, Vol. 1 by R. W. G. Wyckoff, *Acta Crystallographica*, Wiley, New York, 1964. Available from: https://doi.org/10.1107/S0365110X65000361.
[38] J. Mei, T. Liao, G.A. Ayoko, J. Bell, Z. Sun, Cobalt oxide-based nanoarchitectures for electrochemical energy applications, Prog. Mate. Sci. 103 (2019) 596–677. Available from: https://doi.org/10.1016/j.pmatsci.2019.03.001.
[39] Y.-G. Guo, J.-S. Hu, L.-J. Wan, Nanostructured materials for electrochemical energy conversion and storage devices, Adv. Mater. 20 (2008) 2878–2887. Available from: https://doi.org/10.1002/adma.200800627.
[40] X. Xia, Y. Zhang, D. Chao, C. Guan, Y. Zhang, L. Li, et al., Solution synthesis of metal oxides for electrochemical energy storage applications, Nanoscale. 6 (2014) 5008–5048. Available from: https://doi.org/10.1039/C4NR00024B.
[41] Z. Wu, L. Li, T. Liao, X. Chen, W. Jiang, J. Yang, et al., Janus nanoarchtectures: from structural design to catalytic applications, Nano Today 22 (2018) 62–82. Available from: https://doi.org/10.1016/j.nantod.2018.08.009.
[42] W. Sun, X. Rui, D. Yang, Z. Sun, B. Li, W. Zhang, et al., Two-dimensional tin disulfide nanosheets for enhanced sodium storage, ACS Nano 9 (2015) 11371–11381. Available from: https://doi.org/10.1021/acsnano.5b05229.
[43] K.S. Novoselov, A.K. Geim, S.V. Morozov, D. Jiang, Y. Zhang, S.V. Dubonos, et al., Electric field effect in atomically thin carbon films, Science. 306 (2004) 666–669. Available from: https://doi.org/10.1126/science.1102896.
[44] Q. Tang, Z. Zhou, Graphene-analogous low-dimensional materials, Prog. Mater. Sci. 58 (2013) 1244–1315. Available from: https://doi.org/10.1016/j.pmatsci.2013.04.003.
[45] C. Tan, X. Cao, X.-J. Wu, Q. He, J. Yang, X. Zhang, et al., Recent advances in ultrathin two-dimensional nanomaterials, Chem. Rev. 117 (2017) 6225–6331. Available from: https://doi.org/10.1021/acs.chemrev.6b00558.
[46] H. Zhang, Ultrathin two-dimensional nanomaterials, ACS Nano. 9 (2015) 9451–9469. Available from: https://doi.org/10.1021/acsnano.5b05040.

[47] H. Jin, C. Guo, X. Liu, J. Liu, A. Vasileff, Y. Jiao, et al., Emerging two-dimensional nanomaterials for electrocatalysis, Chem. Rev. 118 (2018) 6337–6408. Available from: https://doi.org/10.1021/acs.chemrev.7b00689.
[48] R. Amirante, E. Cassone, E. Distaso, P. Tamburrano, Overview on recent developments in energy storage: mechanical, electrochemical and hydrogen technologies, Energy Convers. Manag. 132 (2017) 372–387. Available from: https://doi.org/10.1016/j.enconman.2016.11.046.
[49] H. Chen, T.N. Cong, W. Yang, C. Tan, Y. Li, Y. Ding, Progress in electrical energy storage system: a critical review, Prog. Nat. Sci. 19 (2009) 291–312. Available from: https://doi.org/10.1016/j.pnsc.2008.07.014.
[50] Q. Lu, Y. Yu, Q. Ma, B. Chen, H. Zhang, 2D transition-metal-dichalcogenide-nanosheet-based composites for photocatalytic and electrocatalytic hydrogen evolution reactions, Adv. Mater. 28 (2016) 1917–1933. Available from: https://doi.org/10.1002/adma.201503270.
[51] L. Schlapbach, A. Züttel, Hydrogen-storage materials for mobile applications, Nature. 414 (2001) 353–358. Available from: https://doi.org/10.1038/35104634.
[52] K. Uosaki, G. Elumalai, H.C. Dinh, A. Lyalin, T. Taketsugu, H. Noguchi, Highly efficient electrochemical hydrogen evolution reaction at insulating boron nitride nanosheet on inert gold substrate, Sci. Rep. 6 (2016) 32217. Available from: https://doi.org/10.1038/srep32217.
[53] L. Chen, X. Dong, Y. Wang, Y. Xia, Separating hydrogen and oxygen evolution in alkaline water electrolysis using nickel hydroxide, Nat. Commun. 7 (2016) 11741. Available from: https://doi.org/10.1038/ncomms11741.
[54] V.R. Stamenkovic, D. Strmcnik, P.P. Lopes, N.M. Markovic, Energy and fuels from electrochemical interfaces, Nat. Mater. 16 (2017) 57–69. Available from: https://doi.org/10.1038/nmat4738.
[55] A.P. van Troostwyk, J.R. Deiman, Sur une manière de décomposer l'eau en air inflammable & en air vital, Obs Phys 35 (1789) 369–378.
[56] A. Ursua, L.M. Gandia, P. Sanchis, Hydrogen production from water electrolysis: current status and future trends, Proc. IEEE. 100 (2012) 410–426. Available from: https://doi.org/10.1109/JPROC.2011.2156750.
[57] C. Lamy, From hydrogen production by water electrolysis to its utilization in a PEM fuel cell or in a SO fuel cell: Some considerations on the energy efficiencies, Int. J. Hydrog. Energy. 41 (2016) 15415–15425. Available from: https://doi.org/10.1016/j.ijhydene.2016.04.173.
[58] S. Park, Y. Shao, J. Liu, Y. Wang, Oxygen electrocatalysts for water electrolyzers and reversible fuel cells: status and perspective, Energy Environ. Sci. 5 (2012) 9331–9344. Available from: https://doi.org/10.1039/C2EE22554A.
[59] G. Chisholm, L. Cronin, Chapter 16 – Hydrogen from water electrolysis, in: T.M. Letcher (Ed.), Storing Energy, Elsevier, 2016, pp. 315–343. Available from: https://doi.org/10.1016/B978-0-12-803440-8.00016-6.
[60] V. Vij, S. Sultan, A.M. Harzandi, A. Meena, J.N. Tiwari, W.-G. Lee, et al., , Nickel-based electrocatalysts for energy-related applications: oxygen reduction, oxygen evolution, and hydrogen evolution reactions, ACS Catal. 7 (2017) 7196–7225. Available from: https://doi.org/10.1021/acscatal.7b01800.
[61] Y. Jiao, Y. Zheng, M. Jaroniec, S.Z. Qiao, Design of electrocatalysts for oxygen- and hydrogen-involving energy conversion reactions, Chem. Soc. Rev. 44 (2015) 2060–2086. Available from: https://doi.org/10.1039/C4CS00470A.

[62] K. Zeng, D. Zhang, Recent progress in alkaline water electrolysis for hydrogen production and applications, Prog. Energy Combust. Sci. 36 (2010) 307–326. Available from: https://doi.org/10.1016/j.pecs.2009.11.002.
[63] A.J. Bard, L.R. Faulkner, Electrochemical Methods: Fundamentals and Applications, 2nd edition, Wiley, New York, NY, 2001.
[64] M. Gong, D.-Y. Wang, C.-C. Chen, B.-J. Hwang, H. Dai, A mini review on nickel-based electrocatalysts for alkaline hydrogen evolution reaction, Nano Res. 9 (2016) 28–46. Available from: https://doi.org/10.1007/s12274-015-0965-x.
[65] G. Zhang, Y.-S. Feng, W.-T. Lu, D. He, C.-Y. Wang, Y.-K. Li, et al., Enhanced catalysis of electrochemical overall water splitting in alkaline media by Fe doping in Ni_3S_2 nanosheet arrays, ACS Catal. 8 (2018) 5431–5441. Available from: https://doi.org/10.1021/acscatal.8b00413.
[66] R. Subbaraman, D. Tripkovic, K.-C. Chang, D. Strmcnik, A.P. Paulikas, P. Hirunsit, et al., Trends in activity for the water electrolyser reactions on 3d M(Ni,Co,Fe,Mn) hydr(oxy)oxide catalysts, Nat. Mater. 11 (2012) 550–557. Available from: https://doi.org/10.1038/nmat3313.
[67] A. Fujishima, K. Honda, Electrochemical photolysis of water at a semiconductor electrode, Nature. 238 (1972) 37–38. Available from: https://doi.org/10.1038/238037a0.
[68] A. Fujishima, X. Zhang, D.A. Tryk, TiO_2 photocatalysis and related surface phenomena, Surf. Sci. Rep. 63 (2008) 515–582. Available from: https://doi.org/10.1016/j.surfrep.2008.10.001.
[69] S.U.M. Khan, M. Al-Shahry, W.B. Ingler Jr., Efficient photochemical water splitting by a chemically modified n-TiO_2, Science. 297 (2002) 2243–2245. Available from: https://doi.org/10.1126/science.1075035.
[70] G. Wang, H. Wang, Y. Ling, Y. Tang, X. Yang, R.C. Fitzmorris, et al., Hydrogen-treated TiO2 nanowire arrays for photoelectrochemical water splitting, Nano Lett. 11 (2011) 3026–3033. Available from: https://doi.org/10.1021/nl201766h.
[71] X.-Y. Zhang, H.-P. Li, X.-L. Cui, Y. Lin, Graphene/TiO_2 nanocomposites: synthesis, characterization and application in hydrogen evolution from water photocatalytic splitting, J. Mater. Chem. 20 (2010) 2801–2806. Available from: https://doi.org/10.1039/b917240h.
[72] Y. Gai, J. Li, S.-S. Li, J.-B. Xia, S.-H. Wei, Design of narrow-gap TiO_2: a passivated codoping approach for enhanced photoelectrochemical activity, Phys. Rev. Lett. 102 (2009) 036402. Available from: https://doi.org/10.1103/PhysRevLett.102.036402.
[73] M. Ge, Q. Li, C. Cao, J. Huang, S. Li, S. Zhang, et al., One-dimensional TiO_2 nanotube photocatalysts for solar water splitting, Adv. Sci. 4 (2017) 1600152. Available from: https://doi.org/10.1002/advs.201600152.
[74] S.A. Ansari, M.M. Khan, M.O. Ansari, M.H. Cho, Nitrogen-doped titanium dioxide (N-doped TiO_2) for visible light photocatalysis, New J. Chem. 40 (2016) 3000–3009. Available from: https://doi.org/10.1039/c5nj03478g.
[75] S. Kment, F. Riboni, S. Pausova, L. Wang, L. Wang, H. Han, et al., Photoanodes based on TiO_2 and alpha-Fe_2O_3 for solar water splitting - superior role of 1D nanoarchitectures and of combined heterostructures, Chem. Soc. Rev. 46 (2017) 3716–3769. Available from: https://doi.org/10.1039/c6cs00015k.
[76] H.H. El-Maghrabi, A. Barhoum, A.A. Nada, Y.M. Moustafa, S.M. Seliman, A.M. Youssef, et al., Synthesis of mesoporous core-shell CdS@TiO_2 (0D and 1D) photocatalysts for solar-driven hydrogen fuel production,

J. Photochem. Photobiol. *A:* Chem. 351 (2018) 261–270. Available from: https://doi.org/10.1016/j.jphotochem.2017.10.048.

[77] S. He, Y. Meng, Q. Wu, J. Yang, S. Huang, X. Li, et al., Ta-doped porous TiO_2 nanorod arrays by substrate-assisted synthesis: efficient photoelectrocatalysts for water oxidation, Nanoscale. 10 (2018) 19367–19374. Available from: https://doi.org/10.1039/c8nr04003f.

[78] S. Selcuk, X. Zhao, A. Selloni, Structural evolution of titanium dioxide during reduction in high-pressure hydrogen, Nat. Mater. 17 (2018) 923–928. Available from: https://doi.org/10.1038/s41563-018-0135-0.

[79] M.G. Walter, E.L. Warren, J.R. McKone, S.W. Boettcher, Q. Mi, E.A. Santori, et al., Solar water splitting cells, Chem. Rev. 110 (2010) 6446–6473. Available from: https://doi.org/10.1021/cr1002326.

[80] X. Gan, D. Lei, K.-Y. Wong, Two-dimensional layered nanomaterials for visible-light-driven photocatalytic water splitting, Mater. Today Energy. 10 (2018) 352–367. Available from: https://doi.org/10.1016/j.mtener.2018.10.015.

[81] A.J. Nozik, Photoelectrolysis of water using semiconducting TiO_2 crystals, Nature. 257 (1975) 383–386. Available from: https://doi.org/10.1038/257383a0.

[82] C.A. Grimes, Synthesis and application of highly ordered arrays of TiO2 nanotubes, J. Mater. Chem. 17 (2007) 1451–1457. Available from: https://doi.org/10.1039/b701168g.

[83] J.G. Mavroides, D.I. Tchernev, J.A. Kafalas, D.F. Kolesar, Photoelectrolysis of water in cells with TiO_2 anodes, Mater. Res. Bull. 10 (1975) 1023–1030. Available from: https://doi.org/10.1016/0025-5408(75)90210-X.

[84] S. Hoang, S. Guo, N.T. Hahn, A.J. Bard, C.B. Mullins, Visible light driven photoelectrochemical water oxidation on nitrogen-modified TiO_2 nanowires, Nano Lett. 12 (2012) 26–32. Available from: https://doi.org/10.1021/nl2028188.

[85] M.A. Butler, Photoelectrolysis and physical properties of the semiconducting electrode WO_2, J. Appl. Phys. 48 (1977) 1914–1920. Available from: https://doi.org/10.1063/1.323948.

[86] J. Su, L. Guo, N. Bao, C.A. Grimes, Nanostructured $WO_3/BiVO_4$ heterojunction films for efficient photoelectrochemical water splitting, Nano Lett. 11 (2011) 1928–1933. Available from: https://doi.org/10.1021/nl2000743.

[87] P.M. Rao, L. Cai, C. Liu, I.S. Cho, C.H. Lee, J.M. Weisse, et al., Simultaneously efficient light absorption and charge separation in $WO_3/BiVO_4$ core/shell nanowire photoanode for photoelectrochemical water oxidation, Nano Lett. 14 (2014) 1099–1105. Available from: https://doi.org/10.1021/nl500022z.

[88] Y. Mao, H. Yang, J. Chen, J. Chen, Y. Tong, X. Wang, Significant performance enhancement of ZnO photoanodes from $Ni(OH)_2$ electrocatalyst nanosheets overcoating, Nano Energy. 6 (2014) 10–18. Available from: https://doi.org/10.1016/j.nanoen.2014.02.008.

[89] X. Yang, A. Wolcott, G. Wang, A. Sobo, R.C. Fitzmorris, F. Qian, et al., Nitrogen-doped ZnO nanowire arrays for photoelectrochemical water splitting, Nano Lett. 9 (2009) 2331–2336. Available from: https://doi.org/10.1021/nl900772q.

[90] X. Zhang, Y. Liu, Z. Kang, 3D branched ZnO nanowire arrays decorated with plasmonic Au nanoparticles for high-performance photoelectrochemical water splitting, ACS Appl. Mater. Interfaces. 6 (2014) 4480–4489. Available from: https://doi.org/10.1021/am500234v.

[91] M. Shao, F. Ning, M. Wei, D.G. Evans, X. Duan, Hierarchical nanowire arrays based on ZnO core – layered double hydroxide shell for largely enhanced photoelectrochemical water splitting, Adv. Funct. Mater. 24 (2014) 580–586. Available from: https://doi.org/10.1002/adfm.201301889.
[92] Y.Y. Lee, H.S. Jung, Y.T. Kang, A review: effect of nanostructures on photocatalytic CO_2 conversion over metal oxides and compound semiconductors, J. CO_2 Util. 20 (2017) 163–177. Available from: https://doi.org/10.1016/j.jcou.2017.05.019.
[93] T. Hisatomi, K. Domen, Progress in the demonstration and understanding of water splitting using particulate photocatalysts, Curr. Opin. Electrochem. 2 (2017) 148–154. Available from: https://doi.org/10.1016/j.coelec.2017.04.005.
[94] Z. Wang, C. Li, K. Domen, Recent developments in heterogeneous photocatalysts for solar-driven overall water splitting, Chem. Soc. Rev. 48 (2019) 2109–2125. Available from: https://doi.org/10.1039/C8CS00542G.
[95] G. Liu, C. Zhen, Y. Kang, L. Wang, H.-M. Cheng, Unique physicochemical properties of two-dimensional light absorbers facilitating photocatalysis, Chem. Soc. Rev. 47 (2018) 6410–6444. Available from: https://doi.org/10.1039/C8CS00396C.
[96] T. Su, Q. Shao, Z. Qin, Z. Guo, Z. Wu, Role of interfaces in two-dimensional photocatalyst for water splitting, ACS Catal. 8 (2018) 2253–2276. Available from: https://doi.org/10.1021/acscatal.7b03437.
[97] M. Carmo, D.L. Fritz, J. Mergel, D. Stolten, A comprehensive review on PEM water electrolysis, Int. J. Hydrog. Energy. 38 (2013) 4901–4934. Available from: https://doi.org/10.1016/j.ijhydene.2013.01.151.
[98] R.M. Bullock, Reaction: earth-abundant metal catalysts for energy conversions, Chem. 2 (2017) 444–446. Available from: https://doi.org/10.1016/j.chempr.2017.03.019.
[99] D.F. van der Vliet, C. Wang, D. Tripkovic, D. Strmcnik, X.F. Zhang, M.K. Debe, et al., Mesostructured thin films as electrocatalysts with tunable composition and surface morphology, Nat. Mater. 11 (2012) 1051–1058. Available from: https://doi.org/10.1038/nmat3457.
[100] E. Fabbri, A. Habereder, K. Waltar, R. Kötz, T.J. Schmidt, Developments and perspectives of oxide-based catalysts for the oxygen evolution reaction, Catal. Sci. Technol. 4 (2014) 3800–3821. Available from: https://doi.org/10.1039/C4CY00669K.
[101] E. Gülzow, Alkaline fuel cells: a critical view, J. Power Sources. 61 (1996) 99–104. Available from: https://doi.org/10.1016/S0378-7753(96)02344-0.
[102] J. Duan, S. Chen, M. Jaroniec, S.Z. Qiao, Heteroatom-doped graphene-based materials for energy-relevant electrocatalytic processes, ACS Catal. 5 (2015) 5207–5234. Available from: https://doi.org/10.1021/acscatal.5b00991.
[103] S. Chen, S.-Z. Qiao, Hierarchically porous nitrogen-doped graphene-$NiCo_2O_4$ hybrid paper as an advanced electrocatalytic water-splitting material, Acs Nano. 7 (2013) 10190–10196. Available from: https://doi.org/10.1021/nn404444r.
[104] H. Liang, F. Meng, M. Caban-Acevedo, L. Li, A. Forticaux, L. Xiu, et al., Hydrothermal continuous flow synthesis and exfoliation of NiCo layered double hydroxide nanosheets for enhanced oxygen evolution catalysis, Nano Lett. 15 (2015) 1421–1427. Available from: https://doi.org/10.1021/nl504872s.

[105] Y. Li, P. Hasin, Y. Wu, NixCo$_{3-x}$O$_4$ nanowire arrays for electrocatalytic oxygen evolution, Adv. Mater. 22 (2010) 1926–1929. Available from: https://doi.org/10.1002/adma.200903896.
[106] P. Du, R. Eisenberg, Catalysts made of earth-abundant elements (Co, Ni, Fe) for water splitting: recent progress and future challenges, Energy Environ. Sci. 5 (2012) 6012–6021. Available from: https://doi.org/10.1039/c2ee03250c.
[107] M. Oezaslan, F. Hasché, P. Strasser, Pt-based core–shell catalyst architectures for oxygen fuel cell electrodes, J. Phys. Chem. Lett. 4 (2013) 3273–3291. Available from: https://doi.org/10.1021/jz4014135.
[108] P. Paunović, I. Radev, A.T. Dimitrov, O. Popovski, E. Lefterova, E. Slavcheva, et al., New nano-structured and interactive supported composite electrocatalysts for hydrogen evolution with partially replaced platinum loading, Int. J. Hydrog. Energy. 34 (2009) 2866–2873. Available from: https://doi.org/10.1016/j.ijhydene.2009.01.024.
[109] Y. Luo, D. Huang, M. Li, X. Xiao, W. Shi, M. Wang, et al., MoS_2 nanosheet decorated with trace loads of Pt as highly active electrocatalyst for hydrogen evolution reaction, Electrochim. Acta. 219 (2016) 187–193. Available from: https://doi.org/10.1016/j.electacta.2016.09.151.
[110] T.F. Jaramillo, K.P. Jorgensen, J. Bonde, J.H. Nielsen, S. Horch, I. Chorkendorff, Identification of active edge sites for electrochemical H_2 evolution from MoS_2 nanocatalysts, Science. 317 (2007) 100–102. Available from: https://doi.org/10.1126/science.1141483.
[111] D. Voiry, H. Yamaguchi, J. Li, R. Silva, D.C.B. Alves, T. Fujita, et al., Enhanced catalytic activity in strained chemically exfoliated WS_2 nanosheets for hydrogen evolution, Nat. Mater. 12 (2013) 850–855. Available from: https://doi.org/10.1038/nmat3700.
[112] D. Vikraman, S. Hussain, K. Akbar, L. Truong, A. Kathalingam, S.-H. Chun, et al., Improved hydrogen evolution reaction performance using MoS_2–WS_2 heterostructures by physicochemical process, ACS Sustain. Chem. Eng. 6 (2018) 8400–8409. Available from: https://doi.org/10.1021/acssuschemeng.8b00524.
[113] D. Voiry, M. Salehi, R. Silva, T. Fujita, M. Chen, T. Asefa, et al., Conducting MoS_2 nanosheets as catalysts for hydrogen evolution reaction, Nano Lett. 13 (2013) 6222–6227. Available from: https://doi.org/10.1021/nl403661s.
[114] B. Hinnemann, P.G. Moses, J. Bonde, K.P. Jørgensen, J.H. Nielsen, S. Horch, et al., Biomimetic hydrogen evolution: MoS2 nanoparticles as catalyst for hydrogen evolution, J. Am. Chem. Soc. 127 (2005) 5308–5309. Available from: https://doi.org/10.1021/ja0504690.
[115] X. Hong, J. Mei, L. Wen, Y. Tong, A. Vadileff, L. Wang, et al., Non-lithium metal-sulfur batteries: steps towards a leap, Adv. Mater. 31 (2019) 1802822. Available from: https://doi.org/10.1002/adma.201802822.
[116] Z. Ren, J. Li, Y. Ren, S. Wang, Y. Qiu, J. Yu, Large-scale synthesis of hybrid metal oxides through metal redox mechanism for high-performance pseudocapacitors, Sci. Rep. 6 (2016) 20021. Available from: https://doi.org/10.1038/srep20021.
[117] H. Wang, H.-W. Lee, Y. Deng, Z. Lu, P.-C. Hsu, Y. Liu, et al., Bifunctional non-noble metal oxide nanoparticle electrocatalysts through lithium-induced conversion for overall water splitting, Nat. Commun. 6 (2015) 7261. Available from: https://doi.org/10.1038/ncomms8261.
[118] R.D.L. Smith, M.S. Prévot, R.D. Fagan, Z. Zhang, P.A. Sedach, M.K.J. Siu, et al., Photochemical route for accessing amorphous metal oxide

materials for water oxidation catalysis, Science. 340 (2013) 60–63. Available from: https://doi.org/10.1126/science.1233638.
[119] M. Gong, W. Zhou, M.-C. Tsai, J. Zhou, M. Guan, M.-C. Lin, et al., Nanoscale nickel oxide/nickel heterostructures for active hydrogen evolution electrocatalysis, Nat. Commun. 5 (2014) 4695. Available from: https://doi.org/10.1038/ncomms5695.
[120] F. Liu, X. Wang, J. Hao, S. Han, J. Lian, Q. Jiang, High density arrayed Ni/NiO core-shell nanospheres evenly distributed on graphene for ultrahigh performance supercapacitor, Sci. Rep. 7 (2017) 17709. Available from: https://doi.org/10.1038/s41598-017-17899-6.
[121] J.Y. Hwang, M.F. El-Kady, Y. Wang, L. Wang, Y. Shao, K. Marsh, et al., Direct preparation and processing of graphene/RuO_2 nanocomposite electrodes for high-performance capacitive energy storage, Nano Energy. 18 (2015) 57–70. Available from: https://doi.org/10.1016/j.nanoen.2015.09.009.
[122] H.B. Li, M.H. Yu, F.X. Wang, P. Liu, Y. Liang, J. Xiao, et al., Amorphous nickel hydroxide nanospheres with ultrahigh capacitance and energy density as electrochemical pseudocapacitor materials, Nat. Commun. 4 (2013) 1894. Available from: https://doi.org/10.1038/ncomms2932.
[123] D.F. Abbott, M. Meier, G.R. Meseck, E. Fabbri, S. Seeger, T.J. Schmidt, Silicone nanofilament-supported mixed nickel-metal oxides for alkaline water electrolysis, J. Electrochem. Soc. 164 (2017) F203–F208. Available from: https://doi.org/10.1149/2.0201704jes.
[124] J. Chi, H. Yu, G. Li, L. Fu, J. Jia, X. Gao, et al., Nickel/cobalt oxide as a highly efficient OER electrocatalyst in an alkaline polymer electrolyte water electrolyzer, RSC Adv. 6 (2016) 90397–90400. Available from: https://doi.org/10.1039/C6RA19615B.
[125] X. Long, Z. Ma, H. Yu, X. Gao, X. Pan, X. Chen, et al., Porous FeNi oxide nanosheets as advanced electrochemical catalysts for sustained water oxidation, J. Mater. Chem. A. 4 (2016) 14939–14943. Available from: https://doi.org/10.1039/C6TA05907D.
[126] L. Sheng, T. Liao, L. Kou, Z. Sun, Single-crystalline ultrathin 2D TiO_2 nanosheets: a bridge towards superior photovoltaic devices, Mater. Today Energy 3 (2017) 32–39. Available from: https://doi.org/10.1016/j.mtener.2016.12.004.
[127] Z. Sun, T. Liao, L.Y. Sheng, L.Z. Kou, J.H. Kim, S.X. Dou, Deliberate design of TiO_2 nanostructures towards superior photovoltaic cells, Chem. Eur. J. 22 (2016) 11357–11364. Available from: https://doi.org/10.1002/chem.201601546.
[128] T. Liao, Z. Sun, C.H. Sun, S.X. Dou, D.J. Searles, Electronic coupling and catalytic effect on H_2 evolution of MoS_2/graphene nanocatalyst, Sci. Rep. 4 (2014) 6256. Available from: https://doi.org/10.1038/srep06256.
[129] T. Liao, C.H. Sun, A.J. Du, Z.Q. Sun, D.H. Jurcakova, S. Smith, Charge carrier exchange at chemically modified graphene edges: a density functional theory study, J. Mater. Chem. 22 (2012) 8321–8326. Available from: https://doi.org/10.1039/C2JM30387F.
[130] T. Liao, C.H. Sun, Z.Q. Sun, A.J. Du, S. Smith, Chemically modified ribbon edge stimulated H_2 dissociation: a first-principles computational study, Phys. Chem. Chem. Phys. 15 (2013) 8054–8057. Available from: https://doi.org/10.1039/C3CP50654A.
[131] T. Liao, C.H. Sun, Z.Q. Sun, A.J. Du, D.H. Jurcakova, S. Smith, How to achieve maximum charge carrier loading on heteroatom-substituted graphene nanoribbon edges: density functional theory study, J. Mater.

Chem. 22 (2012) 13751–13755. Available from: https://doi.org/10.1039/C2JM31445B.
[132] T. Liao, Z.Q. Sun, J.H. Kim, S.X. Dou, Theoretically designed metal-welded carbon nanotubes: extraordinary electronic properties and promoted catalytic performance, Nano Energy 32 (2017) 209–215. Available from: https://doi.org/10.1016/j.nanoen.2016.12.033.
[133] T. Liao, L.Z. Kou, A.J. Du, L.N. Cheng, C. Cao, Z.Q. Sun, H_2S sensing and splitting on atom-functionalized carbon nanotubes: a theoretical study, Adv. Theory Simul. 1 (2018) 1700033. Available from: https://doi.org/10.1002/adts.201700033.
[134] T. Liao, Z.Q. Sun, S.X. Dou, Theoretically manipulating quantum dots on 2D TiO_2 monolayer for effective visible light absorption, ACS Appl. Mater. Interfaces 9 (2017) 8255–8262. Available from: https://doi.org/10.1021/acsami.6b15741.
[135] T. Liao, L.Z. Kou, A.J. Du, Z.Q. Sun, Simplest MOF units for effective photodriven hydrogen evolution reaction, J. Am. Chem. Soc. 140 (2018) 9159–9166. Available from: https://doi.org/10.1021/jacs.8b04599.
[136] M.M.Y.A. Alsaif, A.F. Chrimes, T. Daeneke, S. Balendhran, D.O. Bellisario, Y. Son, et al., High-performance field effect transistors using electronic inks of 2D molybdenum oxide nanoflakes, Adv. Funct. Mater. 26 (2016) 91–100. Available from: https://doi.org/10.1002/adfm.201503698.
[137] M.T. Greiner, L. Chai, M.G. Helander, W.-M. Tang, Z.-H. Lu, Transition metal oxide work functions: the influence of cation oxidation state and oxygen vacancies, Adv. Funct. Mater. 22 (2012) 4557–4568. Available from: https://doi.org/10.1002/adfm.201200615.
[138] B. Seger, J. McCray, A. Mukherji, X. Zong, Z. Xing, L. Wang, An n-type to p-type switchable photoelectrode assembled from alternating exfoliated titania nanosheets and polyaniline layers, Angew. Chem. Int. Ed. 52 (2013) 6400–6403. Available from: https://doi.org/10.1002/anie.201302062.
[139] A.K. Geim, I.V. Grigorieva, Van der Waals heterostructures, Nature. 499 (2013) 419–425. Available from: https://doi.org/10.1038/nature12385.
[140] S.J. Haigh, A. Gholinia, R. Jalil, S. Romani, L. Britnell, D.C. Elias, et al., Cross-sectional imaging of individual layers and buried interfaces of graphene-based heterostructures and superlattices, Nat. Mater. 11 (2012) 764–767. Available from: https://doi.org/10.1038/nmat3386.
[141] H. Tao, Y. Gao, N. Talreja, F. Guo, J. Texter, C. Yan, et al., Two-dimensional nanosheets for electrocatalysis in energy generation and conversion, J. Mater. Chem. A. 5 (2017) 7257–7284. Available from: https://doi.org/10.1039/C7TA00075H.
[142] Y. Sun, Q. Liu, S. Gao, H. Cheng, F. Lei, Z. Sun, et al., Pits confined in ultrathin cerium(IV) oxide for studying catalytic centers in carbon monoxide oxidation, Nat. Commun. 4 (2013) 2899. Available from: https://doi.org/10.1038/ncomms3899.
[143] Y. Sun, S. Gao, F. Lei, Y. Xie, Atomically-thin two-dimensional sheets for understanding active sites in catalysis, Chem. Soc. Rev. 44 (2015) 623–636. Available from: https://doi.org/10.1039/C4CS00236A.
[144] Y. Wang, J. Li, Z. Wei, Transition-metal-oxide-based catalysts for the oxygen reduction reaction, J. Mater. Chem. A 6 (2018) 8194–8209. Available from: https://doi.org/10.1039/C8TA01321G.
[145] Y. Zhu, X. Liu, S. Jin, H. Chen, W. Lee, M. Liu, et al., Anionic defect engineering of transition metal oxides for oxygen reduction and evolution

reactions, J. Mater. Chem. A. 7 (2019) 5875–5897. Available from: https://doi.org/10.1039/C8TA12477A.
[146] Z. Lin, B.R. Carvalho, E. Kahn, R. Lv, R. Rao, H. Terrones, et al., Defect engineering of two-dimensional transition metal dichalcogenides, 2D Mater. 3 (2016) 022002. Available from: https://doi.org/10.1088/2053-1583/3/2/022002.
[147] G. Wang, Y. Ling, Y. Li, Oxygen-deficient metal oxide nanostructures for photoelectrochemical water oxidation and other applications, Nanoscale. 4 (2012) 6682-6691. Available from: https://doi.org/10.1039/c2nr32222f.
[148] H.-S. Kim, J.B. Cook, H. Lin, J.S. Ko, S.H. Tolbert, V. Ozolins, et al., Oxygen vacancies enhance pseudocapacitive charge storage properties of MoO_{3-x}, Nat. Mater. 16 (2017) 454–460. Available from: https://doi.org/10.1038/nmat4810.
[149] N. Mahmood, I.A. De Castro, K. Pramoda, K. Khoshmanesh, S.K. Bhargava, K. Kalantar-Zadeh, Atomically thin two-dimensional metal oxide nanosheets and their heterostructures for energy storage, Energy Storage Mater. 16 (2019) 455–480. Available from: https://doi.org/10.1016/j.ensm.2018.10.013.
[150] Y. Dou, T. Liao, Z. Ma, D. Tian, Q. Liu, F. Xiao, et al., Graphene-like holey Co_3O_4 nanosheets as a highly efficient catalyst for oxygen evolution reaction, Nano Energy. 30 (2016) 267–275. Available from: https://doi.org/10.1016/j.nanoen.2016.10.020.
[151] Y. Dou, Y. Wang, D. Tian, J. Xu, Z. Zhang, Q. Liu, et al., Atomically thin Co_3O_4 nanosheet-coated stainless steel mesh with enhanced capacitive Na^+ storage for high-performance sodium-ion batteries, 2D Mater. 4 (2016) 015022. Available from: https://doi.org/10.1088/2053-1583/4/1/015022.
[152] J. Mei, T. Liao, G.A. Ayoko, Z. Sun, Two-dimensional bismuth oxide heterostructured nanosheets for lithium- and sodium-ion storages, ACS Appl. Mater. Interfaces 11 (2019) 28205–28212. Available from: https://doi.org/10.1021/acsami.9b09882.
[153] Y. Dou, J. Xu, B. Ruan, Q. Liu, Y. Pan, Z. Sun, et al., Atomic layer-by-layer Co_3O_4 /graphene composite for high performance lithium-ion batteries, Adv. Energy Mater. 6 (2016) 1501835. Available from: https://doi.org/10.1002/aenm.201501835.
[154] K. Kalantar-zadeh, J.Z. Ou, T. Daeneke, A. Mitchell, T. Sasaki, M.S. Fuhrer, Two dimensional and layered transition metal oxides, Appl. Mater. Today. 5 (2016) 73–89. Available from: https://doi.org/10.1016/j.apmt.2016.09.012.
[155] J.Y. Hwang, M.F. El-Kady, Y. Wang, L. Wang, Y. Shao, K. Marsh, et al., Direct preparation and processing of graphene/RuO_2 nanocomposite electrodes for high-performance capacitive energy storage, Nano Energy 18 (2015) 57–70. Available from: https://doi.org/10.1016/j.nanoen.2015.09.009.
[156] J. Mei, Y. Zhang, T. Liao, Z.Q. Sun, S.X. Dou, Strategies for improving the lithium storage performance of 2D nanomaterials, Natl. Sci. Rev. 5 (2018) 389–416. Available from: https://doi.org/10.1093/nsr/nwx077.
[157] J. Mei, Y. Zhang, T. Liao, X. Peng, G.A. Ayoko, Z. Sun, Black phosphorus nanosheets promoted 2D-TiO_2-2D heterostructured anode for high-performance lithium storage, Energy Storage Mater. 19 (2019) 424–431. Available from: https://doi.org/10.1016/j.ensm.2019.03.010.
[158] Y. Zhang, J. Mei, C. Yan, T. Liao, J. Bell, Z.Q. Sun, Bioinspired 2D nanomaterials for sustainable applications, Adv. Mater. 31 (2019) 1902806. Available from: https://doi.org/10.1002/adma.201902806.

[159] Z.Q. Sun, T. Liao, W. Li, Y. Qiao, K. Ostrikov, Beyond Seashells: bio-inspired 2D photonic and photoelectronic devices, Adv. Funct. Mater. 29 (2019) 1901460. Available from: https://doi.org/10.1002/adfm.201901460.
[160] J. Mei, T. Liao, H. Spratt, G.A. Ayoko, X.S. Zhao, Z. Sun, Honeycomb-inspired heterogeneous bimetallic Co–Mo oxide nanoarchitectures for high-rate electrochemical lithium storage, Small Methods. 3 (2019) 1900055. Available from: https://doi.org/10.1002/smtd.201900055.

Further reading

A.H. Khan, S. Ghosh, B. Pradhan, A. Dalui, L.K. Shrestha, S. Acharya, et al., Two-dimensional (2D) nanomaterials towards electrochemical nanoarchitectonics in energy-related applications, Bull. Chem. Soc. Jpn. 90 (2017) 627–648. Available from: https://doi.org/10.1246/bcsj.20170043.

3

Graphene-based hybrid materials for advanced batteries

Zhongkan Ren, Santanu Mukherjee and Gurpreet Singh
Mechanical and Nuclear Engineering Department, Kansas State University, Manhattan, KS, United States

3.1 Introduction

Lithium-ion batteries (LIBs), comprising of a graphite anode and lithium cobalt oxide cathode, have become the overwhelmingly popular electrochemical energy storage devices currently [1]. They have propelled the modern portable and commercial electronics industry and have revolutionized the way humans connect with personal electronic items, for example, cellular phones, google glass, etc. [2]. Li poses some inherent advantages which has made this possible: its rather low ionic radii of 0.76 Å, a high working potential of about 3.6 V in which a majority of electrolytes are stable, elevated gravimetric energy densities of $\sim$120–150 Wh kg^{-1} and high theoretical capacity (3861 mAh g^{-1}) to enumerate a few [2–4]. These beneficial properties make the current LIB systems the undisputed gold standard for small-scale storage [5]. It is estimated that the global market for rechargeable LIB systems now stands at 10 billion dollars [6]. However, with an ever-increasing need for greater energy storage, especially for medium and large scale, to meet our bourgeoning demands, a time has come to develop alternatives electrochemical energy storage with even more superior characteristics. As such increasing scrutiny has been drawn to some issues of LIBs. Firstly, Li metal's rarity in the earth's crust and consequent high cost ($5000/ton vis-à-vis $150/ton for Na resources) makes it almost economically impractical for large grid-scale storage [4]. Limited capacity of the graphite anode ($\sim$370 mAh g^{-1} for Li^+ ion) is another practical issue. To overcome this and to obtain higher capacities, studies have focused on utilizing electrodes that

2D Nanomaterials for Energy Applications. DOI: https://doi.org/10.1016/B978-0-12-816723-6.00003-4

operate at high voltages (≥4.2 V), for example, the spinel $LiMn_{1.5}Ni_{0.5}O_4$; however, the organic electrolyte starts to dissociate at such voltages [7]. Li plating on the anode when operated below 1.7 V making the battery potentially prone to an explosion and the use of toxic organic electrolytes are other important challenges of LIB systems that are driving greater research toward alternative electrochemical energy storage systems [8,9].

To meet these rapidly expanding energy storage necessities, newer and more advanced battery chemistries are being proposed. Lithium-sulfur (Li-S) batteries use sulfur as the cathode and take advantage of sulfur's large theoretical capacity for Li (1672 mAh g^{-1}) and sulfur's abundance in the earth's crust [10]. However, Li usually tends to undergo a conversion reaction with S (forming predominantly Li_2S and other intermediate polysulfides) which produces almost an 80% volume expansion and seriously impedes the lifetime of these batteries [10]. Also, sulfur and the intermediate products formed during its discharge demonstrates poor electronic and ionic conductivities which considerably lowers the energy efficiency of the system [11]. Other alkali-metal ion systems such as sodium, potassium, calcium, and even magnesium systems (SIB, KIB, CIB, and MIB) have received notable attention [12–15]. Sodium and potassium's advantage lies in their geologically greater availability and consequent low cost ($150/ton for Na and $216/ton for K) which makes them possible candidates for large-scale grid storage [4,16]. Magnesium has the added advantage of being divalent thereby being able to contribute twice as much charge per atom as compared to Li, Na, and K atoms [17]. Ca is the fifth most abundant element in the earth and it demonstrates a standard reduction potential only 0.17 V lower than Li, thereby theoretically being able to provide higher overall cell voltages when compared to MIB systems [18,19]. Regardless, issues in these alkali systems persist as well. Na and K are relatively large ions which makes intercalation into host systems an issue, coupled with low working voltages and energy densities [9,20]. Particularly for SIB systems, graphite cannot be used as an anode because of thermodynamic considerations [21]. Low energy density and insufficient Coulombic efficiency have hindered the progress of CIB systems [22]. For MIB systems, the absence of a suitable stable electrolyte that will aid the efficient reversible transport of Mg^{2+} ions remains a cause of concern, along with a lack of clear understanding of the insertion of the divalent species in the host [23]. Therefore it becomes clear from this discussion that the prudent choice of electrode material is very important in the efficient running of the battery system under consideration. *Thus it is to be noted that for this discussion, Li-S, SIB, KIB, MIB, and CIB systems have been considered as advanced battery systems.*

Graphene, essentially a single layer of graphite consisting of a monolayer of sp^2 carbon atoms arranged in a hexagonal pattern and a C–C bond length of 1.42 Å, have attracted a lot of attention lately as an electrode material for electrochemical energy storage [24–26]. This is primarily because of the rather large theoretical specific surface area that they present ($2630\,m^2\,g^{-1}$), elevated intrinsic carrier mobility ($200{,}000\,cm^2\,V^{-1}\,s^{-1}$) and excellent mechanical properties (a breaking strength of $42\,N\,m^{-1}$ for a defect-free sheet) [27–29]. Harnessing its favorable properties as discussed, graphene has been studied in detail for electrode applications in electrochemical storage systems [30,31]. For example, Wen et al. used expanded graphite nanosheets as anodes in SIBs and were able to demonstrate a reversible capacity of $284\,mAh\,g^{-1}$ at a current density of $20\,mA\,g^{-1}$ with a capacity retention of 73.92% after 2000 cycles of operation [32]. However, the use of standalone graphene poses its own set of challenges. The techniques used to fabricate graphene (liquid-phase exfoliation, chemical vapor deposition of hydrocarbons, and Hummer's method) are energy-intensive, involve large quantities of corrosive acids, environmentally unfriendly (liberating considerable amounts of CO_2 to the atmosphere), and are not very scalable [33,34]. Ionic repulsion on individual graphene layers has also been observed to reduce the working capacity [35]. These issues make obtaining large quantities of graphene for electrochemical energy storage purposes much costlier than graphite.

To overcome these challenges posed by standalone graphene, graphene-based hybrids where graphene or reduced graphene oxide (rGO) has been combined with other materials have been proposed and studied. *For this chapter, only those hybrids where graphene has been combined with other 2D Li-active materials such as transition metal dichalcogenides (TMDs), phosphorene, or MXenes are considered.* There are several ways in which graphene-based 2D material hybrids have been synthesized for electrochemical energy storage applications. Graphene as a conducting agent is one such strategy, where it substitutes for the more commonly used carbon black and it greatly enhances the electrochemical kinetics. Theoretical studies have proven this concept, with the composite providing a low diffusion barrier, along with a high stiffness which maintains the overall structural integrity of the electrode [36,37]. Also, as only 5 wt.% (w.r.t to total weight of active electrode material) of conducting agent is needed, cost does not become an important concern in such cases. Strategies also involve using graphite as component of a composite and its application as a free-standing electrode [38]. This has played a synergistic role in improving the overall capacity and cycle life stability [38,39]. The subsequent sections will discuss the application and usefulness of graphene-based hybrids for the advanced battery systems listed above.

3.2 Graphene hybrid electrodes in advanced batteries

The key feature of graphene-based hybrid materials discussed in this chapter is that graphene stays largely inactive while other 2D materials are responsible to electrochemical energy storage [40]. In these hybrids, graphene works both as conducting material and a binder which not only enhances the electrical and thermal conductivity but also provide a support matrix for 2D nanoparticles [40]. There are three kinds of graphene/2D materials hybrid mechanics reported in this chapter (shown in Fig. 3.1).

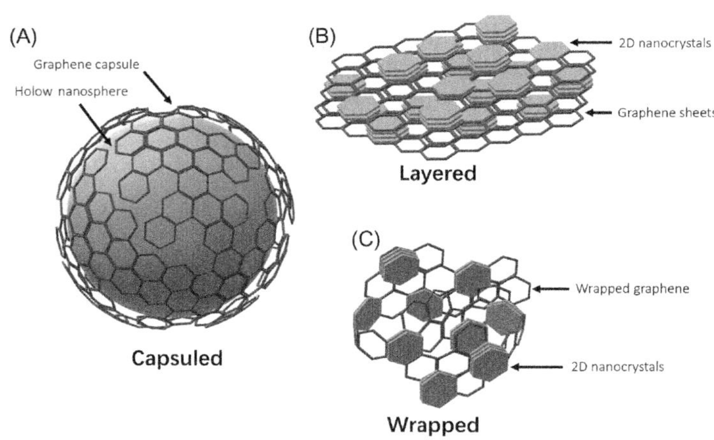

Figure 3.1 Hybrid mechanics of graphene/2D nanomaterials. (A) Encapsulated structure. 2D active materials are in the form of several-layer-thin multiwalled (similar to multiwalled carbon nanotube structure) hollow sphere and encapsulated by graphene shell on the outer surface. The hollowed structure is beneficial in controlling volume expansion during intercalation. (B) Layered structure. Sandwich-like structures of graphene sheets or flakes and single or few-layered active nanocrystals. (C) Wrapped structure. Mixed status of wrapped graphene sheets and active nanocrystals. *Source*: R. Raccichini, A. Varzi, S. Passerini, B. Scrosati, The role of graphene for electrochemical energy storage, Nat. Mater. 14 (2015) 271–297.

It is understood that graphene can significantly improve the electrochemical behavior of lithium active 2D particle (e.g., TMD, MXene, or phosphorene) in a three-pronged way [41–45]. Firstly, exfoliated 2D nanoparticles are known to restack into their bulk counterpart during electrode assembly thereby compromising characteristic properties leading to a loss of Li-reversibility and fast electrode degradation. Here incorporation of graphene (or reduced graphene oxide) prevents agglomeration of active 2D nanoparticles [41]. Secondly, graphene sheets also provide a shorter electron path to current collector for coated nanoparticles thereby improving rate capability [41]. Thirdly, large free interstitial space in a composite electrode system can accommodate volumetric changes during intercalation/deintercalation of ions and enhance the cyclability (Fig. 3.2) [41].

The contributions by different research groups in "2D materials/graphene hybrid electrodes in advanced batteries" over the past decade will be discussed in sections 3.3–3.7. The chapter will be arranged in the order of various 2D active materials/graphene applications in different types of advanced batteries, that is, TMDs, SnS_2, and graphene applications in SIB, KIB, and Mg batteries.

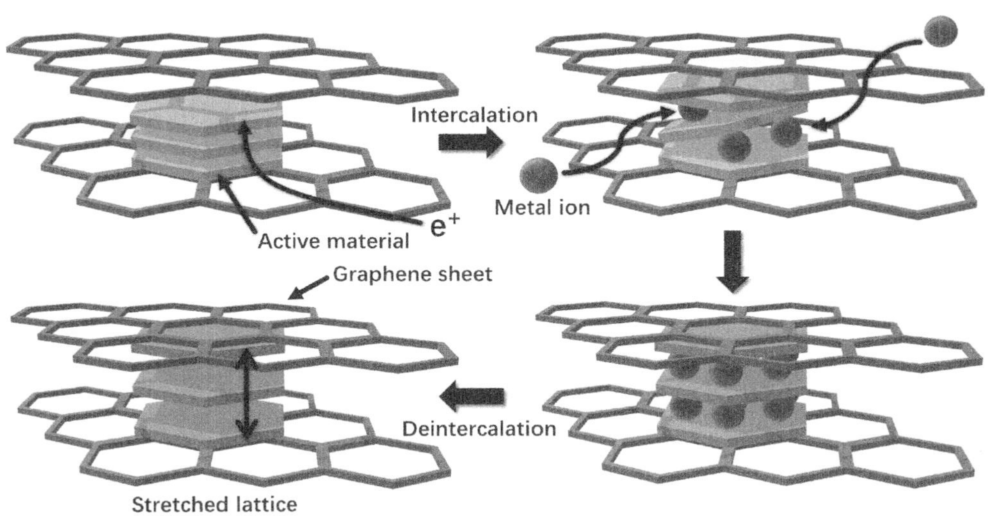

Figure 3.2 Intercalation/deintercalation of hybrid electrode. Schematics of graphene enhancing mechanism in hybrid electrode while intercalation and deintercalation: improved electron conductivity, controlled volume expansion after charge and discharge.

3.3 Transition metal dichalcogenides/graphene composite in sodium ion battery

Na-ion batteries have been considered one of the alternatives to Li-ion batteries owing to the plentiful and evenly distributed sodium resources [39]. However, the issues such as large volume changes and low diffusion kinetics that occur with SIB electrodes have hindered their commercialization [45]. Graphite, as an anode material, is largely electrochemically inactive to Na-ion intercalation. To overcome this problem, and to allow for intercalation of Na-ions, exfoliated graphene sheets have been proposed instead [45]. However, the charge capacity improved only marginally [45]. Hence, incorporation of other 2D materials as active sodium storing site is an important step to next-generation SIB.

A MoS_2/reduced graphene oxide composite was synthesized by Qin et al. Reduction of graphene oxide was carried out in MoS_2 solution via microwave irradiation followed by annealing at 800°C in N_2/H_2 for 2 h [46]. The composite performed better than pure MoS2 as an anode in a SIB system, both in terms of specific capacity and cyclical stability (Fig. 3.3M) [46]. A specific capacity 305 $mAh\,g^{-1}$ was recovered after 50 cycles when the current density raised back to 100 $mA\,g^{-1}$ (Fig. 3.3L) with rate performance of 214 $mAh\,g^{-1}$ (70% of initial) at 1 $A\,g^{-1}$ (Fig. 3.3L) [46]. Highly porous structure (Fig. 3.3B–D), reduced electrical resistance by incorporation of graphene, and well-controlled volume expansion of MoS_2/rGO composite attribute to the elevated electrochemical performance (Fig. 3.3G–M) [46].

Park et al. proposed a series of hybrid materials from layered $MoSe_2$, graphene, and carbon nanotube (CNT) fabricated by a spray pyrolysis method [47]. Ammonium molybdate, graphene oxide, and multiwall CNT were used as precursors to produce hybrid powders [47]. The GO nanosheets prevent the growth of $MoSe_2$ nanocrystals that minimizes the stacking of $MoSe_2$ [47]. Incorporation of CNT backbone into the composite can effectively improve the porosity and further optimize the stacking issue of $MoSe_2$ and rGO layers [47]. The electrochemical performance (sodium storage) tests conducted as anode materials in SIB (Fig. 3.4H–O, Q) [47]. The results showed best performance found in $MoSe_2$-rGO-CNT composite (vs Bare $MoSe_2$, $MoSe_2$-rGO, and $MoSe_2$-CNT) which confirmed enhancement effect of rGO and CNT [47]. The reported values are an initial capacity of 501.6 $mAh\,g^{-1}$ at a current density of 1.0 $A\,g^{-1}$ (Figs. 3.4N),

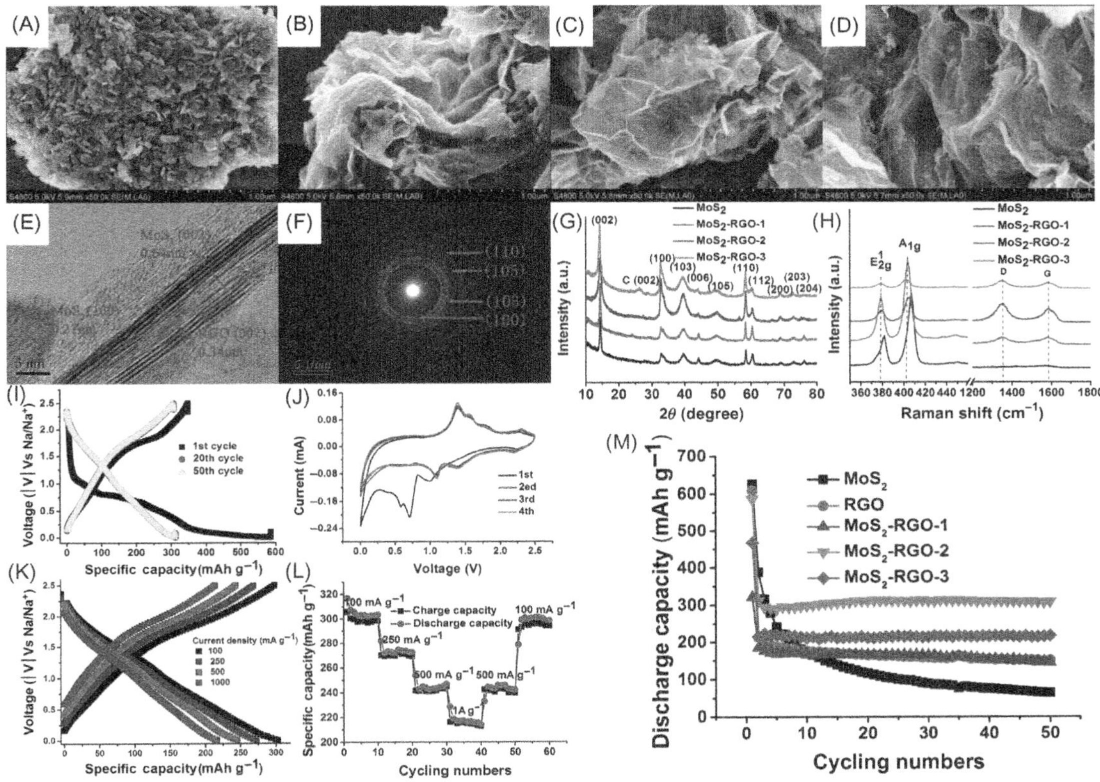

Figure 3.3 Structural characterization and electrochemical performance analysis of MoS_2/rGO hybrid electrode for SIB. FEM images of (A) MoS_2, and (B, C, and D) different MoS_2/rGO with 0.2, 0.4, and 0.6 g rGO content. (E) HRTEM image of MoS_2/0.4 g rGO. (F) SAED pattern of MoS_2/0.4 g rGO. (G) XRD patterns of MoS_2, MoS_2/rGO with 0.2, 0.4, and 0.6 g rGO content. (H) Raman spectra of MoS_2, MoS_2/rGO with 0.2, 0.4, and 0.6 g rGO content. (I) GCD profiles of MoS_2/0.4 g rGO. (J) CV curves of MoS_2/0.4 g rGO. (K) GCD profiles of MoS_2/0.4 g rGO at different current densities. (L) Rate performance of MoS_2/0.4 g rGO. (M) Cycling performances of MoS_2, rGO, MoS_2/rGO with 0.2, 0.4, and 0.6 g rGO content at 100 mA g^{-1} [46]. *Source*: W. Qin, T. Chen, L. Pan, L. Niu, B. Hu, D.L.J. Li, Z. Sun, MoS2-reduced graphene oxide composites via microwave assisted synthesis for sodium ion battery anode with improved capacity and cycling performance, Electrochim. Acta 153 (2015) 55–61.

393 mAh g^{-1} retained capacity after 200 cycles (Fig. 3.4L), as well as 411 mAh g^{-1} at a current density of 30 A g^{-1} (high charge rate) (Fig. 3.4M) [47].

Another recent work performed by David et al. proposed acid-treated MoS_2/rGO composite paper application in SIB as anode material [39]. Composite electrodes were prepared through thermal reduction of GO with sonication-assisted acid-exfoliated MoS_2 dispersion [39]. The 2D structure of MoS_2 was confirmed by TEM, XRD, and Raman (Fig. 3.5C–H) [39].

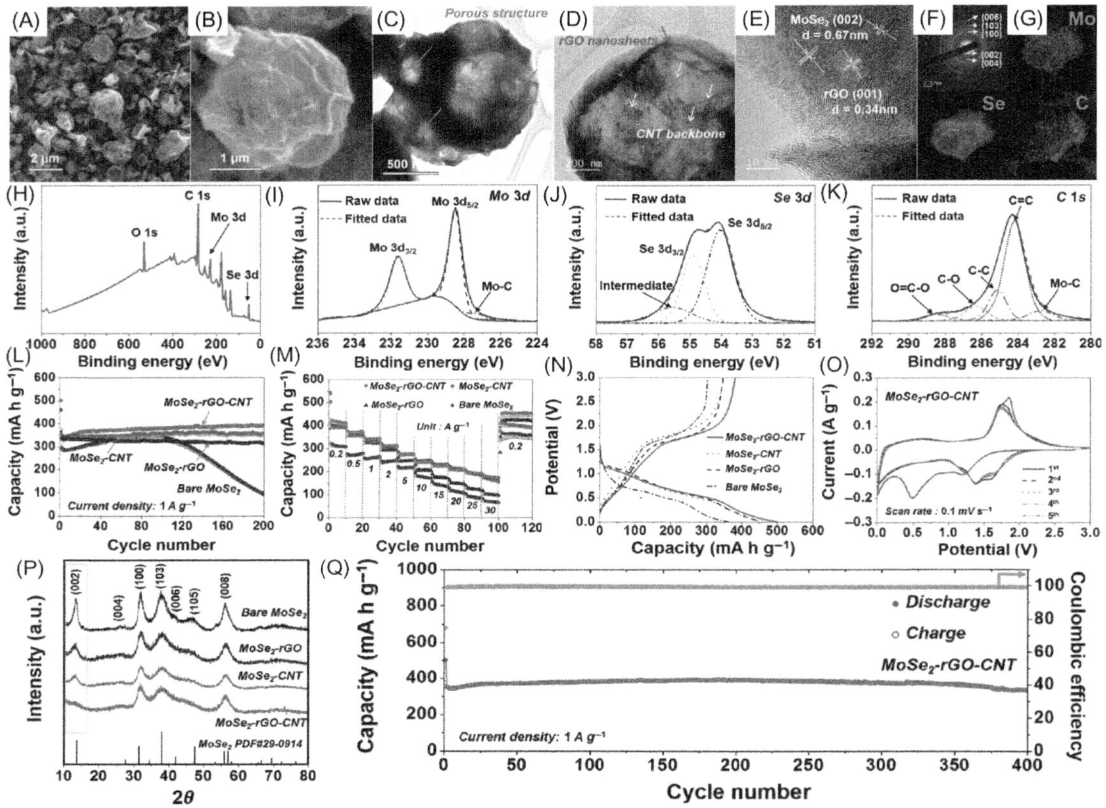

Figure 3.4 Structural characterization and electrochemical performance analysis of $MoSe_2$/rGO/CNT hybrid electrode for SIB. (A, B) SEM images, (C, D) TEM images, (E) HRTEM image, and (F, G) SEAD and element mapping of MoS_2/rGO/CNT composite. XPS spectrum of (H) wide scan, (I) Mo3d, (J) Se3d, and (K) C1s. (L) Cycling performance, (M) rate performance, and (N) initial charge/discharge of $MoSe_2$/rGO/CNT, $MoSe_2$/CNT, $MoSe_2$/rGO, and bare $MoSe_2$ at 1.0 A g^{-1}. (O) CV curves of initial 5 cycles of $MoSe_2$/rGO/CNT at 0.1 mV s^{-1}. (P) XRD patterns of $MoSe_2$/rGO/CNT, $MoSe_2$/CNT, $MoSe_2$/rGO, and bare $MoSe_2$. (Q) Cyclability of $MoSe_2$/rGO/CNT at 1.0 A g^{-1}. *Source*: G.D. Park, J.H. Kim, S.K. Park, Y.C. Kang, $MoSe_2$ Embedded CNT-reduced graphene oxide composite microsphere with superior sodium ion storage and electrocatalytic hydrogen evolution performances, Appl. Mater. Interfaces, 9 (2017) 10673–10683.

Charging capacity of the half-cell reached 230 mAh g^{-1} with a Coulombic efficiency of about 99% at a current density of 25 mA g^{-1} (Fig. 3.5N–Q) [39].

WS_2/graphene composite anode application in SIB was studied by Su et al. with 594 mAh g^{-1} of sodium storage capacity (Fig. 3.6I and L) [48]. Nanocomposites, synthesized via hydrothermal method, exhibited symmetric hexagonal unit cell structure which was indicated by XRD (Fig. 3.6C) and SAED pattern (Fig. 3.6E) [48]. Cell cycling study found enhanced capacity

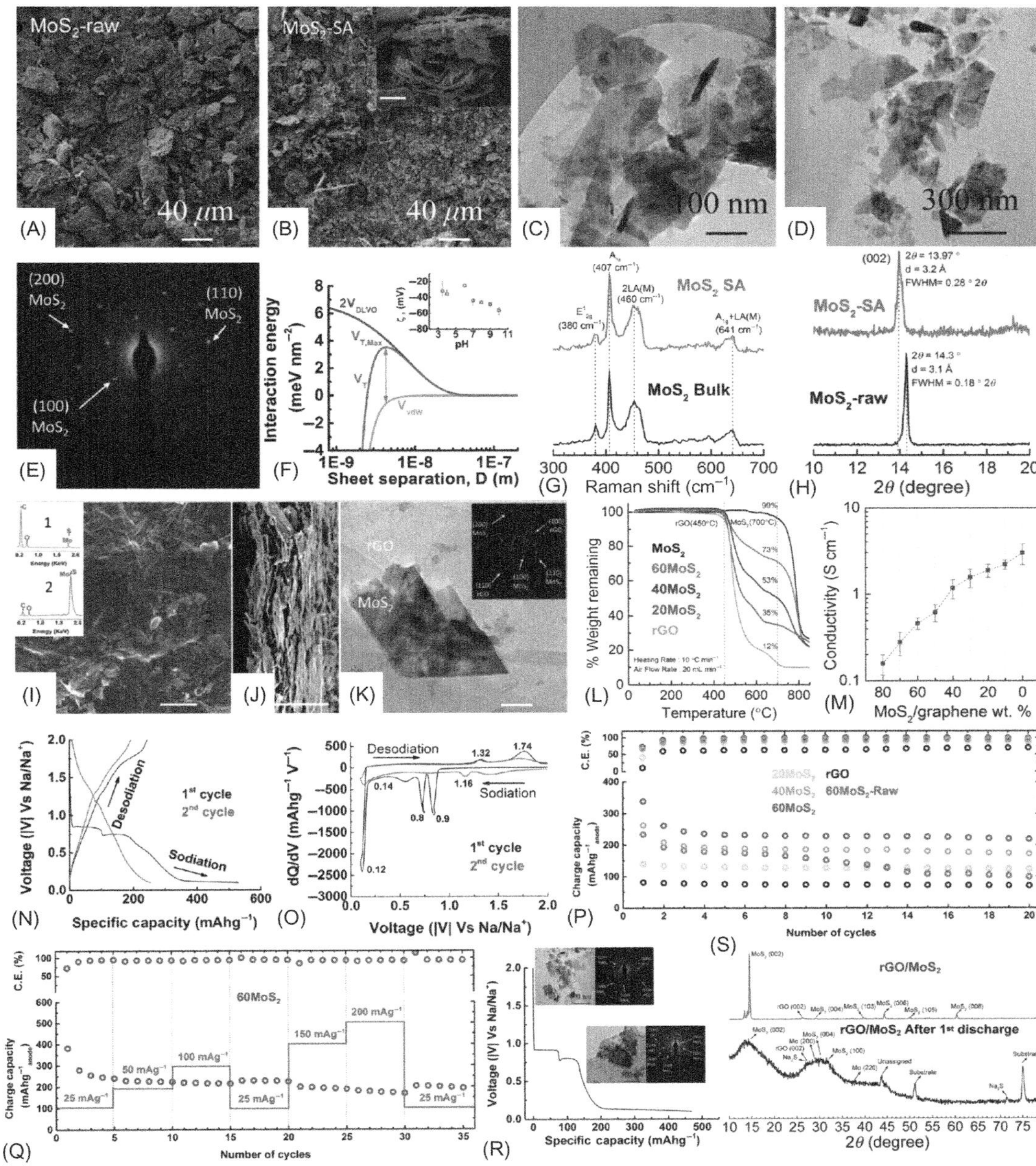

Figure 3.5 Structural characterization and electrochemical performance analysis of MoS_2/graphene hybrid electrode for SIB. SEM images of MoS_2 (A) raw and (B) after superacid treatment. (C, D) HRTEM images of superacid treated MoS_2. (E) Corresponding SAED pattern. (F) Calculated intercalation energies (potential, repulsion, and attraction) versus MoS_2 sheet separation distance. (G) Raman spectra of bulk MoS_2 and acid-treated MoS_2. (H) XRD patterns of bulk MoS_2 and acid-treated MoS_2. SEM images of MoS_2/graphene paper (I) top view with EDX spectra of corresponding spots (J) cross-sectional view. (K) TEM image of MoS_2/graphene with corresponding SAED pattern. (L) TGA results of acid-treated MoS_2, MoS_2/graphene, and rGO. (M) Electrical conductivity with different MoS_2 loading amount. (N) GCD curves of MoS_2/graphene. (O) CV curves of MoS_2/graphene. (P) Cycling performance and Coulombic efficiency of MoS_2/graphene at current density of 20 mA g^{-1}. (Q) Rate performance and Coulombic efficiency of MoS_2/graphene at different current densities. (R) TEM images with corresponding SAED patterns, and (S) XRD patterns of MoS_2/graphene before and after first cycle at 0.01 V [39]. *Source*: L. David, R. Bhandavat, G. Singh, MoS_2/graphene composite paper for sodium-ion battery electrodes, ACS Nano 8 (2014) 1759–1770.

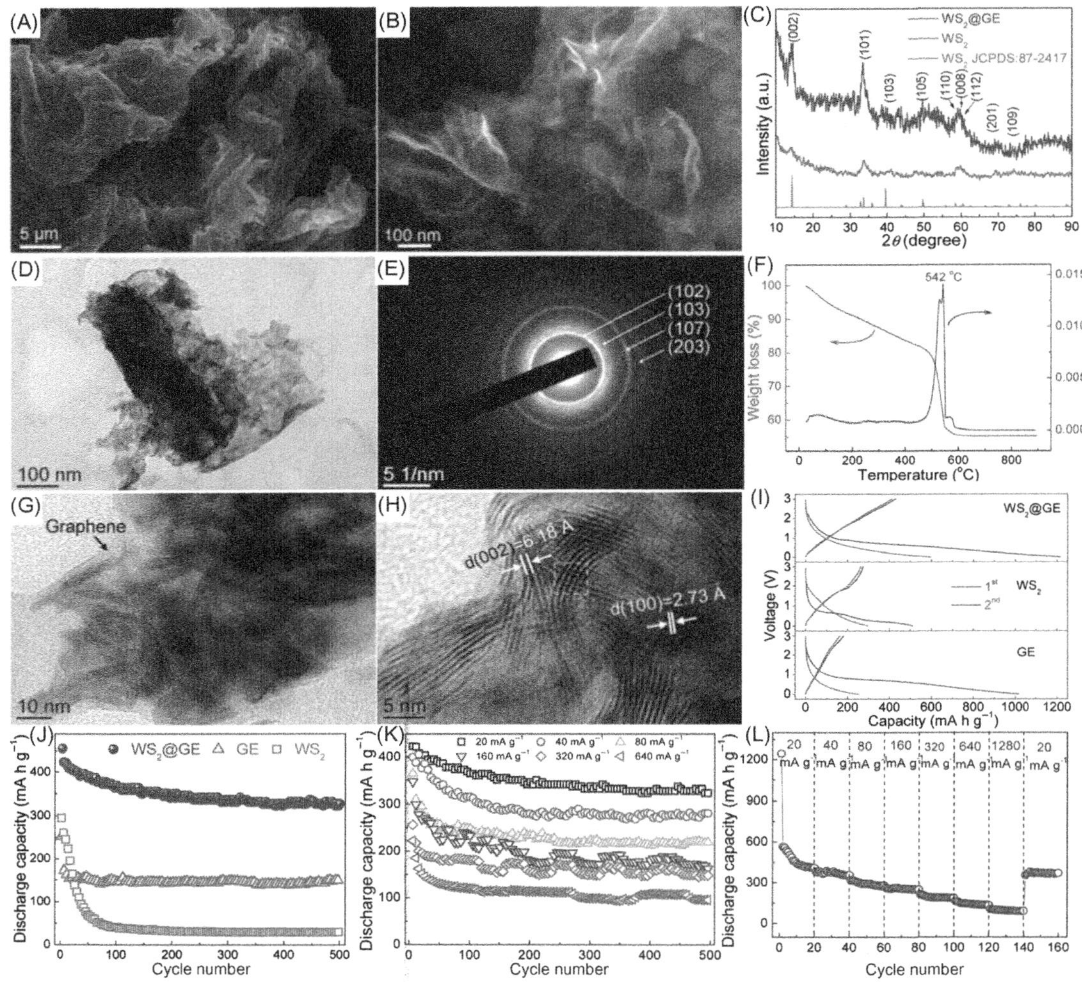

Figure 3.6 Structural characterization and electrochemical performance analysis of WS_2/graphene hybrid electrode for SIB. (A, B) FESEM images of WS_2/graphene. (C) XRD patterns of WS_2 and WS_2/graphene. (D, G, H) TEM and HRTEM images of WS_2/graphene and (E) matching SAED pattern of (D). (F) TG and DTA results of WS_2/graphene. (I) Initial two cycles GCD profile of WS_2/graphene, WS_2, and graphene. (J) Discharge performances of WS_2/graphene, WS_2, and graphene over cycle number. (K) Discharge performance of WS_2/graphene at different current densities. (L) Rate performance of WS_2/graphene at different current densities [48]. *Source*: D. Su, S. Dou, G. Wang, WS_2@graphene nanocomposites as anode materials for Na-ion batteries with enhanced electrochemical performances, Chem. Commun., 50 (2014) 4192–4195.

(329 mAh g^{-1} after incorporation of graphene versus 32 mAh g^{-1} for bare WS_2) after 500 cycles (Fig. 3.6J) [48].

A spherical WS_2/rGO composite material was synthesized by Choi et al. for SIB anode application [42]. Unlike direct "one-pot" hydrothermal method by Su et al., Choi et al. applied two-step syntheses to achieve the layered WS_2 decorated graphene oxide microsphere structure [49]. WO_3/rGO spheres were initially produced by ultrasonic spray pyrolysis method, then WO_3 to WS_2 sulfidation was completed by thiourea [49]. The internal phase transformation from bulk WO_3 to layered WS_2 was revealed by HRTEM and XRD patterns (Fig. 3.7E–G) [49]. They reported 356 mAh g^{-1} initial capacity with 56% Coulombic efficiency and 334 mAh g^{-1} retention after 200 cycles at 200 mA g^{-1} (Fig. 3.7K and L) [49].

There also exist notable investigations of 2D TMDs/graphene hybrid materials as electrode applications in SIB. Sahu et al. studied sonication-assisted MoS_2/graphene in SIB which reported reversible capacity as high as 575 mAh g^{-1} at 100 mA g^{-1} under potential window of 0.01–2.6 V and this value reduced to 218 mAh g^{-1} at 50 mA g^{-1} when potential range changed to 0.4–2.6 V [44]. For hydrothermally synthesized MoS_2/graphene hybrid, Wang et al. and Xie et al. sequentially announced capacity, respectively, 313 mAh g^{-1} at 100 mA g^{-1} after 200 cycles and 254 mAh g^{-1} at 80 mA g^{-1} after 300 cycles [43,50]. While Zhang et al. applied hydrothermal method to $MoSe_2$ and obtained stable and even higher reversible capacity of 430 mAh g^{-1} over 200 cycles at 500 mA g^{-1} high charge rate [51]. Choi et al. employed a similar spray pyrolysis strategy as introduced above to WS_2 and graphene. They successfully applied in SIB as an anode material with 334 mAh g^{-1} recovered capacity at 200 mA g^{-1} after 200 cycles [49].

3.4 SnS_2/graphene composite in sodium ion battery

The investigation of electrochemical behavior of exfoliated SnS_2/graphene composite electrode material was conducted by Liu et al. in SIB systems [52]. Fast hydrolysis and sonication led to the formation of H_2 at $LiSnS_2$ interlayer space which pushed SnS_2 layers to separate from each other [52]. As-formed two- to five-layered hexagonal SnS_2 particles were restacked on graphene nanosheets via hydrothermal method (Fig. 3.8A, B, F–H) [52]. A specific capacity of 650 mAh g^{-1} at a current density of 200 mA g^{-1} was achieved by the composite anodes and maintained at 610 mAh g^{-1} after 300 cycles. A capacity of 326 mAh g^{-1} was obtained at a current density 4000 mA g^{-1} and showed remarkable rate performance as well (Fig. 3.8N and O) [52].

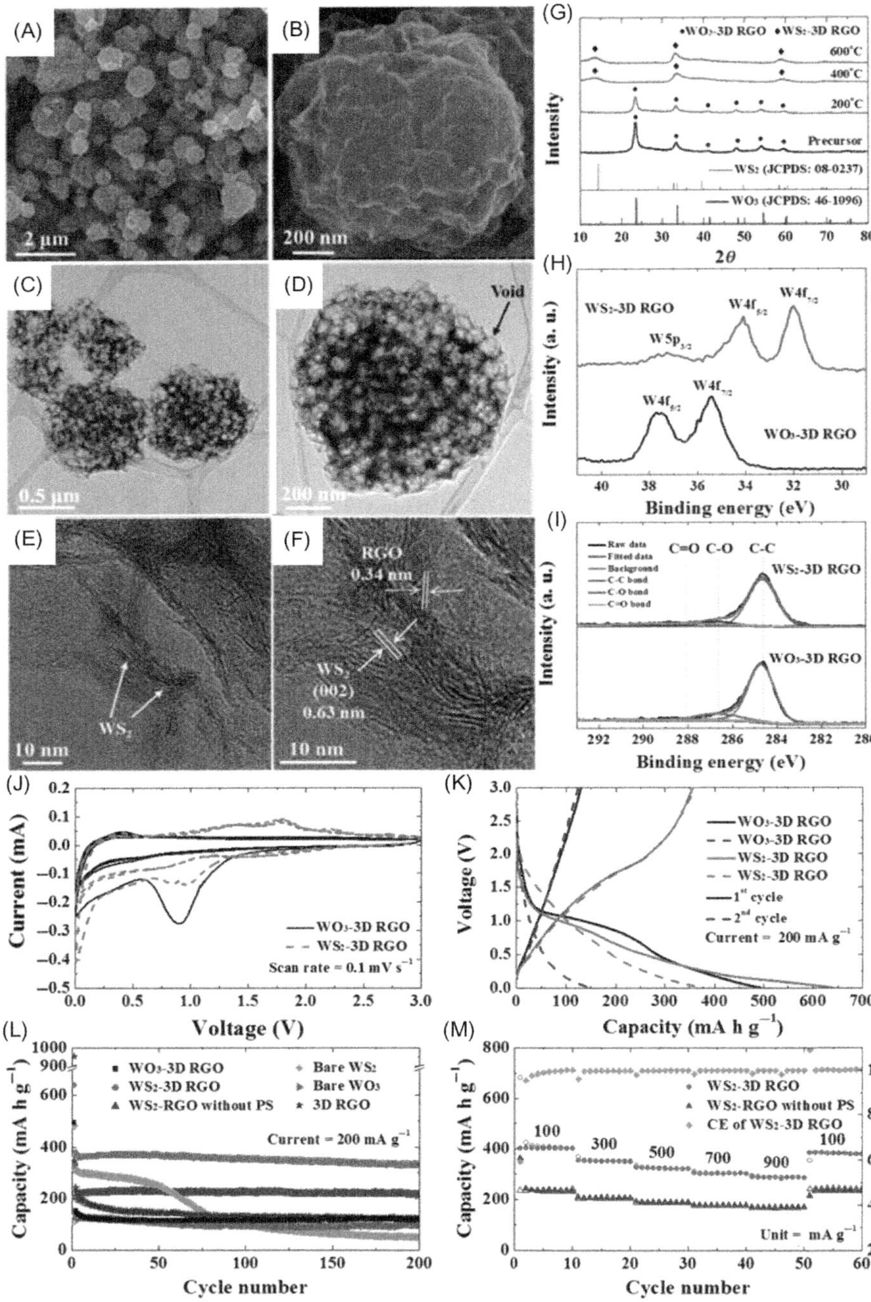

Figure 3.7 Structural characterization and electrochemical performance analysis of WS_2/3D-rGO hybrid electrode for SIB. (A, B) SEM images (C–F) TEM and HRTEM images of WS_2/rGO (G) XRD pattern. XPS of (H) W4f and (I) C1s. (J) CV curves, (K) initial charge/discharge curve, (L) cyclability, and (M) rate performance of WS_2/rGO nanocomposite. *Source*: S.H. Choi, Y.C. Kang, Sodium ion storage properties of WS_2-decorated three-dimensional reduced graphene oxide microspheres, Nanoscale, 7 (2015) 3956–3970.

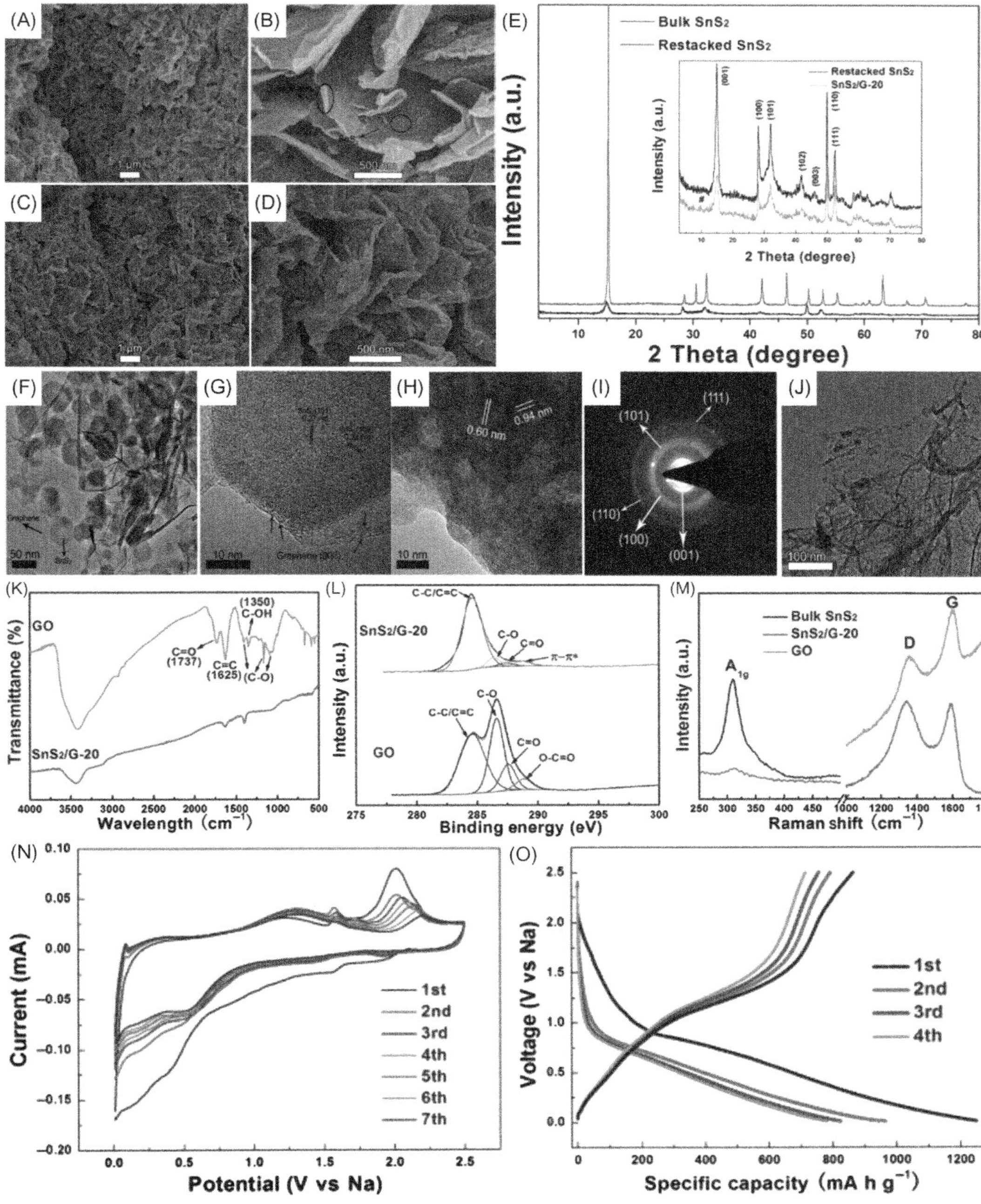

Figure 3.8 Structural characterization and electrochemical performance analysis of SnS_2/graphene hybrid electrode for SIB. SEM images of SnS_2/graphene (A, B) before cycles and (C, D) after 100 cycles. (E) XRD patterns of bulk, restacked SnS_2 and SnS_2/graphene. TEM and HRTEM images of SnS_2/graphene (F, G, H) before cycles and (J) after 100 cycles. (I) SAED pattern of SnS_2/graphene. (K) FT-IR spectra of SnS_2/graphene and GO. (L) XPS spectra of SnS_2/graphene and GO (C 1s). (M) Raman spectra of GO, SnS_2/graphene, and bulk SnS_2. (N) CV curves of SnS_2/graphene at 0.1 mV s^{-1} between potential of 0.01–2.5 V. (O) GCD curves of SnS_2/graphene at 200 mA g^{-1} current density [52]. *Source*: Y. Liu, H. Kang, L. Jiao, C. Chen, K. Cao, Y. Wang, H. Yuan, Exfoliated-SnS_2 restacked on graphene as a high-capacity, high-rate, and long-cycle life anode for sodium ion batteries, Nanoscale, 7 (2015) 1325–1332.

Another comparable application of layered SnS_2/rGO composite anode for SIB was reported by Qu et al. with a slightly different synthesis technique [41]. A hydrothermal route was applied to $SnCl_4$, thioacetamide, and GO mixture [41]. The composite electrode was able to deliver specific charge capacity of 630 mAh g^{-1} at current density of 0.2 A g^{-1}, rate performance of 544 mAh g^{-1} at 2 A g^{-1} as well as cyclability of 500 mAh g^{-1} after 400 cycles at 1 A g^{-1} (Fig. 3.9J) [41].

3.5 Phosphorene/graphene composite in SIB

Phosphorene/graphene hybrid anodes were prepared via liquid-phase exfoliation by Sun et al. for SIB application [45]. Na-ion intercalation mechanism through black phosphorus was emphasized in their study to explain the specific capacity of 2440 mAh g^{-1} achieved by phosphorene/graphene composite [45]. Formation of Na_3P alloy enabled enormous Na-ion storage in phosphorene however about 92% of volumetric expansion is inevitable [45]. Therefore incorporation of graphene sheet greatly controlled expansion and provided a buffer layer [45]. Thin layered phosphorene optimized diffusions for Na-ions and electrons as well [45]. At a current density of 50 mA g^{-1}, up to 2440 mAh g^{-1} initial capacity was achieved while 2080 mAh g^{-1} was recovered after 100 cycles [45].

3.6 SnS_2/graphene composite in KIB

K-ion batteries have gained rather less attention compared to SIBs, even though potassium resources are also abundant [53]. The main research in this area involves exploring applicable electrode materials [53]. Although graphite has been reported as an anode for K^+ storage, which takes place in the form of multiple intermediate intercalation stages; however, large volume changes during the charge/discharge process ruin long-term cycling performance [53].

An application of layered hybrid material in potassium-ion battery was reported recently by Lakshmi et al. demonstrating improved electrochemical properties (350 mAh g^{-1} reported vs 278 mAh g^{-1} theoretical potassium capacity of graphite) (Fig. 3.10J) [53]. SnS_2/rGO was fabricated by H_2S treated peroxostannate-GO composite [53]. Two-dimensional hexagonal platelet morphology of coated SnS_2 nanoparticles was confirmed by various characterization techniques namely,

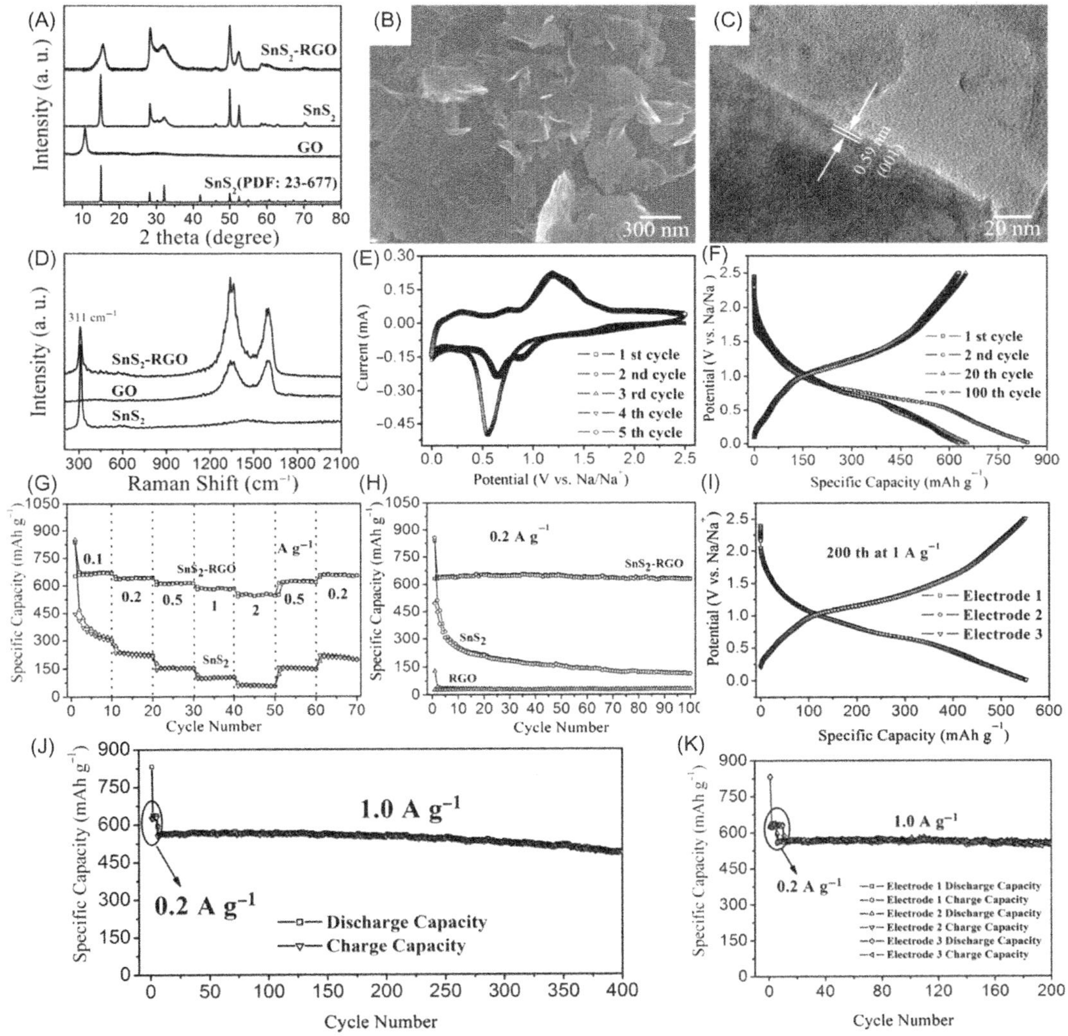

Figure 3.9 Structural characterization and electrochemical performance analysis of SnS_2/rGO hybrid electrode for SIB. (A) XRD patterns of SnS_2/rGO, SnS_2, and GO. (B) FESEM image of SnS_2/rGO. (C) HRTEM image of SnS_2/rGO. (D) Raman spectra of SnS_2/rGO, GO, and SnS_2. (E) CV curves of SnS_2/rGO at scan rate of 0.1 mV s^{-1}. (F) GCD profiles of SnS_2/rGO at different cycles. (G) Rate performance of SnS_2/rGO versus SnS_2. (H) Cycling performance of SnS_2/rGO, SnS_2, and rGO. (I) GCD profiles of different SnS_2/rGO at 200 cycles. (J) Cycling performance of SnS_2/rGO for 400 cycles. (K) Cycling performance of different SnS_2/rGO [41]. *Source*: B. Qu, C. Ma, G. Ji, C. Xu, J. Xu, Y.S. Meng, T. Wang, J.Y. Lee, Layered SnS_2-reduced graphene oxide composite – a high-capacity, high-rate, and long-cycle life sodium-ion battery anode material, Adv. Mater., 26 (2014) 3854–3859.

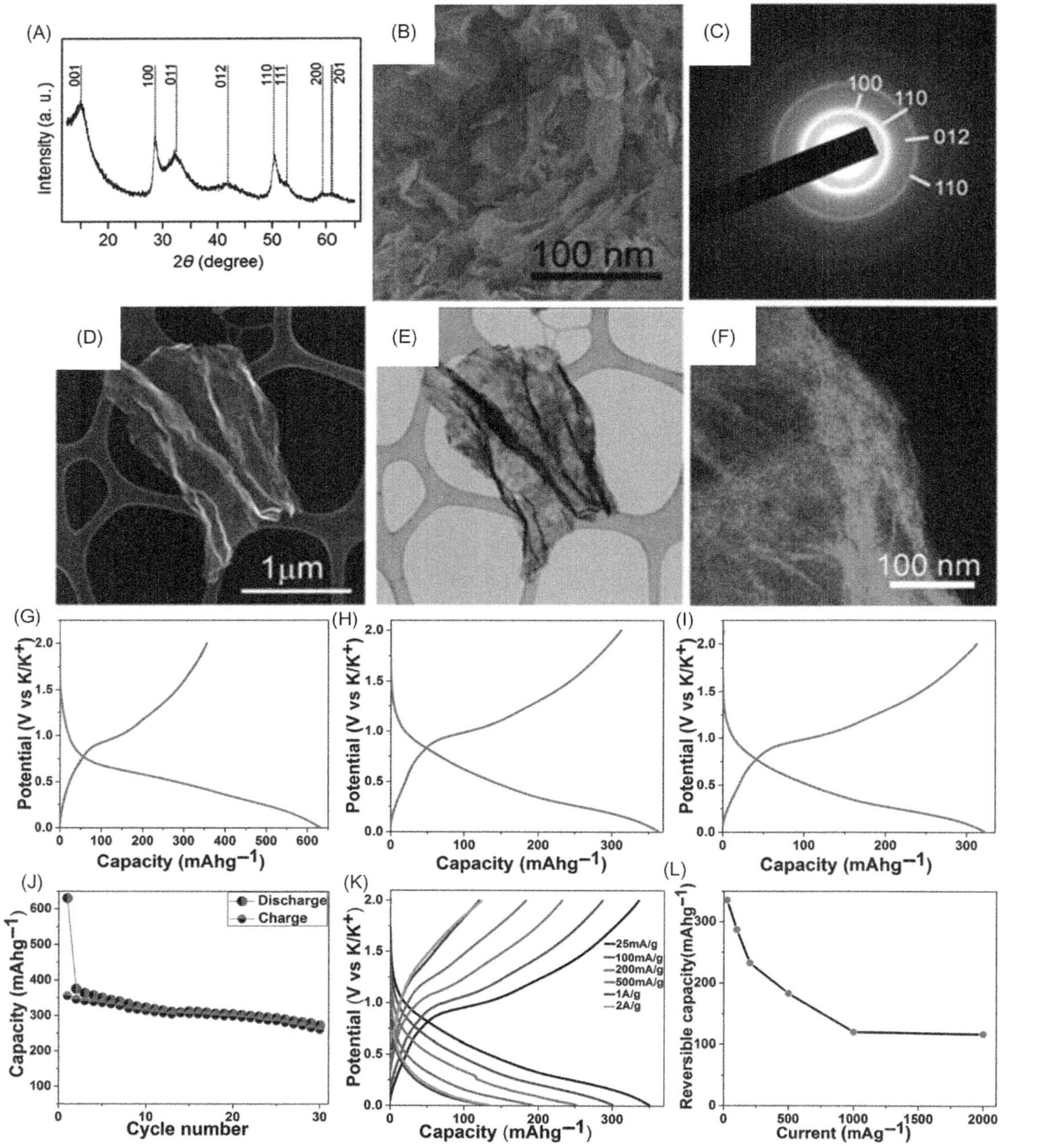

Figure 3.10 Structural characterization and electrochemical performance analysis of SnS_2/rGO hybrid electrode for KIB. (A) XRD pattern, (B) SEM image, (C) SAED pattern, (D, E) SEM and STEM, and (F) EFTEM image of SnS_2/rGO demonstrating of sulfur distribution on the surface of rGO nanosheets. GCD profiles of SnS_2/rGO in the (G) 1st, (H) 3rd, and (I) 10th cycles. (J) Cycling performance of SnS_2/rGO. (K) GCD profiles of SnS_2/rGO at different current density. (L) Reversible capacity SnS_2/rGO with different current density. *Source*: V. Lakshmi, Y. Chen, A.A. Mikhaylov, A. G. Medvedev, I. Sultana, M.M. Rahman, O. Lev, P.V. Prikhodchenko, A.M. Glushenkov, Nanocrystalline SnS_2 coated onto reduced graphene oxide: demonstrating the feasibility of a nongraphitic anode with sulfide chemistry for potassium-ion batteries, Chem. Commun., 53 (2017) 8272–8275.

XRD, XPS, and STEM (Fig. 3.10A–E) [53]. Rate capability test showed 120 mAh g^{-1} of remaining capacity at 2 A g^{-1} (Fig. 3.10L) [53].

3.7 Transition metal dichalcogenides/graphene composite in Mg battery

Mg-based storage systems have the same benefits as previously introduced SIB and KIB systems, such as low cost, easy availability, and so on. While bivalent Mg^{2+} ions are superior in transferring charges over other systems by its nature. Yet the development of Mg batteries is limited by the poor kinetics and low reversibility [54].

Batteries that can reversibly store Mg ions were proposed as an alternative to LIB due to the better cost-effectiveness in large-scale energy storage [55]. Hsu et al. illustrated effective Mg battery system with MoS_2/graphene as its cathodes and Li-ion containing all-phenyl-complex (APC) as its electrolyte [55]. Hexagonal MoS_2 plates exhibited a "2H-to-1T" phase transformation by initial Li^+ intercalation then incorporated with graphene (Fig. 3.11C) [55]. Specific capacity of 225 mAh g^{-1} at 25 mA g^{-1} current density was obtained while 150 mAh g^{-1} was recovered at a current density of 1000 mA g^{-1} (Fig. 3.11J) [55]. Cyclability test of the cell showed 90% of retention at 100 mA g^{-1} after 200 cycles [55].

Liu et al. announced the first hydrothermally synthesized MoS_2/graphene hybrid electrode application in Mg battery [56]. As prepared rechargeable Mg battery reached enhanced performance of 104 mAh g^{-1} after 30 cycles at 20 mA g^{-1} when compared to bulk MoS_2 and standalone graphene [56]. And later, Liu and his group reported a different MoS_2/graphene composite synthesis with hydrolysis sonication and freeze-drying technique [57]. The electrochemical test showed stable cycling performance of 82.5 mAh g^{-1} after 50 cycles at 20 mA g^{-1} [57] (Table 3.1).

3.8 Conclusion

It has been more than a decade since the first successful exfoliation of multilayered graphene samples with thickness of just several nanometers was reported. Not surprisingly, graphene rapidly found its applications in almost all areas of science and technology including, but not limited to chemistry,

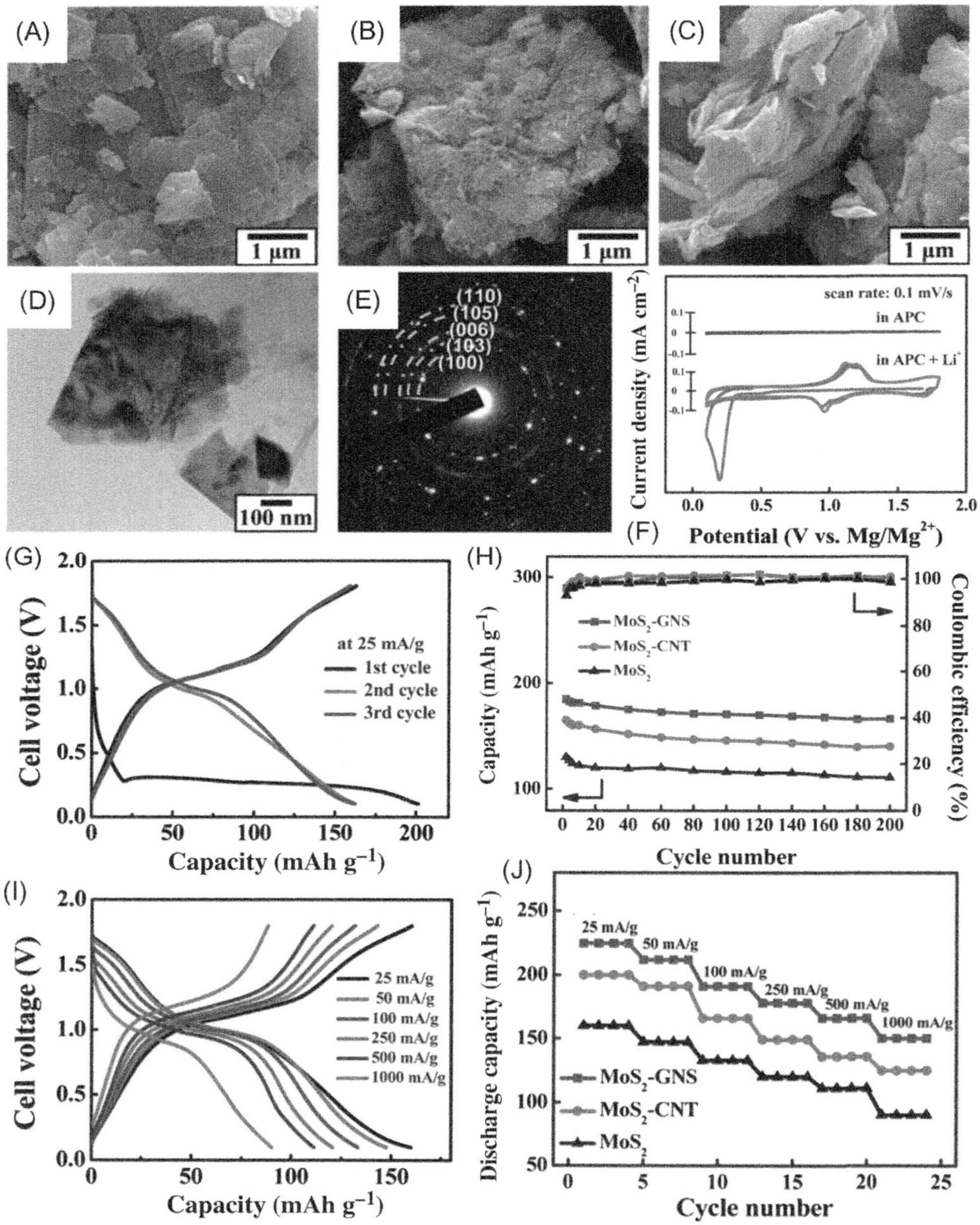

Figure 3.11 Structural characterization and electrochemical performance analysis of MoS_2/graphene hybrid electrode in Mg battery. SEM images of (A) MoS_2, (B) MoS_2/CNT, and (C) MoS_2/graphene. (D, E) TEM and electron diffraction pattern of MoS_2/graphene composite. (F) CV curve and (G) initial three charge/discharge of MoS_2 in APC with 0.5 M Li^+. Charge/discharge curves in different charge rate of (H) MoS_2 and (I) transformed MoS_2 in APC. (J) Rate performance of MoS_2, MoS_2/CNT, and MoS_2/graphene electrodes. *Source*: C. Hsu, C. Chou, C. Yang, T. Lee, J. Chang, MoS_2/graphene cathodes for reversibly storing Mg^{2+} and Mg^{2+}/Li^+ in rechargeable magnesium-anode batteries, Chem. Commun., 52 (2016) 1701–1704.

Table 3.1 Performance of hybrid electrodes with only layered active materials and graphene composite in different beyond Li-ion rechargeable system.

Electrode-layered active material	Fabrication techniques	Electrolyte chemistry	Battery performance	Voltage range	Reference
Na-ion					
MoS_2	Microwave-assisted synthesis	1 M $NaClO_4$ in EC:PC:FEC (1:1:0.05)	305/50/100	0.005–2.5	[46]
MoS_2	Acid-exfoliation, vacuum filtration	1 M $NaClO_4$ in EC:DMC (1:1)	230/15/25	0.01–2.25	[39]
MoS_2	Spray pyrolysis	1 M $NaClO_4$ in EC:DMC (1:1)	322/600/1500	0.001–3.0	[42]
MoS_2	Sonication, centrifugation	1 M $NaClO_4$ in PC:EC (7:3)	203/100/50	0.4–2.6	[44]
MoS_2	Hydrothermal synthesis	1 M $NaClO_4$ in EC:PC:FEC (1:1:0.05)	313/200/100	0.01 – 2.5	[50]
MoS_2	Hydrothermal synthesis	1 M $NaClO_4$ in PC:EC (1:1)	254/300/80	0.01–3.0	[43]
$MoSe_2$-CNT	Spray pyrolysis	1 M $NaClO_4$ in EC:DMC:FEC (1:1:0.05)	393/200/1000	0.001–3.0	[47]
$MoSe_2$	Hydrothermal synthesis	0.8 M $NaClO_4$ in EC:DEC:FEC (1:1:0.05)	430/200/500	0.01–3.0	[51]
WS_2	Spray pyrolysis	1 M $NaClO_4$ in EC:DMC:FEC (1:1:0.05)	334/200/200	0.001–3.0	[49]
WS_2	Hydrothermal synthesis	1 M $NaClO_4$ in PC:EC (1:1)	329/500/20	0.01–3.0	[48]
SnS_2	Hydrothermal synthesis	1 M $NaClO_4$ in EC:DEC (1:1)	619/100/200	0.01–2.5	[52]
SnS_2	Hydrothermal synthesis	1 M $NaClO_4$ in EC:DEC (1:1)	500/400/1000	0.01–2.5	[41]
Phosphorene	Self-assembly	1 M $NaPF_6$ in EC:DEC:FEC (1:1:0.1)	2080/100/50	0.02–1.5	[45]
K-ion					
SnS_2	Sonication, centrifugation	0.75 M KPF_6 in EC:DEC	280/25/25	0.01–2.0	[53]
Mg					
MoS_2	Self-assembly	0.4 M PhMgCl, 0.2 M $AlCl_3$ in THF: LiCl	225/25	0.1–1.8	[55]
MoS_2	Sonication, freeze-drying	0.4 M PhMgCl, $AlCl_3$ in THF:LiCl	83/50/20	0–2.2	[57]
MoS_2	Hydrothermal synthesis	0.4 M PhMgCl, 0.2 M $AlCl_3$ in THF	104/30/20	<2.3	[56]

physics, biology, and other engineering. Meanwhile, the advantageous properties of graphene also suggest a tremendous potential of these special group of materials with heterostructures: 2D materials. This new era demonstrates that the major materials engineering potential markets, such as semiconductor, electronics, composites, and energy storage can now go beyond graphene and still be able to harness these favorable properties.

Energy storage has always been one of the most studied applications of graphene owing to its structure and related unique electrochemical and mechanical properties. Still, massive production and application of graphene products have not yet been realized due to the challenges and issues that lie with standalone graphene, that is, large energy consumption in producing graphene, and ionic repulsion between graphene layers that weakens working capacity.

Hence, hybrid of graphene and other 2D materials have been proposed to raise the stability and performance of electrode systems. Large interlayer spacing of 2D materials can accommodate the intercalation of relatively larger (Na, K, Mg, etc.) alkali-metal ions. Considerable amount of intercalation active sites makes 2D materials as ideal electrodes for energy storage. Yet those 2D materials have not been widely applied due to their limitations, such as poor capacity, reduced kinetics, etc. However, recent investigations suggest enhancement of 2D electrodes by incorporation of graphene nanosheets. The formation of these nanocomposites tailored the structure to be perfect for storing ions by increasing conductivity, better accommodation of volume expansion as well as decreasing diffusion barrier. The research works discussed in this chapter indicate there still are ways to further improve the electrochemical performances of 2D materials. However, it is believed that the alternative to non-LIB batteries which are safer and reliable, eco-friendly, and inexpensive will soon be a reality. To achieve this, additional communications are needed between laboratories and industries to reduce the cost and issues to scale the productions.

Acknowledgments

Authors would like to thank by National Science Foundation grant no. 1454151 for financial support.

References

[1] V. Etacheri, R. Marom, R. Elazari, G. Salitra, D. Aurbach, Challenges in the development of advanced Li-ion batteries: a review, Energy Environ. Sci. 4 (2011) 3243–3262.

[2] J.M. Tarascon, M. Armand, Issues and challenges facing rechargeable lithium batteries, Nature 414 (2001) 359–367.

[3] R.D. Shannon, Revised effective ionic radii and systematic studies of interatomic distances in halides and chalcogenides, Acta Crystall. A32 (1976) 751–767.

[4] L. Peng, Y. Zhu, D. Chen, R.S. Ruoff, G. Yu, Two-dimensional materials for beyond-lithium-ion batteries, Adv. Energy Mater. 6 (2016) 1–21.

[5] J.B. Goodenough, K.-S. Park, The Li-ion rechargeable battery: a perspective, JACS 135 (2013) 1167–1176.

[6] P.G. Bruce, B. Scrosati, J.-M. Tarascon, Nanomaterials for rechargeable lithium batteries, Angew. Chem. 47 (2008) 2930–2946.

[7] A. Manthiram, Materials challenges and opportunities of lithium ion batteries, J. Phys. Chem. Lett. 2 (2011) 176–184.

[8] J.C. Pramudita, D. Pontiroli, G. Magnani, M. Gaboardi, M. Riccý, C. Milanese, et al., Graphene and selected derivatives as negative electrodes in sodium- and lithium-ion batteries, ChemElectroChem 2 (2015) 600–610.

[9] J.-Y. Hwang, S.-T. Myung, Y.-K. Sun, Sodium-ion batteries: present and future, Chem. Soc. Rev. 46 (2017) 3529–3614.

[10] A. Manthiram, Y. Fu, S.-H. Chung, C. Zu, Y.-S. Su, Rechargeable lithium – sulfur batteries, Chem. Rev. 114 (2014) 11751–11787.

[11] Y.-X. Yin, S. Xin, Y.-G. Guo, L.-J. Wan, Lithium–sulfur batteries: electrochemistry, materials, and prospects, Angew. Chem. 52 (2013) 13186–13200.

[12] S. Chen, C. Wu, L. Shen, C. Zhu, Y. Huang, K. Xi, et al., Challenges and perspectives for NASICON-type electrode materials for advanced sodium-ion batteries, Adv. Mater. 29 (2017).

[13] B. Ji, F. Zhang, X. Song, Y. Tang, Novel potassium-ion-based dual-ion battery, Adv. Mater. 29 (2017).

[14] P. Novak, R. Imhof, O. Haas, Magnesium insertion electrodes for rechargeable nonaqueous batteries - a competitive alternative to lithium? Electrochim. Acta 45 (1999) 351–367.

[15] D. Datta, J. Li, V.B. Shenoy, Defective graphene as a high-capacity anode material for Na- and Ca-ion batteries, Appl. Mater. Interfaces 6 (2014) 1788–1795.

[16] B.L. Ellis, L.F. Nazar, Sodium and sodium-ion energy storage batteries, Curr. Opin. Solid State Mater. Sci. 16 (2012) 168–177.

[17] T. Ichitsubo, T. Adachi, S. Yagi, T. Doi, Potential positive electrodes for high-voltage magnesium-ion batteries, J. Mater. Chem. 21 (2011) 11764–11772.

[18] A.L. Lipson, B. Pan, S.H. Lapidus, C. Liao, J.T. Vaughey, B.J. Ingram, Rechargeable Ca-ion batteries: a new energy storage system, Chem. Mater. 27 (2015) 8442–8447.

[19] A. Ponrouch, C. Frontera, F. Bardé, M.R. Palacín, Towards a calcium-based rechargeable battery, Nat. Mater. 15 (2015) 169–172.

[20] P. Adelhelm, P. Hartmann, C.L. Bender, M. Busche, C. Eufinger, J. Janek, From lithium to sodium: cell chemistry of room temperature sodium-air and sodium-sulfur batteries, Beilstein J. Nanotechnol. 6 (2015) 1016–1055.

[21] F. Legrain, J. Sottmann, K. Kotsis, S. Gorantla, S. Sartori, S. Manzhos, Amorphous (glassy) carbon, a promising material for sodium ion battery

anodes: a combined first-principles and experimental study, J. Phys. Chem. C 119 (2015) 13496–13501.
[22] D. Aurbach, R. Skaletsky, Y. Gofer, The electrochemical behavior of calcium electrodes in a few organic electrolytes, J. Electrochem. Soc. 138 (1991) 3536–3545.
[23] M.M. Huie, D.C. Bock, E.S. Takeuchi, A.C. Marschilok, K.J. Takeuchi, Cathode materials for magnesium and magnesium-ion based batteries, Coord. Chem. Rev. 287 (2015) 15–27.
[24] S. Hu, S.R. Das, H. Monshat, 2D materials: applications for electrochemical energy storage devices, in: Springer (Ed.), Advances in Nanomaterials, 2017, pp. 61–75.
[25] B.L. Dasari, J.M. Nouri, D. Brabazon, S. Naher, Graphene and derivatives: synthesis techniques, properties and their energy applications, Energy 140 (2017) 766–778.
[26] K.S. Novoselov, A.K. Geim, S.V. Morozov, D. Jiang, Y. Zhang, S.V. Dubonos, et al., Electric field effect in atomically thin carbon films, Science 306 (2004) 666–669.
[27] J.C. Meyer, A.K. Geim, M.I. Katsnelson, K.S. Novoselov, T.J. Booth, S. Roth, The structure of suspended graphene sheets, Nature 446 (2007) 60–63.
[28] A.K. Geim, K.S. Novoselov, The rise of graphene, Nat. Mater. 6 (2007) 183–191.
[29] C. Lee, X. Wei, J.W. Kysar, J. Hone, Measurement of the elastic properties and intrinsic strength of monolayer graphene, Science 321 (2008) 385–388.
[30] J. Ding, H. Wang, Z. Li, A. Kohandehghan, K. Cui, Z. Xu, et al., Carbon nanosheet frameworks derived from peat moss as high performance sodium ion battery anodes, ACS Nano 7 (2013) 11004–11015.
[31] J. Zhu, D. Yang, Z. Yin, Q. Yan, H. Zhang, Graphene and graphene-based materials for energy storage applications, Small 10 (2014) 3480–3498.
[32] Y. Wen, K. He, Y. Zhu, F. Han, Y. Xu, I. Matsuda, et al., Expanded graphite as superior anode for sodium-ion batteries, Nat. Commun. 5 (2014) 1–10.
[33] J.H. Warner, F. Schaffel, M. Rummeli, A. Bachmatiuk, Graphene: Fundamentals and Emergent Applications, Newnes, 2012.
[34] K.R. Paton, E. Varrla, J.N. Coleman, Scalable production of large quantities of defect-free few-layer graphene by shear exfoliation in liquids, Nat. Mater. 13 (2014) 624–630.
[35] W. Lv, Z. Li, Y. Deng, Q.-H. Yang, F. Kang, Graphene-based materials for electrochemical energy storage devices: opportunities and challenges, Energy Storage Mater. 2 (2016) 107–138.
[36] L. Wang, Z. Jiang, W. Li, X. Gu, L. Huang, Hybrid phosphorene/graphene nanocomposite as an anode material for Na-ion batteries: a first-principles study, J. Phys. D: Appl. Phys. 50 (2017) 165501–165508.
[37] G.-C. Guo, D. Wang, X.-L. Wei, Q. Zhang, H. Liu, W.-M. Lau, et al., First-principles study of phosphorene and graphene heterostructure as anode materials for rechargeable Li batteries, J. Phys. Chem. Lett. 6 (2015) 5002–5008.
[38] B.H. Qu, C.Z. Ma, G. Ji, C.H. Xu, J. Xu, Y.S. Meng, et al., Layered SnS_2-reduced graphene oxide composite: a high-capacity, high-rate, and long-cycle life sodium-ion battery anode material, Adv. Mater. 26 (2014) 3854–3859.
[39] L. David, R. Bhandavat, G. Singh, MoS_2/graphene composite paper for sodium-ion battery electrodes, ACS Nano 8 (2014) 1759–1770.
[40] R. Raccichini, A. Varzi, S. Passerini, B. Scrosati, The role of graphene for electrochemical energy storage, Nat. Mater. 14 (2015) 271–279.
[41] B. Qu, C. Ma, G. Ji, C. Xu, J. Xu, Y.S. Meng, et al., Layered SnS_2-reduced graphene oxide composite – a high-capacity, high-rate, and long-cycle life sodium-ion battery anode material, Adv. Mater. 26 (2014) 3854–3859.

[42] S.H. Choi, Y.N. Ko, J.K. Lee, Y.C. Kang, 3D MoS_2-graphene microspheres consisting of multiple nanospheres with superior sodium ion storage properties, Adv. Funct. Mater. 25 (2015) 1780–1788.
[43] X.Q. Xie, Z.M. Ao, D.W. Su, J.Q. Zhang, G.X. Wang, MoS_2/graphene composite anodes with enhanced performance for sodium-ion batteries: the role of the two-dimensional heterointerface, Adv. Funct. Mater. 25 (2015) 1393–1403.
[44] T.S. Sahu, S. Mitra, Exfoliated MoS_2 sheets and reduced graphene oxide: an excellent and fast anode for sodium-ion battery, Sci. Rep. 5 (2015).
[45] J. Sun, H.W. Lee, M. Pasta, H.T. Yuan, G.Y. Zheng, Y.M. Sun, et al., A phosphorene-graphene hybrid material as a high-capacity anode for sodium-ion batteries, Nat. Nanotechnol. 10 (2015) 980–U184.
[46] W. Qin, T. Chen, L. Pan, L. Niu, B. Hu, D.L.J. Li, et al., MoS_2-reduced graphene oxide composites via microwave assisted synthesis for sodium ion battery anode with improved capacity and cycling performance, Electrochim. Acta 153 (2015) 55–61.
[47] G.D. Park, J.H. Kim, S.K. Park, Y.C. Kang, $MoSe_2$ embedded CNT-reduced graphene oxide composite microsphere with superior sodium ion storage and electrocatalytic hydrogen evolution performances, ACS Appl. Mater. Interfaces 9 (2017) 10673–10683.
[48] D. Su, S. Dou, G. Wang, WS2@graphene nanocomposites as anode materials for Na-ion batteries with enhanced electrochemical performances, Chem. Commun. 50 (2014) 4192–4195.
[49] S.H. Choi, Y.C. Kang, Sodium ion storage properties of WS_2-decorated three-dimensional reduced graphene oxide microspheres, Nanoscale 7 (2015) 3965–3970.
[50] Y.X. Wang, S.L. Chou, D. Wexler, H.K. Liu, S.X. Dou, High-performance sodium-ion batteries and sodium-ion pseudocapacitors based on MoS_2/graphene composites, Chem.: Eur. J. 20 (2014) 9607–9612.
[51] Z.A. Zhang, Y. Fu, X. Yang, Y.H. Qu, Z.Y. Zhang, Hierarchical $MoSe_2$ nanosheets/reduced graphene oxide composites as anodes for lithium-ion and sodium-ion batteries with enhanced electrochemical performance, ChemNanoMat 1 (2015) 409–414.
[52] Y. Liu, H. Kang, L. Jiao, C. Chen, K. Cao, Y. Wang, et al., Exfoliated-SnS_2 restacked on graphene as a high-capacity, high-rate, and long-cycle life anode for sodium ion batteries, Nanoscale 7 (2015) 1325–1332.
[53] V. Lakshmi, Y. Chen, A.A. Mikhaylov, A.G. Medvedev, I. Sultana, M.M. Rahman, et al., Nanocrystalline SnS_2 coated onto reduced graphene oxide: demonstrating the feasibility of a non-graphitic anode with sulfide chemistry for potassium-ion batteries, Chem. Commun. 53 (2017) 8272–8275.
[54] S. Mukherjee, Z. Ren, G. Singh, Beyond graphene anode materials for emerging metal ion batteries and supercapacitors, Nano-Micro Lett. 10 (2018).
[55] C.J. Hsu, C.Y. Chou, C.H. Yang, T.C. Lee, J.K. Chang, MoS_2/graphene cathodes for reversibly storing Mg2 + and Mg2 + /Li + in rechargeable magnesium-anode batteries, Chem. Commun. 52 (2016) 1701–1704.
[56] Y.C. Liu, L.F. Jiao, Q. Wu, Y.P. Zhao, K.Z. Cao, H.Q. Liu, et al., Synthesis of rGO-supported layered MoS_2 for high-performance rechargeable Mg batteries, Nanoscale 5 (2013) 9562–9567.
[57] Y.C. Liu, L.Z. Fan, L.F. Jiao, Graphene intercalated in graphene-like MoS_2: a promising cathode for rechargeable Mg batteries, J. Power Sources 340 (2017) 104–110.

4

2D materials as the basis of supercapacitor devices

Michael P. Down and Craig E. Banks
Faculty of Science and Engineering, Manchester Fuel Cell Innovation Center, Manchester Metropolitan University, Manchester, United Kingdom

4.1 Introducing supercapacitors

This section will introduce supercapacitors in the spectrum of energy storage devices, their relative performance, mechanisms and characteristics, and their relative standing in the field of energy storage. It will compare supercapacitors to the competing technologies, most notably batteries and film capacitors, and identify the systems and technologies that look to enhance the performance by the exploitation of nanotechnology and nanomaterials [1–29].

Supercapacitors are devices capable of managing high-power rates compared to batteries, and higher energy densities compared to solid-state/film capacitors. Although, supercapacitors provide up to many thousand times higher power in the same volume as batteries they are not able to store the same amounts of charge as batteries do, typically of the order of 3–10 times lower [1]. This makes supercapacitors suitable for those applications in which bursts of power are needed, but high energy storage capacity is not essential. Supercapacitors, however, can also be included within a hybridized energy storage system, utilizing both supercapacitors and batteries, to decouple the power and energy characteristics of the system, thus improving the sizing and practicality while fulfilling the power and energy requirements and elongating the lifetime. Such systems can be used to provide either high-power discharging, where a burst of power is required for the driven system, that is, starting motors or the reverse situation where energy for charging is provided sporadically and quick collection is required, that is, quick charging circuits. The power output of supercapacitors is lower than that of electrolytic capacitors but can reach about 10 kW kg^{-1}.

2D Nanomaterials for Energy Applications. DOI: https://doi.org/10.1016/B978-0-12-816723-6.00004-6

However, their specific energy is several orders of magnitude higher than the one of solid-state and film capacitors [2]. These devices are of significant interest because they occupy a region of the power and energy storage spectrums between aluminum electrolytic capacitors and batteries. This can be graphically illustrated in a Ragone plot, in which the plot of the energy density and power density in horizontal and vertical axes, respectively [26–29]. This can further be related to the discharge time, t, of the devices in diagonal lines ($E = Pt$). Different storage technologies are represented in a Ragone plot in Fig. 4.1.

The Ragone plot is an incredibly useful tool to illustrate the relative performance of an experimental technology in the spectrum of energy and power performances. However, the Ragone plot does not reflect many other performance parameters such as cost, safety, weight, and cycling lifetime. Therefore it should be considered separately for a complete understanding of advantages and limitations of a particular energy storage technology.

It is extremely important to note that, given their high-power characteristics, supercapacitors cannot only be discharged in a matter of seconds but also be charged in such a short time period. This is an important benefit for energy recovery systems, for example, for dynamic braking of transport systems or piezoelectric energy harvesters, where the energy is generated

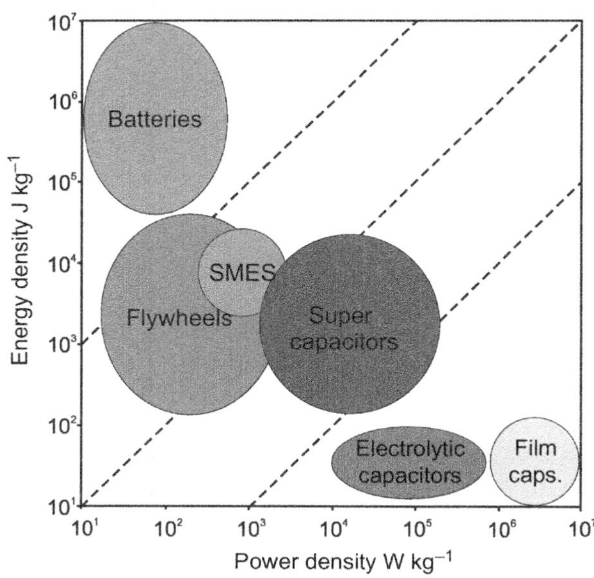

Figure 4.1 A Ragone plot showing the relative performance of supercapacitors and other energy storage devices.

for a brief period. For example, capacitors have a typical specific energy capacity of up to 0.1 W h kg^{-1}, which limits the potential applications outside of discreet circuit power handling and signal filtering applications. This compared to 1–10 W h kg^{-1} for supercapacitors and 10–100 W h kg^{-1} for batteries, which represent the ranges for larger scale energy storage applications. Furthermore, there is a reverse trend in the powers of the devices, from capacitors which demonstrate significantly higher powers due to the remarkably faster response times, with powers well in excess of 10,000 W kg^{-1}, which is a significant characteristic lending to their potential applications in RF signal handling and filtering. Whereas batteries demonstrate much more modest power handling with powers well below 1000 W kg^{-1}, which is largely considered one of the main limiting factors for batteries in fast-charging or high-power circuits [1]. This can be addressed by simply wiring a number of batteries in parallel to increase the current handling capabilities, and hence increase the available power range according to $P = VI$, where V is the potential provided by the battery and I, its applied current. This is however a significantly more expensive solution and introduces its own problems in Coulombic efficiency as batteries demonstrate relatively poor efficiency and significantly reduced lifetimes. As such, supercapacitors represent a vital component in the battle for improved power capabilities of energy storage architectures. Double-layer capacitor cells do not rely on metals chemistries and do not thus run the risk of metal plating, which is an important battery degradation and failure mechanism as well as a safety concern that can lead to short circuits and uncontrollably energetic chemical reactions.

Another significant advantage of supercapacitors is their cycle life being capable of withstanding millions of cycles thanks to their charge storage mechanism, which does not involve irreversible chemical reactions, storing charges physically at the surface of the electrodes in an electric double layer. This allows exceeding the cycle life of batteries, which are at best capable of withstanding a few thousand cycles. The highly reversible electrostatic storage does not produce changes in the electrode physical structure or volume, eliminating the swelling occurring in typical redox reactions in the bulk of a battery's active material during charge and discharge cycles. A supercapacitor electrode has no such rate limitations as those of redox battery electrodes due to electrochemical kinetics through a polarization resistance [3]. The most significant limiting factor related to the energy storage mechanism for supercapacitors is the operating voltage of a supercapacitor cell, which should be kept low to avoid the chemical decomposition of electrolytes. The working temperature range is another feature highlighted, as high-power performance down to $-40°C$ can be achieved with supercapacitors [3], which is not currently possible with batteries [26–28,30].

A supercapacitor cell comprises two electrodes, either identical in a symmetric cell or different for an asymmetric cell with a separator between them. The separator is soaked in electrolyte and mechanically prevents any shorting between the electrodes that occur when an electrical contact is made. The separator material should be ion-permeable to allow the ionic charge transfer, while at the same time having a high electrical resistance to prevent shorting, high ionic conductance, and low thickness in order to achieve the best performance. Usually porous or fibrous polymer or paper separators are used together with organic electrolytes while ceramic or glass fiber separators are usually coupled with aqueous electrolytes [4]. The electrolyte breakdown potential at one of the electrodes limits the attainable cell voltage whereas the equivalent series resistance (ESR), the resistance of all of the system as if connected in series, of the cell will depend strongly on the electrolyte conductivity. Aqueous electrolytes typically have a breakdown voltage of around 1 V, which is significantly lower than the one achievable with organic electrolytes (around 3 V) but the conductivity of aqueous electrolytes is higher than that of organic electrolytes, which is desirable for high-power devices. Aqueous electrolytes also feature such important assets as low cost and easiness in handling [31,32].

Depending on the storage mechanism or cell configuration, the *type* of supercapacitor can be distinguished between electric double-layer capacitors (EDLCs), pseudocapacitors, and hybrid capacitors. EDLCs are based on high specific surface area (41,000 $m^2\,g^{-1}$) nanoporous materials as active electrode materials, leading to a huge capacitance in comparison with electrostatic capacitors. The electrodes are usually made of nanoporous carbon materials thanks to their availability, existing industrial production, and comparatively low cost. Pseudocapacitors are based on conducting polymer or metal oxide–based electrodes, and sometimes functionalized, modified, doped, or enhanced porous carbons, combining electrostatic and pseudocapacitive charge storage mechanisms. These materials can hold much higher specific capacitance values as compared to EDLCs, with the charge storage mechanism relying on fast redox reactions occurring on the electrode surface but not in the bulk like in batteries. However, like in the case of batteries, redox reactions can lead to mechanical changes making the electrodes swell and shrink, giving rise to poor mechanical stability. Consequently, lower cycle life is an important deficiency of pseudocapacitive materials. Finally, hybrid capacitors

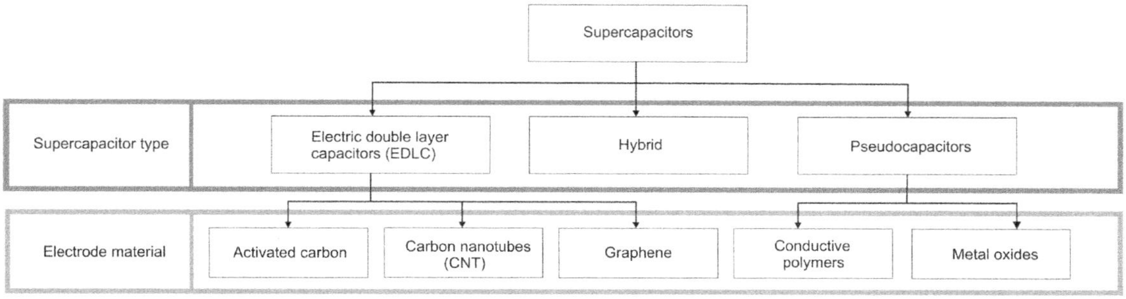

Figure 4.2 The classifications and typical electrode materials for different supercapacitor types.

are composed of an EDLC electrode and a pseudocapacitive or battery type electrode, combining the properties of both systems and leading to an intermediate performance in some cases. A good example of such a system is the lithium-ion capacitors. Different types of supercapacitors are classified in Fig. 4.2.

Here two main research lines can be distinguished concerning pseudocapacitors and EDLCs [2]. As macroscopically these devices work like capacitors, the capacitance, C, will depend on the dielectric constant of the electrolyte, ε, the effective thickness of the double layer, d (separation between charges), and the accessible surface, A, as follows:

$$C = \frac{\varepsilon_r \varepsilon_0 A}{d} \tag{4.1}$$

where ε_0 is the dielectric constant of the vacuum. The capacitance for an electric double layer on carbon surface varies usually from 5 to 20 μF^{-2} depending on the electrolyte [11,21], although much higher values are sometimes reported for edge carbon atoms. The energy stored within a supercapacitor, E, is the equivalent for any capacitor:

$$E = \frac{1}{2} CV^2 \tag{4.2}$$

where V is the cell voltage. Eq. (4.2) shows that the stored energy is proportional to both the capacitance of the device and the square of the cell voltage. Therefore increasing both of them is a general strategy to increase the energy density of the cell.

The maximum instantaneous power that a supercapacitor is able to deliver, P_{Max}, depends on the voltage and the internal resistance, R, as follows:

$$P_{Max} = \frac{V^2}{4R} \tag{4.3}$$

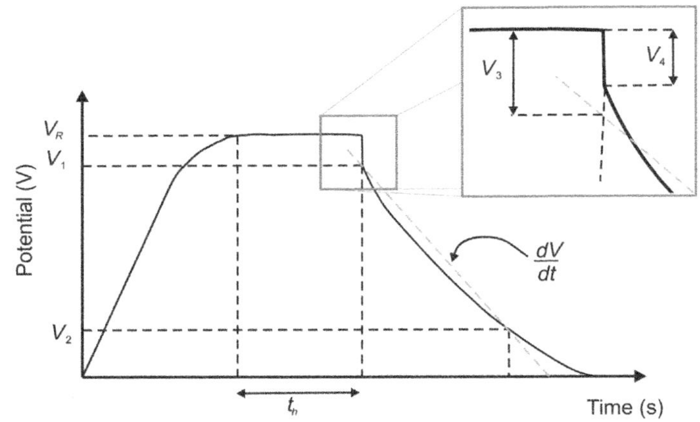

Figure 4.3 The typical potential behavior of a galvanostatic charge–discharge cycle. The dotted line labelled dV/dt represents the line fit to the characteristic discharge after the initial IR drop.

Also the current across the supercapacitor will be

$$I = C\frac{dV}{dt} \tag{4.4}$$

In industry, galvanostatic tests, or constant current tests, are performed to determine the main characteristics of devices. This is where a constant current is applied to the capacitor until either a potential limit or time limit is reached, this is repeated for charging and discharging. Fig. 4.3 shows the typical charge–discharge characteristic of a supercapacitor under galvanostatic charging conditions. This includes capacitance calculation (integral of the area contained during the discharge) and resistances associated to the cell such as ESR and equivalent distributed resistance (EDR), which represents ESR and the resistance in the pores (part of the discharge with curvature). These are calculated using the notation in Fig. 4.5 with the following expressions:

$$C = \frac{I_{\text{discharge}}t_{\text{discharge}}}{V_1 - V_2} = \frac{I_{\text{discharge}}}{\left(\frac{dV}{dt}\right)} \tag{4.5}$$

$$\text{ESR} = \frac{V_4}{I_{\text{discharge}}} \tag{4.6}$$

$$\text{EDR} = \frac{V_3}{I_{\text{discharge}}} \tag{4.7}$$

The electrical properties of a supercapacitor are mainly determined by components such as electrode materials, electrolytes, separators, and current collectors. Electrode fabrication is made through coating a metallic current collector with an about 100 μm thin layer of high-surface-area material. This active material is mixed with a binder so as to form slurry. The thickness of the slurry should be controlled for making the coated layer of active material sufficiently thin to be conductive throughout the material.

Since the ESR of supercapacitor cells must be very low, special attention must be paid to the contact resistance between the active material and the current collector. The surface of current collectors should be treated before coating it with active materials. Surface treatments decrease the Ohmic drop at the current collector/active material interface [5]. For supercapacitors designed to work with organic electrolytes, treated aluminum foils or grid current collectors are used. Using nanostructured current collectors with increased contact area is a way to control the current collector/active material interface. A widely used measurement is the specific capacitance, which is the intrinsic capacitance of an electrode material expressed in F g^{-1}. Although this is a very useful characteristic of the material, a higher specific capacitance does not necessarily mean that the material will be a highly performing supercapacitor electrode. There are other factors substantially affecting capacitance such as electrical conductivity (both that of materials and between electrode particles), which governs electron and ion transfer into the layer [6].

4.2 Electric double layer

An electric double layer is a structure appearing when a charged object is placed into a liquid. The balancing counter charge for this charged surface will form in the liquid, concentrated at the electrode–liquid interface. For the formation of the interface between the liquid and solid there are a number of models. In Fig. 4.4 the Helmhotlz, the Gouy–Chapman, and the Stern model are all illustrated, where Ψ is the electric potential as a function of position with respect to the electrode surface and Ψ_0 is the electric potential at the electrode. IHP refers to the inner Helmholtz plane and the OHP refers to the outer Helmholtz plane, which is defined by the Stern model.

4.2.1 The Helmholtz model

This is the simplest approximation for the modeling of the spatial charge distribution of double-layer interfaces. The charge of the solid electronic conductor is neutralized by opposite sign ions at a distance from the solid. This is the distance

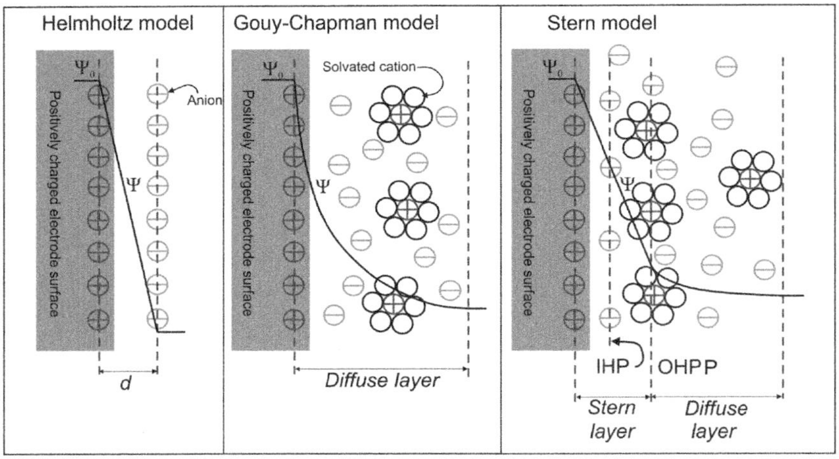

Figure 4.4 An illustration of the performance and key dimensional characteristics and definitions of the three key electric double-layer capacitance models: the Helmholtz model, the Gouy–Chapman model, and the Stern model.

from the surface to the center of the ions. This theory considers rigid layers of opposite charge ions from the solid electrode, analogous to that of a film capacitor with two planar electrodes separated by a distance, d. Therefore the capacitance per unit surface area (or specific capacitance) of the Helmholtz double layer denoted by C_{Hs} and expressed in F m^{-2} is given by

$$C_{Hs} = \left\{ \frac{\varepsilon_0 \varepsilon_r}{d} \right. \quad \text{for planar electrodes} \tag{4.8a}$$

$$C_{Hs} = \left\{ \frac{\varepsilon_0 \varepsilon_r}{R_0 \log\left(1 + \frac{d}{R_0}\right)} \right. \quad \text{for cylindrical electrodes of radius } R_0 \tag{4.8b}$$

$$C_{Hs} = \left\{ \frac{\varepsilon_0 \varepsilon_r}{d} \left(1 + \frac{d}{R_0}\right) \right. \quad \text{for spherical electrode of radius } R_0 \tag{4.8c}$$

where ε_0 and ε_r and the free space permittivity and the relative permittivity of the electrolyte solutions, respectively. The thickness of the Helmholtz layer, d, can be approximated as the radius of the solvated ions. It should be noted, especially when considering particle size and nanomaterials, that the Eqs. (4.8b) and (4.8c) reduce to asymptotic expression given by Eq. (4.8a) when the electrode radius is significantly larger than the thickness of the Helmholtz layer, that is, $R_0 \gg H$. As such, the particle size and dimensions have a profound and important impact on the capacitive performance of the material.

The resulting distribution is that of a linearly distributed potential with distance from the electrode surface. For example, in the case of the example illustrated in Fig. 4.4, we can see a layer of anions forming parallel to the electrode surface, perfectly counterbalancing the charge at the electrode surface. The model essentially behaves as an ideal parallel plate capacitor, with a separation of d, with a linear potential distribution between the two charges. As of today, this represents the simplest theory providing a basic approximation for the complexities of the EDLC and does not adequately explain what occurs in nature [7].

4.2.2 The Gouy–Chapman model

Gouy proposed that the amount of opposite ionic charge appears in a liquid surrounding a charged solid in a fluid manner influenced by thermodynamics and the ions are not rigidly attached to the surface [7]. These ions in the solution tend to diffuse into the liquid phase until the counter potential set up by their displacement restricts this tendency. The kinetic energy and thermal properties of the ions in the solution will partially determine the thickness of the diffuse layer.

Gouy and Chapman developed theories of this diffuse layer in which the ion concentration on the solution near the surface follows the Boltzmann distribution:

$$c_i = c_{i\infty}\left(\frac{-z_i e\Psi}{k_B T}\right) \tag{4.9}$$

where z_i and $c_{1\infty}$ are the valency and bulk molar concentration of ion species, i, respectively. The absolute temperature is denoted by T, e is the elementary charge, and k_B is the Boltzmann constant. In the Gouy–Chapman model, the local electric potential Ψ in the diffuse layer is determined by Poisson–Boltzmann (PB) equation assuming constant electrolyte permittivity. For binary symmetric electrolytes, $z_1 = -z_2 = z$ and $c_{1\infty} = c_{2\infty} = c_\infty$. Then the PB equation is expressed as

$$\nabla \cdot (\varepsilon_0 \varepsilon_r \nabla\Psi) = 2zeN_A c_\infty \sinh\left(\frac{ze\Psi}{k_B T}\right) \tag{4.10}$$

where N_A is the Avogadro's number.

For planar electrodes and constant electrolyte properties, the exact solution of Eq. (4.10) exists subject to the following boundary conditions: (1) $\Psi(0) = \Psi_D$ and (2) $\Psi(\infty) = 0$. Then, the specific capacitance corresponding to the diffuse layer C^D_s is given by

$$C_s^D = \frac{q_s}{\Psi_D} = \frac{4zeN_A c_\infty \lambda_D}{\Psi} \sinh\left(\frac{ze\Psi_D}{2k_B T}\right) \quad (4.11)$$

where q_s is the surface charge density and λ_D is the Debye length for symmetric electrolyte defined as $\lambda_D = [(\varepsilon_0 \varepsilon_r k_B T / 2e^2 z^2 N_A c_\infty)^{1/2}]$.

For spherical electrodes the exact solution of Eq. (4.10) exists provided that the Debye–Hückel approximation requiring $z_i e\Psi / k_B T \ll 1$ is satisfied. Then, the diffuse layer–specific capacitance is given by

$$C_s^D = \frac{\varepsilon_0 \varepsilon_r}{\lambda_D}\left(1 + \frac{\lambda_D}{R_0}\right) \quad (4.12)$$

Eqs. (4.11) and (4.12) both reduce to the asymptotic expression of $C_S^D = \varepsilon_0\, \varepsilon_r / \lambda_D$ when the Debye–Hückel approximation and the thin double-layer approximation are satisfied.

This model fails for highly charged double layers. Experimentally, the measured thickness of the double layers is greater than those calculated. As such, additional effects are required to be considered.

4.2.3 Stern modification of the diffuse double layer

The Gouy–Chapman model makes a much better approach to the reality that the Helmholtz model; however, it still has significant limitations and, as such, limited quantitative applications. It assumes that ions are point charges and that they can approach the surface with no limits, which is not true. Stern modified the Gouy–Chapman model stating that the ions have a finite size, so limiting their approach to the surface. In order to resolve the shortcomings of the Gouy–Chapman model for the diffuse layer, Stern suggested the combination of both previous models, giving an internal layer with a linear potential decline with distance from the electrode surface (e.g., the Helmholtz layer) and an outer diffuse layer with a nonlinear decline (e.g., the Gouy–Chapman layer). The first ions in the Gouy–Chapman model are at the distance away from the surface, δ, but the Stern model assumes that there can be specifically surface-adsorbed ions in plane δ, this is known as the Stern layer. Ions are strongly adsorbed by the electrode within this so-called compact layer. In the compact layer there are specifically adsorbed ions (forming the inner Helmholtz plane), and nonspecifically adsorbed counterions (forming the outer Helmholtz plane) [11]. It should be noted that there are no free charges within the Stern layer.

The total electric double-layer capacitance consists of the Stern layer contributions and the diffuse layer capacitances in series. Mathematically, the Gouy–Chapman–Stern model for symmetric electrodes is expressed as

$$\nabla\cdot(\varepsilon_0\varepsilon_r\nabla\Psi) = \{\, 0 \quad \text{in the Stern layer} \tag{4.13a}$$

$$\nabla\cdot(\varepsilon_0\varepsilon_r\nabla\Psi) = \left\{\, 2zeN_A c_\infty \sinh\left(\frac{ze\Psi}{k_B T}\right) \quad \text{in the diffuse layer} \right. \tag{4.13b}$$

The point-charge assumption associated the PB equation (as utilized in Eqs. 4.10 and 4.13b) is only valid for very low ion concentration c_∞ and low electric potential. In reality, the ions have finite size and thus a maximum ion concentration, $c_{\max} = 1/(N_A a^3)$, which corresponds to the closed packing of ions. Therefore the ion concentration given by the Boltzmann distribution in Eq. (4.9) should not exceed $c_{\max}$. This corresponds to a maximum surface potential $\Psi_{\max}$ given by

$$\Psi_{\max} = -\frac{k_B T}{ze}\log\left(N_A a^3 c_\infty\right) \tag{4.14}$$

The magnitude of the electric potential in the diffuse layer should never exceed this value in either the Gouy–Chapman model or the Stern modification for these to be valid. This result can have a profound effect on electrodes with restricted dimensions, and this is where the application of nanomaterials can be significantly beneficial.

Numerous studies have been reported throughout literature to account for the effect of finite ion size in the electrolyte solution. Among them, the modified Poisson–Boltzmann (MPB) models based on the local-density and mean-field approximations are relatively convenient both mathematically and numerically. Among the different MPB models, Bikerman's model is the simplest form for applications in symmetric electrolytes and is expressed as

$$\nabla\cdot(\varepsilon_0\varepsilon_r\nabla\Psi) = \frac{\left(2zeN_A c_\infty \sinh\left(\frac{ze\Psi}{k_B T}\right)\right)}{1 + 2\upsilon\sinh^2\left(\frac{ze\Psi}{2k_B T}\right)} \tag{4.15}$$

where the packing parameter is defined as $\upsilon = 2a^3 N_A c_\infty = 2c_\infty / c_{\max}$. When $\upsilon = 0$, the MPB model reduces to the Gouy–Chapman model. For planar electrode and constant electrolyte properties, the surface charge density and diffuse layer–specific capacitance are given by

$$C_{Ds} = \frac{q_s}{\Psi_D} = \frac{2zeN_A c_\infty \lambda_D}{\Psi_D}\sqrt{\frac{2}{\upsilon}\log\left[1 + 2\upsilon \sinh^2\left(\frac{ze\Psi_D}{2k_B T}\right)\right]} \tag{4.16}$$

Note that when $\Psi_D = 0$, Eq. (4.16) predicts an extremum for the diffuse capacitance, that is, $C_{Ds} = \varepsilon_0 \varepsilon_r / \lambda_D$. The capacitance could be either a maximum or minimum value depending on the packing parameter.

4.2.4 Electric double layer in supercapacitors

These models provide the simplistic basis of the mechanisms that drive supercapacitors and their characteristics in charging and discharging, although they fall short of describing the real charge distribution in nanoporous and nanomaterial-based electrodes regularly employed in supercapacitors. There are a number of unique behaviors with ion electrosorption in porous media which make the process of charge storage extremely difficult; there is still a lack of complete understanding of ions behaviors in nanopores. When a supercapacitor is charged, electrons are forced to go from the positive electrode to the negative electrode through an external circuit. As a consequence, cations within the electrolyte concentrate in the negative electrode and anions in the positive electrode forming an electric double layer that compensates the external charge unbalance. During the discharge, electrons travel from the negative electrode to the positive electrode through an external circuit, and both kinds of ions in the pores become mixed again until the cell is discharged.

Ions do not move in the bulk electrolyte the same way as they do within the pores of an electrode material. The mobility of ions into the pores is greatly influenced by the pore size, which if too small makes the pores inaccessible, not contributing to double-layer capacitance [4].

Since not all the pores are accessible to the ions, there is no linear relation between the capacitance exhibited by a material and its specific surface area [10,11,19,20] measured with a small gas molecule probe such as N_2 or Ar. Various studies suggest that pore size below 0.5 nm is not accessible to hydrated ions [12,19] and pores smaller than 1 nm can be too small for organic electrolytes [20]. Generally, there is a controversy regarding the effect of pore size on capacitance. Chmiola et al. claimed that pores with sizes below 1 nm greatly contribute to the capacitance [21]. This increase was explained with the distortion of the solvation shell, thus reducing the distance between charges and enhancing the capacitance [22]. However, constant capacitance in the micropores was measured in organic electrolyte on the basis of a detailed assessment of pore size using complementary adsorption techniques. The decrease in the distance between the electronic and ionic charges is counterbalanced by the corresponding decrease in the effective dielectric permittivity inside the pores, which occurs due to gradual ion desolvation [24,33]. Leaving the controversy of capacitance versus pore size behind, it is worth mentioning that industrially important values are also calculated on a volumetric basis. It then becomes clear that too wide pores contain free space, which is not used for capacitive charge storage but decreases the density of electrodes. This effect is detrimental to volume-based capacitance as well as the existence of narrow electrolyte-inaccessible pores. Thus tuning pore size is anyway needed to have carbon materials with narrow, short, and electrolyte-accessible pores [21].

Apart from it, there is a general agreement that the power capability of a supercapacitor can be enhanced by the presence of a small amount of mesopores (pores wider than 2 nm) for a rapid supply of electrolyte to the micropore surface where main charge storage takes place [34,35]. There have been numerous attempts to properly describe the capacitance of carbon materials depending on the pore shape and size and the specific character of their interaction with electrolytes. For porous carbons with cylindrical pores, the traditional model is used [35]:

$$\frac{C}{A} = \frac{\varepsilon_r \varepsilon_0}{b \ln\left(\frac{b}{b-d}\right)} \tag{4.17}$$

where b is the pore radius and d the distance between the ion and the carbon surface. But for submicropores, it is assumed that the ions line up in the center of a cylindrical pore, so the capacitance is calculated from Ref. [35]:

$$\frac{C}{A} = \frac{\varepsilon_r \varepsilon_0}{b \ln\left(\frac{b}{a_0}\right)} \tag{4.18}$$

where a_0 is the effective size of the ion. This ionic radius was found to be close to the bare ion size, which means that the ions could be fully desolvated. However, as the more realistic approximation to the pore shape in carbons is a slit, not a cylinder, a sandwich capacitance model was later proposed [35].

$$\frac{C_{\text{total}}}{2A} = \frac{C_s}{A}\frac{\varepsilon_r\varepsilon_0}{b\ln\left(\frac{b}{b-a_0}\right)} \tag{4.19}$$

4.3 Electric double-layer capacitance and the influence of scale

The discussed models have been used extensively within the literature to investigate the electric double layer near nanoscale electrodes despite their inherent limitations, and have been used to demonstrate the influence of electrode and materials scale on the capacitive performance. Huang et al. used the Helmholtz model to predict the specific capacitance of EDLCs based on single spherical and cylindrical electrode carbon particles as well as single cylindrical pores, indicating a beneficial impact of both nanomaterials and nanoporous materials [35–38]. Their application of the established models predicted that the total specific capacitance, for a finite system, decreases with increasing electrode radius larger than 2 nm. Fig. 4.5 shows there results, in which a consistent decrease in the areal capacitance of a material would occur based as a function of particles size. The insets are illustrations of the particle geometries used in the approximations: an exohedral electric doubles-sphere capacitor and a exohedral electric double-cylinder capacitor. Despite the apparent correlation between their approximation and the experimental values, it should be noted that the electrolyte permittivity was used as an empirical parameter to match the specific capacitance predicted by Eqs. (4.8a) and (4.8c) with experimental data. In this approach, they proposed alterations to the electrolyte's established permittivity's to match the mathematical approach [35–38].

Further to this, Compton and et al. investigated the effect of the electrode curvature and dimensional geometry on the diffuse layer formed at the surface of single hemispherical and cylindrical nanoelectrodes [39–41]. They went on to solve the Gouy–Chapman model, as shown in Eq. (4.10), numerically for electrode radii ranging from 2 nm to 100 μm. The surface electric potential Ψ_s was less than 0.25 V, and the electrolyte concentration c_∞ was less than 0.1 mol L^{-1}. It was observed that a significantly enhanced surface charge density for a sphere radius is less than 50 nm and attributed this phenomenon to a

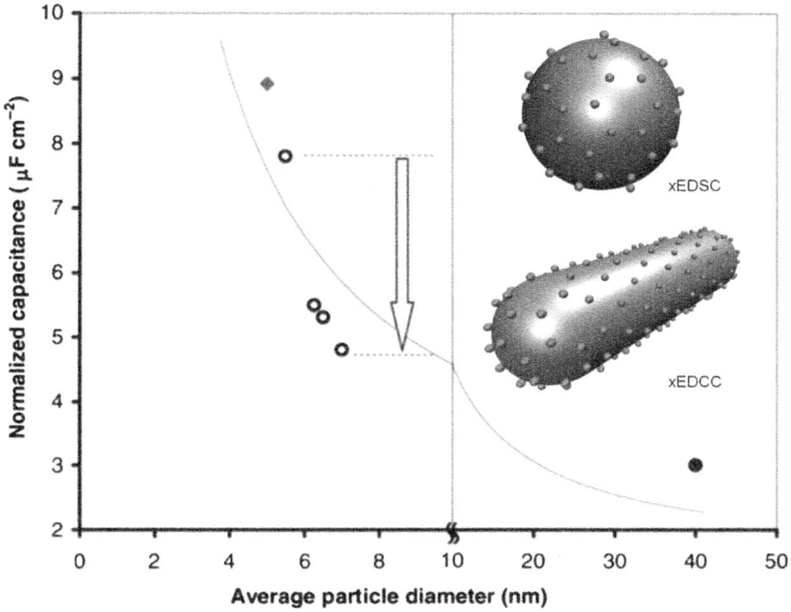

Figure 4.5 Huang et al. application of the Helmholtz model demonstrated that for increasing particle sizes the normalized areal capacity of the nanoparticles would significantly reduce, as represented by the solid line. The data points represent experimental values for nanodiamond soot (gray diamond), carbon onions (hollow circles), and carbon black (solid circles) as a function of average particle size. The insets are illustrations of the geometries used in the model. *Source*: Reproduced from J. Huang, B.G. Sumpter, V. Meunier. Theoretical model for nanoporous carbon supercapacitors. Angew. Chem. Int. Ed. 47 (2008) 520–524; J. Huang, R. Qiao, B. Sumpter, V. Meunier, Effect of diffuse layer and pore shapes in mesoporous carbon supercapacitors. J. Mater. Res. 25 (2010) 1469; J. Huang, B. Sumpter, V. Meunier, A universal model for nanoporous carbon supercapacitors applicable to diverse pore regimes, carbon materials, and electrolytes. Chem. Eur. J. 14 (2008) 6614; J. Huang, B. Sumpter, V. Meunier, G. Yushin, C. Portet, Y. Gogotsi, Curvature effects in carbon nanomaterials: exohedral versus endohedral supercapacitors. J. Mater. Res. 25 (2010) 1525.

"nonclassical behavior" caused by the large electrode curvature [39–41]. However, it should be noted that this trend can also be readily predicted by the exact solution given by Eq. (4.12), and the Gouy–Chapman model used in [39–41] is limited to very low ion concentrations and electric potentials (Fig. 4.6).

Huang et al. [38] used the Gouy–Chapman–Stern model to investigate the effects of the shape and geometry of a single nanopore on the specific capacitance. They explored "slit" and cylindrical pores with diameters ranging from 2 to 16 nm. The dielectric permittivity $\varepsilon_r = 9.73$ was imposed in the Stern layer based on the values previously fitted in the literature [37,38], while $\varepsilon_r = 36$ in the diffuse layer [42] The electrolyte concentration was $c_\infty = 1.0\ \mathrm{mol\,L^{-1}}$ and the electrode surface potential

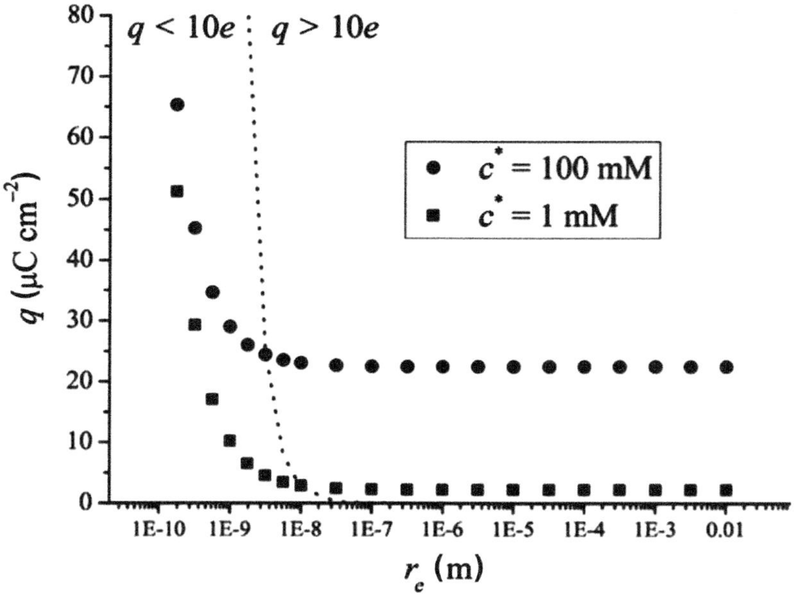

Figure 4.6 Compton's analysis of the modified Helmholtz model provided analysis of the areal charge stored per unit area of the particles of increasing radii from 0.1 nm to 0.01 m for the following conditions: $\epsilon_S = 78.54$, $T = 298$ K, $c^* = 1$, and 100 mM. *Source*: From S. Belding, F. Campbell, E. Dickinson, R. Compton, Nanoparticle-modified electrodes. Phys. Chem. Chem. Phys. 12 (2010) 11208; E. Dickinson, R. Compton, Diffuse double layer at nanoelectrodes. J. Phys. Chem. C 113 (2009) 17585; M. Henstridge, E. Dickinson, R. Compton, On the estimation of the diffuse double layer of carbon nanotubes using classical theory: curvature effects on the Gouy–Chapman limit, Chem. Phys. Lett. 485 (2010) 167.

was $\Psi_s = 1$ V. However, the Stern modified Gouy–Chapman model breaks down for such a concentration and potential as ions can no longer be treated as point charges [43,44].

Hamou et al. numerically investigated the electric potential profile near the probe apex in SECPMs by simulating planar and spherical electrodes [45,46]. Their model accounted for both the Stern layer and the finite size of ions and solved the MPB model given by Eq. (4.16). The electrolyte concentration c_∞ ranged from 10^{-5} to 10^{-2} mol L^{-1}, while the surface potential Ψ_s varied from 0.2 to 0.4 V. The authors investigated the effects of the shape and size of the probe on the electric potential profile in the electrolyte. They found that a sharper apex resulted in (1) higher electric field and surface charge density at the probe vertex and (2) better resolution of the electric potential in the lateral direction [45,46].

In nearly all the abovementioned references, the electrolyte dielectric permittivity was assumed to be constant and sometimes was treated as a fitting parameter. However, the relative permittivity ε_r of polar electrolytes is established to dramatically decrease as the electric field increases [42,47–53]. In fact, the individual electrolyte molecules can be influenced and become highly oriented under a large electric field. Therefore further orientation of the molecules can hardly provide more polarization and the relative permittivity decreases [48–50]. Booth derived the following model to account for the dependency of electrolyte dielectric permittivity on the local electric field larger than $10^7\ \mathrm{V\,m^{-1}}$ [42,47,48]:

$$\varepsilon_r(E) = n^2 + \left(\varepsilon_r(0) - n^2\right)\frac{3}{\beta E}\left[\coth(\beta E) - \frac{1}{\beta E}\right] \tag{4.20}$$

while ε_r can be treated as constant for $E < 10^7\ \mathrm{V\,m^{-1}}$

$$\varepsilon_r(E) = \varepsilon_r(0\Big) \tag{4.21}$$

Here, E is the norm of the local electrical field vector such that $E = |-\nabla\Psi|$, n is the index of refraction of the electrolyte at zero electric field frequency, and $\varepsilon_r(0)$ is the relative permittivity at zero electric field. Results of molecular dynamics simulations for different electrolytes [50–52] have verified that the Booth model accurately predicts the electrolyte permittivity for high electric fields up to $4\ \mathrm{V\,nm^{-1}}$ typically encountered in EDLCs [52,54]. Furthermore, the Booth model has been combined with the Poisson equation in [32,55,56] to investigate the repulsion between two charged planar surface electrodes due to hydration forces in aqueous electrolyte solutions. Hamou et al. investigated the effect of field-dependent electrolyte permittivity using the Booth model in the simulations of SECPM [45,46]. However, they did not observe significant changes in the dielectric permittivity and in electric potential profiles [45,46].

Wang et al. went on to utilize the MPB and Booth models to predict the specific capacitance of closely packed monodispersed nanosphere arrays with different packing morphologies and sphere diameters [57]. The numerical results established that the diffuse layer–specific capacitance of the sphere arrays significantly decreased when the field-dependent electrolyte permittivity was accounted for [57]. However, the Stern layer capacitance was predicted using the Helmholtz model (Eq. 4.8a) rather than by simulating the complete electric double-layer structure consisting of both the Stern and diffuse layers.

An equilibrium model based on continuum theory was developed to predict the specific capacitance of a single spherical electrode particle with various radii. It was considered the first simulation of the electric double-layer capacitance of ultramicroelectrodes by simultaneously accounting for (1) the Stern and diffuse layers, (2) the finite size of ions, and (3) field-dependent dielectric permittivity. They went further to account for the fact that the double-layer capacitances also significantly depend on surface electric potential [43,44], electrolyte concentration [58–60], and temperature (Eq. 4.16).

Wang and Pilot developed a simulation of the capacitance around spherical bodies, which demonstrated a fundamentally pragmatic approach which demonstrated the influence of the particle size on field distribution, capacitance, and how the models should be used to accurately make assumptions at the submicroscale [23]. Fig. 4.7 shows the schematic of the computational domain simulated in the present study. A spherical electrode, with radius R_0, defining the point of the solid–liquid interface, was immersed in an electrolyte solution. The region of electrolyte solution was assumed to consist of two discreet layers corresponding to (1) a Stern layer of thickness H near the electrode surface and with the spherical coordinate from the center of the electrode $R_{Stern} = R_0 + H$ and (2) a diffuse layer beyond. By virtue of symmetry, the problem was one-dimensional in the radial direction and was solved in spherical coordinates. The electric potential was prescribed as positive at the electrode surface and was zero far from the electrode

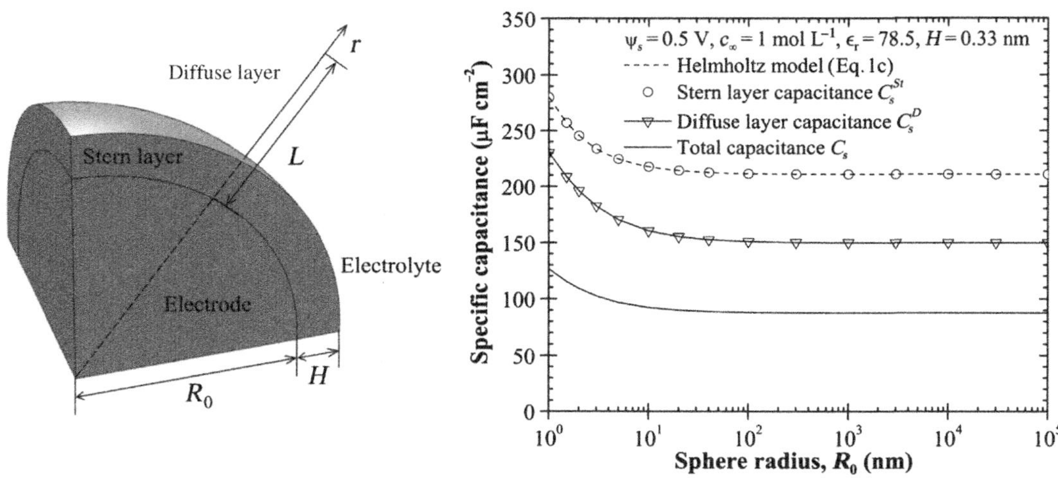

Figure 4.7 Compton's analysis of the modified Helmholtz model provided analysis of the areal charge stored per unit area. *Source*: Modified from H. Wang, J. Varghese, L. Pilon, Simulation of electric double layer capacitors with mesoporous electrodes: effects of morphology and electrolyte permittivity. Electrochim. Acta 56 (2011) 6189; H. Wang, L. Pilon, Accurate simulations of electric double layer capacitance of ultramicroelectrodes, J. Phys. Chem. C 115 (2011) 16711–16719.

surface. The length of the overall computational domain was specified to be $L = 80$ nm for all cases simulated. Increasing this length to 160 nm was found to have no effect on the predicted electric potential profile and on the specific capacitance. To make the problem mathematically tractable, the following assumptions were made: (1) the electric potential and ion concentration were invariant with time (steady state) and reached their equilibrium states, (2) anions and cations had the same and constant effective diameter independent of the electrolyte concentrations [43,44,61], (3) isothermal conditions prevailed throughout the electrode and electrolyte, (4) advection of the electrolyte was assumed to be negligible, and (5) the ions could only accumulate at the electrode surface and could not diffuse into the electrode particle, that is, there was no ion insertion.

The boundary conditions at the submicron scale were further defined by Wang and Pilon [57,62]. The local steady-state and equilibrium electric potential in the electrolyte solution denoted by $\Psi(r)$ was computed by solving, firstly, Eq. (4.13a) in the Stern layer and then Eq. (4.15) in the diffuse layer. It should be noted that the coordinate transformation $R = r - R_0$ was used, in which to simplify the governing equation to account for the electrode radius:

$$\frac{1}{(R+R_0)^2}\frac{d}{dR}\left(\varepsilon_0\varepsilon_r(R+R_0)^2\frac{d\Psi}{dR}\right) \quad 0 \le R \le H \qquad (4.22a)$$

$$= \begin{cases} 0 \\ \dfrac{\left(2zeN_A c_\infty \sinh\left(\dfrac{ze\Psi}{k_BT}\right)\right)}{\left(1 + 2\nu \sinh^2\left(\dfrac{ze\Psi}{k_BT}\right)\right)} \end{cases} \quad R \ge H \qquad (4.22b)$$

where R represents the distance from the electrode surface. The associated boundary conditions were given by [57,62–64]

$$\Psi = \Psi_s \quad \text{when } R = 0 \qquad (4.23a)$$

$$\varepsilon_0\varepsilon_r \frac{d\Psi}{dR}\bigg|_{R=H^-} = \varepsilon_0\varepsilon_r \frac{d\Psi}{dR}\bigg|_{R=H^+} \quad \text{when } R = H \qquad (4.23b)$$

$$\Psi = 0 \quad \text{when } R = L \tag{4.23c}$$

Eq. (4.13b) states that the electric potential and displacement were continuous across the Stern-diffuse layers interface located at $R = H$ [63,64]. Cases when $H = 0$ in Eqs. (4.22a), (4.22b), (4.23a), and (4.23c) correspond to simulations without the Stern layer as performed in Refs. [58–60,65].

In order to solve Eqs. (4.22a), (4.22b), (4.23a), and (4.23c), the electrolyte properties ε_r, z, c_∞, and ion diameter, a, and the temperature, T, are required. Here, the Booth model, which is given by Eqs. (4.20) and (4.21), was used to account for the effects of the electric field on electrolyte relative permittivity [42,47,48]. The study carried out by Wang and Pilon focuses on aqueous binary symmetric electrolyte solution at room temperature ($T = 298$ K) characterized by the following properties: $\varepsilon_r(0) = 78.5$, $n = 1.33$, and $\beta = 1.41\ 108\ \text{V m}^{-1}$ [32,55]. The effective ion diameter was taken as $a = 0.66$ nm and the valency was $z = 1$ corresponding to solvated ions such as K^+, OH, and Cl in aqueous solutions [61], for example. The electrolyte concentration was chosen as $c_\infty = 1.0\ \text{mol L}^{-1}$ corresponding to the typical values in EDLCs. Finally, the Stern layer thickness, H, was approximated as the solvated ion radius, that is, $H = a/2 = 0.33$ nm [66–69]. In reality, the Stern layer thickness may be larger than the solvated ion radius due to the specific adsorption of solvent molecules or anions near the electrode surface [63,64,66–69]. This is typically caused by nonelectrostatic forces [63,64,66–69]. A parametric study was also carried out for different values of Stern layer thickness, $H = 0$, 0.33, and 1.0 nm [57,62]. The specific capacitances of the Stern and diffuse layers were computed by dividing the surface charge density [69–71] $q_s(R) = \varepsilon_0 \varepsilon_r E(R)$ by their respective potential differences as [69–71]

$$C_s^{\text{Stern}} = \frac{q_s(0)}{\Psi_s - \Psi_D} = \frac{\varepsilon_0 \varepsilon_r E(0)}{\Psi_s - \Psi_D} \tag{4.24}$$

$$C_s^D = \frac{q_s(H)}{\Psi_D} = \frac{\varepsilon_0 \varepsilon_r E(H)}{\Psi_D} \tag{4.25}$$

where $E(R) = |d\Psi/dR|(R)$ is the norm of the local electric field at location R while $\Psi_D = \Psi(H)$ is the electric potential computed at the Stern-diffuse layers interface. Then, the total specific capacitance C_s was calculated, utilizing the formula for the capacitance of capacitors wired in series, as [69–71]

$$\frac{1}{C_s} = \frac{1}{C_s^{\text{Stern}}} + \frac{1}{C_s^D} \tag{4.26}$$

Numerical convergence was assessed based on the surface charge densities $q_s(R)$ at $R = 0$ and at $R = H$. The convergence criterion was chosen such that the maximum relative difference in both $q_s(0)$ and $q_s(H)$ was less than 1% when multiplying the total number of finite elements by 2. The total number of finite elements required to obtain a converged solution was less than 400 for all cases simulated in the present study.

The numerical tool was validated against (1) the exact solutions of the Gouy–Chapman model for planar electrodes (Eq. 4.10) and spherical electrodes (Eq. 4.11) for $\varepsilon_r = 78.5$, $c_\infty = 0.01$ mol L^{-1}, and $\Psi_D = 0.01$ V and (2) the numerical results of the modified PB model (Eq. 4.15) for planar electrodes reported for a wide range of packing parameter ν and dimensionless potential $(z_e\Psi D/k_BT)$ [43]. Excellent agreement was found in all cases considered. Fig. 4.8 shows the numerically predicted diffuse layer–specific capacitance, C_s, D as a function of sphere radius R_0 ranging from 1 nm to 100 μm. It was obtained by solving Eqs. (4.22a) and (4.22b) with $H = 0$ assuming constant permittivity $\varepsilon_r = 78.5$, $c_\infty = 0.01$ mol L^{-1}, $a = 0.66$ nm (i.e., $\nu = 0.0035$), and $\Psi_D = 0.01$ V. Fig. 4.8 also shows the exact solutions for planar and spherical electrodes

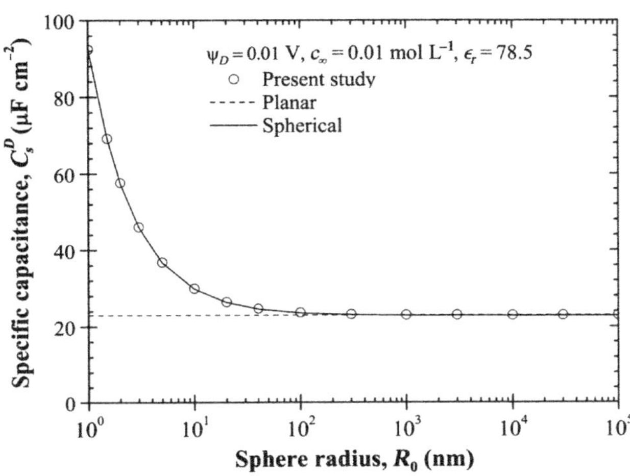

Figure 4.8 The results of the application of the Stern modified Gouy–Chapman model, simulated by Wang and Pilot, according to the constant permittivity of $\varepsilon_r = 78.5$, $c_\infty = 0.01$ mol L^{-1} and $\Psi_D = 0.01$ V along with the exact solutions for the planar and spherical electrodes under the same conditions. *Source*: From H. Wang, L. Pilon, Accurate simulations of electric double layer capacitance of ultramicroelectrodes. J. Phys. Chem. C 115(2011) 16711–16719.

respectively given by Eqs. (4.10) and (4.11). The numerical predictions agreed quite well with the exact solutions for all values of electrode radius considered. It is evident that C_{s}, D decreased with increasing sphere radius. It also reached the asymptotic value of planar electrodes (Eq. 4.10) for sphere radii larger than 100 nm. This can be attributed to the fact that a smaller sphere radius results in a larger surface electric field [71] and thus larger surface charge and specific capacitance. These results are similar to the trend predicted numerically in further literature for spherical and cylindrical nanoelectrodes using $\psi_s = 0.25$ V and $c\infty = 0.1$ mol L^{-1} [59,60]. However, the observed trend is not due to "nonclassical behavior" but is based on classical continuum theory.

4.4 Application of nanostructure electrode materials in electrochemical double-layer capacitance supercapacitors

The two primary attributes associated with an electrochemical capacitor are its energy or capacity and power density, both of which are mostly expressed as a quantity per unit weight, that is, F g^{-1} and W g^{-1}, respectively. Since the energy stored in electrochemical capacitors is related to the charge arising from the potential difference at each interface, especially in the case of nonfaradaic charge storage, they can offer more rapid charge/discharge rates as compared with batteries, whereas their energy density is lower than that of batteries [72]. Many researchers have made an effort to improve the energy density while maintaining high-power density.

In most cases, EDLCs are based on porous carbon electrode materials. EDLCs are often referred to as supercapacitors or ultracapacitors that electrostatically store the charge by using reversible adsorption of the electrolyte ions onto electrochemically stable, high-surface-area carbonaceous electrodes. The specific surface area of EDLCs is enlarged by making the bulk of the carbon material porous [73]. In principle, the main requirements for EDLC electrodes include: (1) fast charge/discharge rate; (2) large potential window; (3) high conductivity; and (4) large effective surface area [74–76].

Pseudocapacitive materials show fast redox reactions during the charge/discharge process at their surface. The fact that charge storage is based on a redox process means that pseudocapacitors have some battery-like behavior (faradaic reaction) in their charge/discharge process [74,77]. Therefore pseudocapacitors show higher capacitances than EDLCs, although they have somewhat slower charge/discharge rates than EDLCs. Pseudocapacitance is typically shown by materials such as conducting polymers and transition metal oxides. The interest is increasing in the development of improved materials for pseudocapacitors. To improve the capacitance of pseudocapacitors, four key factors are required: (1) doping of the conducting polymer to increase the redox state and conductivity; (2) high charge/discharge rate; (3) high surface area for the redox reaction; and (4) a wide potential window [77].

Hybrid capacitors comprising EDLCs and pseudocapacitors combine their advantages, namely high energy and power densities [78–80]. The charge storage mechanisms in such devices are a combination of purely electrostatic adsorption–desorption phenomenon at the nonfaradaic electrode and a reversible faradaic reaction at the electrode. To achieve high energy density, hybrid capacitor systems comprising redox materials have been actively researched and developed in recent years [15]. Both EDLCs and pseudocapacitors are essential for fabricating high-performance hybrid capacitors. Electrochemical capacitors consist of electrolytes, separators, binders, and electrode materials.

4.4.1 Electrochemical double-layer capacitance nanomaterials

Carbon materials are largely considered the most realistic and promising electrode materials for industrialization. The advantages of carbon materials include large specific surface area, good electronic conductivity, relative abundance, and high chemical stability [27,73,81]. The charge storage mechanism of carbon-based electrode materials mostly conforms to that of EDLCs. The important factors influencing their electrochemical performance are specific surface area, pore size distribution, pore shape and structure, electrical conductivity, and surface functionality. Among these, specific surface area and pore size distribution are the two most important factors affecting the performance of EDLCs [82].

4.4.1.1 Porous carbon

In order to increase the performance of the carbonaceous electrode the introduction of pores into the structure can drastically increase surface area. As shown by the application of the

Stern model the size of the pores can be fundamental to the capacitative performance of the electrode material. To realize the enhanced performance that stems from increasing the surface area of electrode materials, the pore size of the materials should be limited to the range of subnanometers to a few hundred nanometers [83]. Extensive research throughout the literature has been conducted on developing porous structures [82]. The small pore sizes of carbonaceous materials result in large specific surface areas. However, at a certain, optimal pore size, the effective surface area of the double layer is maximized while maintaining the optimal ion exchange with the electrolyte [84]. For example, a colloidal crystal templating method was optimized for the synthesis of three-dimensionally highly ordered mesoporous (3DOM) carbon with a well-defined geometry, a three-dimensional (3D) interconnected pore structure, and tunable pore sizes of 8–40 nm [85]. To achieve precise control over the pore sizes in the carbon products, parameters were established for direct syntheses or seed growth of monodispersed silica nanospheres of specific sizes [86]. The 3DOMs exhibited higher capacitance than conventional porous carbon electrodes (Fig. 4.9).

These porous carbon materials have a significant effect when used in conjunction with other carbon materials [73]. A judicious combination of porous carbon and so-called nanocarbons, such as graphene [87], carbon nanotubes (CNTs) [88], and carbon nanofibers (CNFs) [89], have been shown to improve the capacitive performance based purely on the principle of an electrochemical double layer. High-surface-area CNT/microporous carbon composite materials were prepared for EDLC electrodes [26,31,90]. All-carbon-based CNT/microporous carbon core-shell nanocomposites have silica template, silica–polymer composite 3DOM carbon infiltration of carbon precursor, a

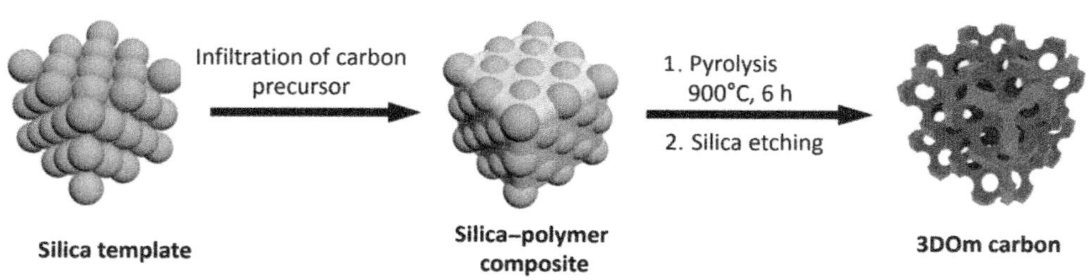

Figure 4.9 Schematic illustration of the synthesis of 3D highly ordered mesoporous carbon. *Source*: From A. Vu, X. Li, J. Phillips, A. Han, W.H. Smyrl, P. Buehlmann, et al., Three-dimensionally ordered mesoporous (3DOm) carbon materials as electrodes for electrochemical double-layer capacitors with ionic liquid electrolytes, Chem. Mater. 25 (2013) 4137–4148.

high-surface-area microporous carbon shell. Therefore these CNT/microporous carbon core-shell nanocomposites are promising electrode materials for EDLCs.

In another case, porous carbon–CNT–graphene ternary all-carbon foams were obtained through multicomponent surface self-assembly of graphene oxide (GO)-dispersed pristine CNTs supported on a commercial sponge [90]. The GO acted as a "surfactant" that dispersed the CNTs, thereby preserving their excellent electronic structure and preventing the aggregation of graphene, which resulted in an overall improvement of the conductivity. The large number of high-surface-area mesopores promotes ion transport/charge storage [23,26,31].

4.4.1.2 Carbon nanofibers, carbon nanotubes, and graphene

Graphene has a maximum theoretical specific surface area of approximately 2600 $m^2\,g^{-1}$, which is twice that of single-walled CNTs and substantially higher than those of most carbon black and activated carbons [91]. Graphene is a unique and attractive electrode material owing to its atom-thick two-dimensional (2D) structure and excellent properties. However, graphene sheets easily form irreversible agglomerates and restack to the graphitic structure. To overcome this problem, graphene can be hybridized with CNTs, CNFs, and porous carbon. Various carbon hybrids have been fabricated by electrospinning followed by heat treatment [92] since electrospinning can readily create multicomponent nanofibers as the carbon precursor [93].

Many studies have shown that highly oriented CNFs have a large surface area and excellent electrical properties [94]. Poly (acrylonitrile) (PAN)-containing GONR nanocomposites also become highly oriented during electrospinning. In addition, GONRs have been converted to an all-carbon material, GNR/CNF, by carbonization [95]. Both the graphene ribbons and CNFs exhibit high performance as electrochemical capacitors; it is expected that the structural synergistic effect would further improve this performance. Various methods of fabricating these graphene fiber materials have been investigated. Graphene-coated nanotube aerogels, a type of graphene fiber material, have been developed by coating the nodes of an isotropic single-wall CNT network within an aerogel with a few layers of graphene. These graphene-based CNT and CNF materials are especially important for textile-enabled materials and devices. They can also act as the building blocks for forming 2D and 3D macroscopic structures for EDLCs [96,97]. The use of CNFs and

CNTs has further enhanced the properties of microelectrochemical capacitors, thereby enabling the fabrication of flexible and adaptable devices [98].

Many researchers have reported on combinations of CNTs, CNFs, metals, and graphene. CNTs and CNFs are connected to the graphene layer through covalent bonds, leading to seamless, high-quality carbon material–graphene–metal interfaces [99]. Ternary hybrid nanostructures consisting of CNFs, manganese oxide (MnO), and graphene were recently fabricated, as shown in Fig. 4.10. MnO-decorated CNFs (MCNFs) were dispersed in an aqueous solution containing isolated GO sheets exfoliated from oxidized graphite. GO sheets are highly dispersible in aqueous solution owing to the oxygen-containing functional groups, and they have high affinity with CNFs because of their similar chemical structures. Thus MCNFs were readily wrapped with GO sheets to yield a 3D nanohybrid architecture. Afterward, GO was converted to reduced graphene oxide (rGO) by using hydrazine. Intercalated MCNFs improved the conductivity of the MCNF/rGO nanocomposite and facilitated ion diffusion by increasing the spacing between the graphene sheets. The rGO sheet is considered to act as a conductive channel in the nanohybrid. Additionally, the intercalation of CNFs between the rGO sheets induced a 3D opened geometry in the electrode, which allowed facile ion and charge transfer. Consequently,

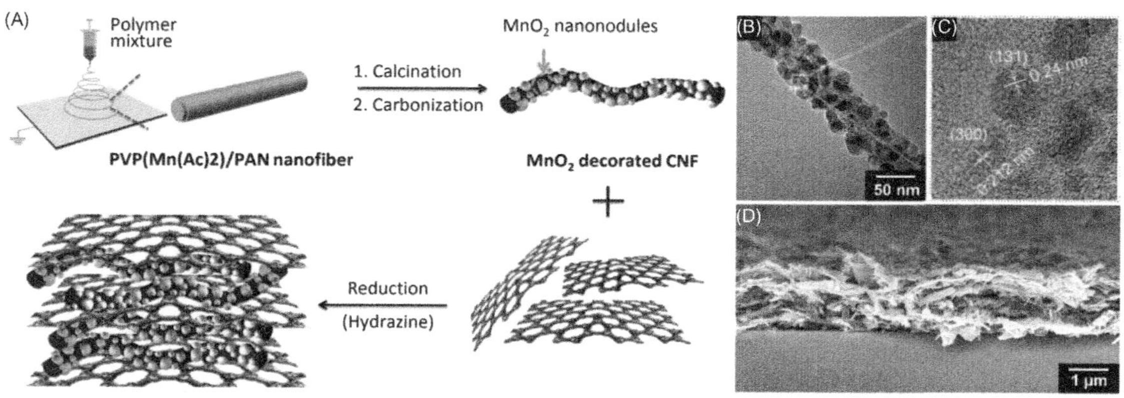

Figure 4.10 (A) The schematic illustration of the synthesis of MnO-decorated CNFs and reduced graphene oxide hybrids, (B) a TEM of the morphology of a MnO decorates CNF, (C) an HR-TEM of an MnO-decorated CNF, and (D) a cross-sectional SEM image of the complete composite deposited onto a silicon wafer. *Source*: Modified from O.S. Kwon, T. Kim, J.S. Lee, S.J. Park, H.-W. Park, M. Kang, et al., Fabrication of graphene sheets intercalated with manganese oxide/carbon nanofibers: toward high-capacity energy storage, Small 9 (2013) 248–254.

high-performance capacitive properties were observed for these new 3D structures [100–103].

CNT-bridged graphene 3D building blocks were synthesized via the Coulombic interaction between positively charged CNTs grafted by cationic surfactants and negatively charged GO sheets [104]. The CNTs were intercalated into the nanoporous graphene layers to build pillared 3D structures, which increased the accessible surface area and allowed fast ion diffusion. Because of this unique 3D porous structure, the electrodes showed remarkable electrochemical performance in ionic liquid electrolytes [87]. However, it is difficult to prepare electrodes by using uniform nanostructured CNTs on special substrates, especially those with porous surface structures. Cellulose is the most abundant and sustainable natural polymer. Cellulose nanofibrils (CeNFs) derived from cellulose have high aspect ratios, excellent mechanical properties, excellent flexibility, and superior hydrophilicity [105,106]. The CeNF-based aerogel possessed a porous structure and an extremely high porosity (resulting in ultralow density and a high specific surface area), as well as excellent electrolyte-absorption properties. Furthermore, the hydrophilicity of the CeNFs in the aerogel improved the contact between the electrodes and electrolytes, and it provided diffusion channels for the electrolyte ions, thus enhancing the performance of the EDLCs.

Various polymer precursors have been used to make new types of carbon nanomaterials. The main characteristics of the resultant carbon nanomaterials depend on the polymer precursors. It is relatively easy to control the morphology and composition of polymers, which offers opportunities for producing carbon nanomaterials with controlled structures and properties [107]. Several researchers have fabricated electrode materials via heat treatment of GO/polymer hybrid precursors, such as PVP(Mn(Ac)2)/PAN nanofiber GO/polypyrrole (PPy) nanowires, GO/PPy nanotubes, and GO/polyaniline. Heat treatment of the GO/polymer hybrids resulted in the formation of graphene-embedded all-carbon nanostructures. These carbonized nanohybrids exhibited good performance as EDLC electrode materials.

4.5 Summary

EDLCs can be applied in the case of stationary and mobile systems requiring high-power pulses. Moreover, owing to their low time constant, they can quickly harvest energy, such as during deceleration or braking of vehicles. Although EDLCs are able to provide higher power with a longer cycle life, they suffer from a relatively low energy density. Therefore the current research is mainly concerned with the optimization of the existing electrode materials and the development of new materials to improve energy density. Many different carbon forms, such as CNTs, CNFs, activated carbon powders, and carbonized materials, can be used as active materials in EDLC electrodes.

References

[1] J.R. Miller, P. Simon, Electrochemical capacitors for energy management, Science 321 (2008) 651–652. Available from: https://doi.org/10.1126/science.1158736.

[2] A.G. Pandolfo, A.F. Hollenkamp, Carbon properties and their role in supercapacitors, J. Power Sources 157 (2006) 11–27. Available from: https://doi.org/10.1016/j.jpowsour.2006.02.065.

[3] R. Kötz, M. Carlen, Principles and applications of electrochemical capacitors, Electrochim. Acta 45 (2000) 2483–2498. Available from: https://doi.org/10.1016/S0013-4686(00)00354-6.

[4] P. Sharma, T.S. Bhatti, A review on electrochemical double-layer capacitors, Energy Convers. Manag. 51 (2010) 2901–2912. Available from: https://doi.org/10.1016/j.enconman.2010.06.031.

[5] C. Portet, P.L. Taberna, P. Simon, C. Laberty-Robert, Modification of Al current collector surface by sol–gel deposit for carbon–carbon supercapacitor applications, Electrochim. Acta 49 (2004) 905–912. Available from: https://doi.org/10.1016/j.electacta.2003.09.043.

[6] G. Wang, L. Zhang, J. Zhang, A review of electrode materials for electrochemical supercapacitors, Chem. Soc. Rev. 41 (2012) 797–828. Available from: https://doi.org/10.1039/C1CS15060J.

[7] M. Endo, T. Takeda, Y. Kim, K. Koshiba, K. Ishii, High Power Electric Double Layer Capacitor (EDLC's): From Operating Principle to Pore Size Control in Advanced Activated Carbons, vol. 1, Carbon Letters, Springer, 2001.

[8] L.L. Zhang, X.S. Zhao, Carbon-based materials as supercapacitor electrodes, Chem. Soc. Rev. 38 (2009) 2520–2531. Available from: https://doi.org/10.1039/b813846j.

[9] O. Barbieri, M. Hahn, A. Herzog, R. Kötz, Capacitance limits of high surface area activated carbons for double layer capacitors, Carbon 43 (2005) 1303–1310. Available from: https://doi.org/10.1016/j.carbon.2005.01.001.

[10] D. Qu, H. Shi, Studies of activated carbons used in double-layer capacitors, J. Power Sources 74 (1998) 99–107. Available from: https://doi.org/10.1016/S0378-7753(98)00038-X.

[11] J. Gamby, P.L. Taberna, P. Simon, J.F. Fauvarque, M. Chesneau, Studies and characterisations of various activated carbons used for carbon/carbon supercapacitors, J. Power Sources 101 (2001) 109–116. Available from: https://doi.org/10.1016/S0378-7753(01)00707-8.

[12] H. Shi, Activated carbons and double layer capacitance, Electrochim. Acta 41 (1996) 1633–1639. Available from: https://doi.org/10.1016/0013-4686(95)00416-5.

[13] H. Jiang, L. Yang, C. Li, C. Yan, P.S. Lee, J. Ma, High-rate electrochemical capacitors from highly graphitic carbon-tipped manganese oxide/ mesoporous carbon/manganese oxide hybrid nanowires, Energy Environ. Sci. 4 (2011) 1813–1819. Available from: https://doi.org/10.1039/ C1EE01032H.
[14] Q. Wang, J. Yan, Z. Fan, Carbon materials for high volumetric performance supercapacitors: design, progress, challenges and opportunities, Energy Environ. Sci. 9 (2016) 729–762. Available from: https://doi.org/10.1039/ C5EE03109E.
[15] J. Liu, G. Cao, Z. Yang, D. Wang, D. Dubois, X. Zhou, et al., Oriented nanostructures for energy conversion and storage, ChemSusChem 1 (2008) 676.
[16] D. Su, R. Schlogl, Nanostructured carbon and carbon nanocomposites for electrochemical energy storage applications, ChemSusChem 3 (2010) 136.
[17] J. Wang, J. Polleux, J. Lim, B. Dunn, Pseudocapacitive contributions to electrochemical energy storage in TiO_2 (anatase) nanoparticles, J. Phys. Chem. C 111 (2007) 14925.
[18] A. Manthiram, A. Murugan, A. Sarkar, T. Muraliganth, Nanostructured electrode materials for electrochemical energy storage and conversion, Energy Environ. Sci. 1 (2008) 621.
[19] D. Qu, Studies of the activated carbons used in double-layer supercapacitors, J. Power Sources 109 (2002) 403–411. Available from: https://doi.org/10.1016/S0378-7753(02)00108-8.
[20] Y.J. Kim, Y. Horie, S. Ozaki, Y. Matsuzawa, H. Suezaki, C. Kim, et al., Correlation between the pore and solvated ion size on capacitance uptake of PVDC-based carbons, Carbon 42 (2004) 1491–1500. Available from: https://doi.org/10.1016/j.carbon.2004.01.049.
[21] J. Chmiola, G. Yushin, Y. Gogotsi, C. Portet, P. Simon, P.L. Taberna, Anomalous increase in carbon capacitance at pore sizes less than 1 nanometer, Science 313 (2006) 1760. Available from: https://doi.org/ 10.1126/science.1132195.
[22] C. Largeot, C. Portet, J. Chmiola, P.-L. Taberna, Y. Gogotsi, P. Simon, Relation between the ion size and pore size for an electric double-layer capacitor, J. Am. Chem. Soc. 130 (2008) 2730–2731. Available from: https:// doi.org/10.1021/ja7106178.
[23] Brownson, C.E. Banks, The Handbook of Graphene Electrochemistry, Springer, 2014.
[24] A. García-Gómez, G. Moreno-Fernández, B. Lobato, T.A. Centeno, Constant capacitance in nanopores of carbon monoliths, Phys. Chem. Chem. Phys. 17 (2015) 15687–15690. Available from: https://doi.org/10.1039/ C5CP01904D.
[25] D.A.C. Brownson, P.J. Kelly, C.E. Banks, In situ electrochemical characterisation of graphene and various carbon-based electrode materials: an internal standard approach, RSC Adv. 5 (2015) 37281–37286. Available from: https://doi.org/10.1039/C5RA03049H.
[26] M.P. Down, C.E. Banks, Freestanding three-dimensional graphene macroporous supercapacitor, ACS Appl. Energy Mater. 1 (2018) 891–899. Available from: https://doi.org/10.1021/acsaem.7b00338.
[27] M.P. Down, C.W. Foster, X. Ji, C.E. Banks, Pencil drawn paper based supercapacitors, RSC Adv. 6 (2016) 81130–81141. Available from: https:// doi.org/10.1039/C6RA18499E.
[28] C.W. Foster, M.P. Down, Y. Zhang, X. Ji, S.J. Rowley-Neale, G.C. Smith, et al., 3D printed graphene based energy storage devices, Sci. Rep. 7 (2017) 42233.

Available from: https://doi.org/10.1038/srep42233. Available from: http://www.nature.com/articles/srep42233#supplementary-information.
[29] A.F. Khan, M.P. Down, G.C. Smith, C.W. Foster, C.E. Banks, Surfactant-exfoliated 2d hexagonal boron nitride (2d-Hbn): role of surfactant upon the electrochemical reduction of oxygen and capacitance applications, J. Mater. Chem. A 5 (2017) 4103–4113. Available from: https://doi.org/10.1039/C6TA09999H.
[30] S. He, W. Chen, 3D graphene nanomaterials for binder-free supercapacitors: scientific design for enhanced performance, Nanoscale 7 (2015) 6957–6990. Available from: https://doi.org/10.1039/C4NR05895J.
[31] D.A.C. Brownson, L.C.S. Figueiredo-Filho, X. Ji, M. Gomez-Mingot, J. Iniesta, O. Fatibello-Filho, et al., Freestanding three-dimensional graphene foam gives rise to beneficial electrochemical signatures within non-aqueous media, J. Mater. Chem. A 1 (2013) 5962–5972. Available from: https://doi.org/10.1039/C3TA10727B.
[32] S. Basu, M. Sharma, Effect of dielectric saturation on disjoining pressure in thin films of aqueous electrolytes, J. Colloid Interface Sci. 165 (1994) 355.
[33] F. Stoeckli, T.A. Centeno, Pore size distribution and capacitance in microporous carbons, Phys. Chem. Chem. Phys. 14 (2012) 11589–11591. Available from: https://doi.org/10.1039/C2CP41545C.
[34] C. Vix-Guterl, E. Frackowiak, K. Jurewicz, M. Friebe, J. Parmentier, F. Béguin, Electrochemical energy storage in ordered porous carbon materials, Carbon 43 (2005) 1293–1302. Available from: https://doi.org/10.1016/j.carbon.2004.12.028.
[35] J. Huang, B.G. Sumpter, V. Meunier, Theoretical model for nanoporous carbon supercapacitors, Angew. Chem. Int. Ed. 47 (2008) 520–524. Available from: https://doi.org/10.1002/anie.200703864.
[36] J. Huang, R. Qiao, B. Sumpter, V. Meunier, Effect of diffuse layer and pore shapes in mesoporous carbon supercapacitors, J. Mater. Res. 25 (2010) 1469.
[37] J. Huang, B. Sumpter, V. Meunier, A universal model for nanoporous carbon supercapacitors applicable to diverse pore regimes, carbon materials, and electrolytes, Chem.: Eur. J 14 (2008) 6614.
[38] J. Huang, B. Sumpter, V. Meunier, G. Yushin, C. Portet, Y. Gogotsi, Curvature effects in carbon nanomaterials: exohedral versus endohedral supercapacitors, J. Mater. Res. 25 (2010) 1525.
[39] S. Belding, F. Campbell, E. Dickinson, R. Compton, Nanoparticle-modified electrodes, Phys. Chem. Chem. Phys. 12 (2010) 11208.
[40] E. Dickinson, R. Compton, Diffuse double layer at nanoelectrodes, J. Phys. Chem. C 113 (2009) 17585.
[41] M. Henstridge, E. Dickinson, R. Compton, On the estimation of the diffuse double layer of carbon nanotubes using classical theory: curvature effects on the Gouy–Chapman limit, Chem. Phys. Lett. 485 (2010) 167.
[42] F. Booth, Dielectric constant of polar liquids at high field strengths, J. Chem. Phys. 23 (1955) 453.
[43] M. Kilic, M. Bazant, A. Ajdari, Steric effects in the dynamics of electrolytes at large applied voltages. I. Double-layer charging, Phys. Rev. E 75 (2007) 021502.
[44] M. Bazant, M. Kilic, B. Storey, A. Ajdari, Towards an understanding of induced-charge electrokinetics at large applied voltages in concentrated solutions, Adv. Colloid Interface Sci. 152 (2009) 48.
[45] R. Hamou, P. Biedermann, A. Erbe, M. Rohwerder, Numerical analysis of Debye screening effect in electrode surface potential mapping by scanning

electrochemical potential microscopy, Electrochem. Commun. 12 (2010) 1391.
[46] R. Hamou, P. Biedermann, A. Erbe, M. Rohwerder, Numerical simulation of probing the electric double layer by scanning electrochemical potential microscopy, Electrochim. Acta 55 (2010) 5210.
[47] F. Booth, The dielectric constant of water and the saturation effect, J. Chem. Phys. 19 (1951) 391.
[48] A. Appleby, B. Conway, C. Vayenas, R. White, M. Gamboa-Adelco, Modern Aspects of Electrochemistry, Springer, 2005.
[49] A. Buckingham, Theory of the dielectric constant at high field strengths, J. Chem. Phys. 25 (1956) 428.
[50] I.C. Yeh, M. Berkowitz, Dielectric constant of water at high electric fields: molecular dynamics study, J. Chem. Phys. 110 (1999) 7935.
[51] R. Fulton, The nonlinear dielectric behavior of water: comparisons of various approaches to the nonlinear dielectric increment, J. Chem. Phys. 130 (2009) 204503.
[52] L. Yang, B. Fishbine, A. Migliori, L. Pratt, Dielectric saturation of liquid propylene carbonate in electrical energy storage applications, J. Chem. Phys. 132 (2010) 044701.
[53] P. Zarzycki, K. Rosso, Molecular dynamics simulation of the AgCl/ electrolyte interfacial capacity, J. Phys. Chem. C 114 (2010) 10019.
[54] B. Conway, Electrochemical Supercapacitors: Scientific Fundamentals and Technological Applications, Springer, 1999.
[55] Y. Gur, I. Ravina, A. Babchin, On the electrical double layer theory. II. The Poisson–Boltzmann equation including hydration forces, J. Colloid Interface Sci. 64 (1978) 333.
[56] V. Paunov, R. Dimova, P. Kralchevsky, G. Broze, A. Mehreteab, The hydration repulsion between charged surfaces as an interplay of volume exclusion and dielectric saturation effects, J. Colloid Interface Sci. 182 (1996) 239.
[57] H. Wang, J. Varghese, L. Pilon, Simulation of electric double layer capacitors with mesoporous electrodes: effects of morphology and electrolyte permittivity, Electrochim. Acta 56 (2011) 6189.
[58] A. Kornyshev, Double-layer in ionic liquids: paradigm change? J. Phys. Chem. B 111 (2007) 5545.
[59] P. Biesheuvel, M. van Soestbergen, Counterion volume effects in mixed electrical double layers, J. Colloid Interface Sci. 316 (2007) 490.
[60] J. Bikerman, Structure and capacity of electrical double layer, Philos. Mag. 33 (1942) 384.
[61] J.N. Israelachvili, Intermolecular and Surface Forces, Academic Press, Elsevier, 2010.
[62] H. Wang, L. Pilon, Accurate simulations of electric double layer capacitance of ultramicroelectrodes, J. Phys. Chem. C 115 (2011) 16711–16719. Available from: https://doi.org/10.1021/jp204498e.
[63] R. Hunter, Foundations of Colloid Science, Oxford University Press, 2001.
[64] J. Lyklema, Fundamentals of Interface and Colloid Science, vol. 2, Academic Press Elsevier, 2001.
[65] V. Vlachy, Ionic effects beyond Poisson–Boltzmann theory, Annu. Rev. Phys. Chem. 50 (1999) 145.
[66] A. Bard, L. Faulkner, Electrochemical Methods: Fundamentals and Applications, Wiley, 2001.
[67] V. Bagotsky, Fundamentals of Electrochemistry, Wiley, 2006.
[68] H.J.Butt, M. Kappl, Surface and Interfacial Forces, Wiley, 2010.

[69] J. Masliyah, S. Bhattacharjee, Electrokinetic and Colloid Transport Phenomena, Wiley, 2006.
[70] D. Griffiths, Introduction to Electrodynamics, Pearson, Cambridge University Press, 1999.
[71] R. Feynman, R. Leighton, M. Sands, The Feynman Lectures on Physics, vol. 2, Reading, Mass. : Addison-Wesley Pub. Co, 2005.
[72] P. Simon, Y. Gogotsi, Materials for electrochemical capacitors, Nat. Mater. 7 (2008) 845.
[73] A.B. Fuertes, M. Sevilla, Hierarchical microporous/mesoporous carbon nanosheets for high-performance supercapacitors, ACS Appl. Mater. Interfaces 7 (2015) 4344–4353. Available from: https://doi.org/10.1021/am508794f.
[74] K. Chen, Y. Yang, K. Li, Z. Ma, Y. Zhou, D. Xue, $CoCl_2$ designed as excellent pseudocapacitor electrode materials, ACS Sustain. Chem. Eng. 2 (2014) 440–444. Available from: https://doi.org/10.1021/sc400338c.
[75] T. Chen, H. Peng, M. Durstock, L. Dai, High-performance transparent and stretchable all-solid supercapacitors based on highly aligned carbon nanotube sheets, Sci. Rep. 4 (2014). Available from: https://doi.org/10.1038/srep03612.
[76] X. Chen, L. Qiu, J. Ren, G. Guan, H. Lin, Z. Zhang, et al., Novel electric double-layer capacitor with a coaxial fiber structure, Adv. Mater. 25 (2013) 6436–6441. Available from: https://doi.org/10.1002/adma.201301519.
[77] W. Chen, R.B. Rakhi, H.N. Alshareef, Morphology-dependent enhancement of the pseudocapacitance of template-guided tunable polyaniline nanostructures, J. Phys. Chem. C 117 (2013) 15009–15019. Available from: https://doi.org/10.1021/jp405300p.
[78] X.J. Li, W. Xing, J. Zhou, G.Q. Wang, S.P. Zhuo, Z.F. Yan, et al., Excellent capacitive performance of a three-dimensional hierarchical porous graphene/carbon composite with a superhigh surface area, Chem.: Eur. J. 20 (2014) 13314–13320. Available from: https://doi.org/10.1002/chem.201402897.
[79] H. Choi, H. Yoon, Nanostructured electrode materials for electrochemical capacitor applications, Nanomaterials 5 (2015) 906–936. Available from: https://doi.org/10.3390/nano5020906.
[80] N. Terasawa, K. Asaka, High-performance hybrid (electrostatic double-layer and faradaic capacitor-based) polymer actuators incorporating nickel oxide and vapor-grown carbon nanofibers, Langmuir 30 (2014) 14343–14351. Available from: https://doi.org/10.1021/la503468z.
[81] J. Zhang, X. Zhang, Y. Zhou, S. Guo, K. Wang, Z. Liang, et al., Nitrogen-doped hierarchical porous carbon nanowhisker ensembles on carbon nanofiber for high-performance supercapacitors, ACS Sustain. Chem. Eng. 2 (2014) 1525–1533. Available from: https://doi.org/10.1021/sc500221s.
[82] X.Y. Chen, C. Chen, Z.J. Zhang, D.H. Xie, Nitrogen-doped porous carbon spheres derived from polyacrylamide, Industr. Eng. Chem. Res. 52 (2013) 12025–12031. Available from: https://doi.org/10.1021/ie4017013.
[83] W.-J. Liu, K. Tian, Y.-R. He, H. Jiang, H.-Q. Yu, High-yield harvest of nanofibers/mesoporous carbon composite by pyrolysis of waste biomass and its application for high durability electrochemical energy storage, Environ. Sci. Technol. 48 (2014) 13951–13959. Available from: https://doi.org/10.1021/es504184c.

[84] J. Chen, C. Li, G. Shi, Graphene materials for electrochemical capacitors, J. Phys. Chem. Lett. 4 (2013) 1244–1253. Available from: https://doi.org/10.1021/jz400160k.
[85] A. Vu, X. Li, J. Phillips, A. Han, W.H. Smyrl, P. Buehlmann, et al., Three-dimensionally ordered mesoporous (3DOm) carbon materials as electrodes for electrochemical double-layer capacitors with ionic liquid electrolytes, Chem. Mater. 25 (2013) 4137–4148. Available from: https://doi.org/10.1021/cm400915p.
[86] K.T. Cho, S.B. Lee, J.W. Lee, Facile synthesis of highly electrocapacitive nitrogen-doped graphitic porous carbons, J. Phys. Chem. C 118 (2014) 9357–9367. Available from: https://doi.org/10.1021/jp501742x.
[87] X. Yang, F. Zhang, L. Zhang, T. Zhang, Y. Huang, Y. Chen, A high-performance graphene oxide-doped ion gel as gel polymer electrolyte for all-solid-state supercapacitor applications, Adv. Funct. Mater. 23 (2013) 3353–3360. Available from: https://doi.org/10.1002/adfm.201203556.
[88] Y. Yao, C. Ma, J. Wang, W. Qiao, L. Ling, D. Long, Rational design of high-surface-area carbon nanotube/microporous carbon core-shell nanocomposites for supercapacitor electrodes, ACS Appl. Mater. Interfaces 7 (2015) 4817–4825. Available from: https://doi.org/10.1021/am5087374.
[89] F.W. Richey, C. Tran, V. Kalra, Y.A. Elabd, Ionic liquid dynamics in nanoporous carbon nanofibers in supercapacitors measured with in operando infrared spectroelectrochernistry, J. Phys. Chem. C 118 (2014) 21846–21855. Available from: https://doi.org/10.1021/jp506903m.
[90] B. You, J. Jiang, S. Fan, Three-dimensional hierarchically porous all-carbon foams for supercapacitor, ACS Appl. Mater. Interfaces 6 (2014) 15302–15308. Available from: https://doi.org/10.1021/am503783t.
[91] Z. Qin, M. Taylor, M. Hwang, K. Bertoldi, M.J. Buehler, Effect of wrinkles on the surface area of graphene: toward the design of nanoelectronics, Nano Lett. 14 (2014) 6520–6525. Available from: https://doi.org/10.1021/nl503097u.
[92] X. Chen, H. Wang, H. Yi, X. Wang, X. Yan, Z. Guo, Anthraquinone on porous carbon nanotubes with improved supercapacitor performance, J. Phys. Chem. C 118 (2014) 8262–8270. Available from: https://doi.org/10.1021/jp5009626.
[93] D. Krishnan, K. Raidongia, J. Shao, J. Huang, Graphene oxide assisted hydrothermal carbonization of carbon hydrates, ACS Nano 8 (2014) 449–457. Available from: https://doi.org/10.1021/nn404805p.
[94] L. Deng, R.J. Young, I.A. Kinloch, A.M. Abdelkader, S.M. Holmes, D.A. De Haro-Del Rio, et al., Supercapacitance from cellulose and carbon nanotube nanocomposite fibers, ACS Appl. Mater. Interfaces 5 (2013) 9983–9990. Available from: https://doi.org/10.1021/am403622v.
[95] J. Zhang, K. Wang, S. Guo, S. Wang, Z. Liang, Z. Chen, et al., One-step carbonization synthesis of hollow carbon nanococoons with multimodal pores and their enhanced electrochemical performance for supercapacitors, ACS Appl. Mater. Interfaces 6 (2014) 2192–2198. Available from: https://doi.org/10.1021/am405375s.
[96] E. Wilson, M.F. Islam, Ultracompressible, high-rate supercapacitors from graphene-coated carbon nanotube aerogels, ACS Appl. Mater. Interfaces 7 (2015) 5612–5618. Available from: https://doi.org/10.1021/acsami.5b01384.
[97] B.R. Stoner, A.S. Raut, B. Brown, C.B. Parker, J.T. Glass, Graphenated carbon nanotubes for enhanced electrochemical double layer capacitor

performance, Appl. Phys. Lett. 99 (2011). Available from: https://doi.org/10.1063/1.3657514.
[98] K. Yu, G. Lu, Z. Bo, S. Mao, J. Chen, Carbon nanotube with chemically bonded graphene leaves for electronic and optoelectronic applications, J. Phys. Chem. Lett. 2 (2011) 1556–1562. Available from: https://doi.org/10.1021/jz200641c.
[99] O.S. Kwon, T. Kim, J.S. Lee, S.J. Park, H.-W. Park, M. Kang, et al., Fabrication of graphene sheets intercalated with manganese oxide/carbon nanofibers: toward high-capacity energy storage, Small 9 (2013) 248–254. Available from: https://doi.org/10.1002/smll.201201754.
[100] Y. Li, Z. Li, P.K. Shen, Simultaneous formation of ultrahigh surface area and three-dimensional hierarchical porous graphene-like networks for fast and highly stable supercapacitors, Adv. Mater. 25 (2013) 2474–2480. Available from: https://doi.org/10.1002/adma.201205332.
[101] J.-S. Lee, S.-I. Kim, J.-C. Yoon, J.-H. Jang, Chemical vapor deposition of mesoporous graphene nanoballs for supercapacitor, ACS Nano 7 (2013) 6047–6055. Available from: https://doi.org/10.1021/nn401850z.
[102] H. Sun, X. You, J. Deng, X. Chen, Z. Yang, J. Ren, et al., Novel graphene/carbon nanotube composite fibers for efficient wire-shaped miniature energy devices, Adv. Mater. 26 (2014) 2868–2873. Available from: https://doi.org/10.1002/adma.201305188.
[103] H. Peng, G. Ma, K. Sun, J. Mu, Z. Zhang, Z. Lei, Facile synthesis of poly(p-phenylenediamine)-derived three-dimensional porous nitrogen-doped carbon networks for high performance supercapacitors, J. Phys. Chem. C 118 (2014) 29507–29516. Available from: https://doi.org/10.1021/jp508684t.
[104] P. Duy Tho, T.H. Lee, D.H. Luong, F. Yao, A. Ghosh, L. Viet Thong, et al., Carbon nanotube-bridged graphene 3D building blocks for ultrafast compact supercapacitors, ACS Nano 9 (2015) 2018–2027. Available from: https://doi.org/10.1021/nn507079x.
[105] G. Zheng, L. Hu, H. Wu, X. Xie, Y. Cui, Paper supercapacitors by a solvent-free drawing method, Energy Environ. Sci. 4 (2011) 3368–3373.
[106] Q. Zheng, Z. Cai, Z. Ma, S. Gong, Cellulose nanofibril/reduced graphene oxide/carbon nanotube hybrid aerogels for highly flexible and all-solid-state supercapacitors, ACS Appl. Mater. Interfaces 7 (2015) 3263–3271. Available from: https://doi.org/10.1021/am507999s.
[107] X. Huang, S. Kim, M.S. Heo, J.E. Kim, H. Suh, I. Kim, Easy synthesis of hierarchical carbon spheres with superior capacitive performance in supercapacitors, Langmuir 29 (2013) 12266–12274. Available from: https://doi.org/10.1021/la4026969.

5

Organometallic hybrid perovskites for humidity and gas sensing applications

Emmanuel Kymakis[1], Apostolos Panagiotopoulos[1,2], Minas M. Stylianakis[1] and Konstantinos Petridis[3]

[1]Department of Electrical and Computer Engineering, Hellenic Mediterranean University (HMU), Heraklion, Greece [2]Department of Materials Science and Technology, University of Crete, Heraklion, Greece [3]Department of Electronic Engineering, Hellenic Mediterranean University (HMU), Chania, Greece

5.1 Introduction

Metal halide perovskite materials have been extensively studied over the past few years. These materials appear as the new rising stars in the optoelectronic sector [1]. They follow a cubic contractual formula of A^{+1} M^{+2} $(X^{-1})_3$, where each "A" (an organic group or an inorganic cation) has twelve neighboring "X" (halide atoms) and each "M" (a metal cation) connects with six adjacent "X" through ionic bonds. In the case, the A cation is organic, such as MA^+ (methyammonium): $CH_3NH_3{}^+$ or FA^+ (formamidinium):$CH(NH_2)_2{}^+$, we have an inorganic–organic metal halide perovskite semiconductor; whereas if the A cation is an inorganic atom such as cesium (Cs^+), we have inorganic metal halide perovskites. Metal halide perovskite materials due to their attractive electronic (bipolar conductivity, long diffusion lengths up to 100 μm), optical (direct bandgap that can be controlled through its constitution, e.g., replacement of lead with tin Sn^{+2}), high absorption coefficients, high quantum emission efficiencies, tunable properties through their morphological and composition, and solution processability, have exploited into various optoelectronic applications from solar cells [2–4], lasers [5], light-emitting diodes [6], photodetectors [7], and sensors [8].

Perovskite solar cells is a technological field that has made spectacular progress in terms of increased photovoltaic

2D Nanomaterials for Energy Applications. DOI: https://doi.org/10.1016/B978-0-12-816723-6.00005-8

efficiency since its launching in 2009 [9] reaching efficiencies over 20% in less than 10 years. The main concern regarding this technology is the inherent chemical and structural instabilities under ambient conditions [10,11]. This is the drawback that can revert into an advantage by applying them as sensors. The ability of metal halide perovskites to transduce any environmental stimulus into an optical or electrical signal is an opportunity in the field of sensing. This functionality, in combination with the aforementioned perovskite's physical properties, provide them with all the attractive characteristics an ideal optical or a chemiresistive sensor should have: room temperature operation, self-power device, high sensitivity, high response and recovery times, reversibility to the prior gas exposure state after the removal of the gas targeted agent, storage and operational stability, selectivity, high surface to bulk ratio and facile and easy fabrication processing.

Physical properties of metal halide perovskites, such as photoluminescence emission intensity and wavelength and electrical conductivity are modulated upon exposure to external stimulus. The most interesting part is that these changes are reversible after the removal of the external stimulus. The perovskite-based sensing elements are extremely attractive and demonstrate number of advantages compared to the traditional metal oxide sensing elements which are employed for gas sensing (e.g., operational in high temperature and limited sensitivity) [12]. Recently, perovskite semiconductors have exhibited their sensing abilities to detect various gases, temperature variations, metal ions and explosives.

This chapter provides the authors' collective and critical review of the most important and recent work regarding the application of inorganic and inorganic–organic metal halide perovskite films, nanocrystals, and nanocubes as humidity and gas sensing elements. All the recently published works in the field have revealed that the targeted gas molecules interact with the defect sites (PbI_2 unpaired ions or dopant ions operate as receptors to bind the target substances) mainly exist more onto the surface rather into the bulk of the perovskite sensing platform. The most important is this interaction of the perovskite platform with external environmental triggering can be engineered and be tailored, which provides a bright future for the sector. Table 5.1 summarizes the most important published results. At the end of the chapter, we discuss the challenges, or the open questions of the field and we propose ways to tackle these problems in the close future.

Table 5.1 Summary of the published results regarding the sensing performance of metal halide perovskite materials.

Target	Material	Sensor type	Performance	Year of publication	Reference
Humidity	$CH_3NH_3PbI_3$ single crystals	Optical—fluorescence	—	2015	Grancini et al. [13]
Humidity	$CH_3NH_3PbI_{3-x}Cl_x$ Films	Chemiresistance—electrical	At 32%, 57%, 75%, 83%, and 97% RH, the resistance decreased to 98%, 87%, 57%, 41%, and 25% of the initial value, respectively	2015	Hu et al. [14]
Humidity	$CH_3NH_3PbBr_3$ (single crystals)	Optical sensor—fluorescence quenching/emission translation	0.03% sensitivity 250 s response time High reversibility to >90% after 10 sensing cycles	2016	Xu et al. [15]
Humidity	$MAPbBr_3$ (single crystals)	Optical sensor—photoluminescence quenching	–	2016	Fang et al. [16]
Humidity	$CH_3NH_3PbI_{3-x}Cl_x$ (nanosheets arrays)	Chemiresistance—electrical resistance reduction	Sensitivity from 0% to 90% Sheet resistance reduction from 1.28×10^8 to 7.39×10^4 Ω (from 30% RH to 90% RH)	2017	Ren et al. [17]
Gas sensing—ammonia	$CH_3NH_3PbI_3$ (films)	Optical sensor—optical bleaching	–	2014	Zhao et al. [18]
Gas sensing—hydrochloric acid (HCl)	$CsPbBr_3$ (nanocrystals)	Blueshift in UV–vis absorption and photoluminescent spectra	Sensitivity down to 5 ppm Response times from 60 min down to 10 min depending on the HCL concentration	2017	Chen et al. [19]
Gas sensing—oxygen gas	$MAPbI_3$ (films)	Electrical resistance variation (decrease of the resistance with oxygen concentration)	3000 fold resistance variation when working with relative oxygen variation from 0% up to 100%/70 ppm sensitivity/response times of the order of 400 ms	2017	Stoeckel et al. [20]

(Continued)

Table 5.1 (Continued)

Target	Material	Sensor type	Performance	Year of publication	Reference
Gas sensing—acetone and NO_2	$MAPbI_{3-x}(SCN)_x$ (films)	Electrical resistance modulation as a function of the targeted gas concentration	20 ppm acetone and 200 ppb NO_2/reaction times of the order of 2–4 min	2017	Zhuang et al. [21]
Gas sensing—NO_2	$MAPbI_3$ (films)	Chemoresistive sensing element—resistance reduction with the concentration of the targeted gas	Reaction times of the order of few seconds/excellent sensitivity to 1 ppm/excellent selectivity to NO_2	2018	Fu et al. [22]
Gas sensing—ozone	$CH_3NH_3PbI_{3-x}Cl_x$ (films)	Electrical resistance reduction	13 s response time at 13 ppb ozone molecules/Sensitivity for ozone concentrations from 2500 to 5 ppb	2018	Kakavelakis et al. [23]
Gas sensing—O_2, acetone, and ethanol	$CsPbBr_3$ (nanocrystals)	Optically activated chemiresistive sensing element	Response and recovery times of 17 and 128 s respectively/responsivity to O_2 0.93/sensitivity to ethanol down to 1 ppm	2018	Chen et al. [24]
Gas sensing—oxygen	$Mn:CsPbCl_3$ (nanocrystals)	Optical sensor	0%–100% of oxygen concentration/response and recovery time of 5 s	2018	Lin et al. [25]
Gas sensing—ozone	$CsPbBr_3$ (nanocubes)	Chemiresistive sensing element	Sensitivity down to 4 ppb of ozone molecules/response and recovery times of the order of few hundreds of seconds	2019	Brintakis et al. [26]

5.2 Humidity sensing elements

Sensing water molecules (humidity) are very important to be measured in various technologies from health monitoring to aerospace exploration. The materials applied as sensing elements include metal oxides, photonic crystals, polymer materials, and graphene composites. However, all these materials "suffer" from (1) complicated synthesis procedures, (2) low sensitivity, and (3) narrow working range. Inorganic, organic–inorganic hybrid perovskite semiconductors, due to their simple fabrication process, high sensitivity to various gas elements, and physical properties (e.g., conductivity, direct band, photoluminescence efficiency, narrow emission bandwidth, and morphology inhomogeneities) appear very attractive and competitive to the other technologies' humidity sensing elements. Key issue in understanding perovskite, as in other semiconductors, sensing capacities is the investigation of the interaction of the surface trap states (especially very dense located at the surface edges of the perovskite single crystals or between the grain boundaries in the case of polycrystalline films [13])—with their environment. The ability of perovskite-based platforms to sense specific gas agents depends on two factors: (1) chemical nature of the gas agent and (2) the local composition of the hybrid perovskite: surface traps. Hu et al. [14] in 2015 reported the reduction of the in situ electrical resistance measurements of solution-processed films based on $CH_3NH_3PbI_{3-x}Cl_x$ when exposed to moisture environment. The solution-processed films were able to change their electric resistance from the 98% to the 25% of their initial value when exposed to relative humidity (RH) of 32–97%, respectively. The process apart that was reversible was very fast as well; within 74 s the initial values of the resistance were recovered with the removal of the humidity. The whole process was reversible when the exposure time in the moisture environment was less than 10 min. The authors attributed the electric resistance lowering to the H_2O molecules induced healing of electron depletion areas on the films' surface (time-resolved XRD measurements confirmed this author's suggestion). Exposures for longer periods and especially under high RH (~85%) resulting in nonreversible situations, due to the decomposition of the films into the more conductive PbI_2 precursor element. Another important factor the authors revealed was the dependence of the sensing elements' sensitivity on the way the film was fabricated: the solution-processed films were the most tolerant whereas the as evaporated ones the more sensitive to the moisture conditions.

Static photoluminescence (PL) and time-resolved photoluminescence (TRPL) measurements showed that the number of the surface trap states increased when the samples exposed to humidity under vacuum conditions rather in ambient conditions. Fang et al. [16] published a very interesting work showing that the majority of the defects in $MAPbBr_3$ single crystals are

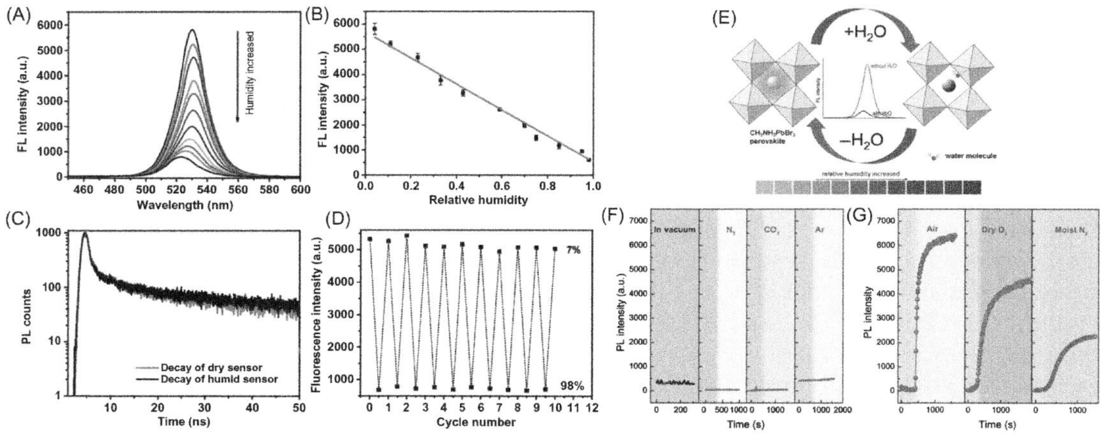

Figure 5.1 (A) PL quenching as a function of the humidity concentration; (B) linear response between PL measured signal and humidity concentration; (C) time-resolved photoluminescent decay as a function of the environmental conditions (within and without moisture conditions); and (D) demonstrated repeatability–reversibility of the $CH_3NH_3PbBr_3$ humidity sensing elements when exposed to 7% and 98% humidity. (E) The sensing mechanism between the water molecules and the perovskite crystals is based on the establishment of a hydrogen bond with the Pb ions. (F)–(G) Selectivity of $MAPbBr_3$ crystals as a function of time when exposed to various gases. *Source*: (E) Reproduced from X. Wu, F. Li, Z. Cai, Y. Wang, F. Luo, X. Chen, An ultrasensitive and reversible fluorescence sensor of humidity using perovskite $CH_3NH_3PbBr_3$, J. Mater. Chem. C 4 (2016) 9651–9655 with permission from the Royal Society of Chemistry; (F–G) Reproduced from H.-H. Fang, S. Adjokatse, H. Wei, J. Jang, G.R. Blake, J. Huang, et al., Ultrahigh sensitivity of methylammonium lead tribromide perovskite single crystals to environmental gases, Sci. Adv. 2 (7) (2016) e1600534 with permission from the American Association for the Advancement of Science.

located onto their surface rather within the bulk. The reversibility to initial PL intensity and electronic stage after the removal of the gas agent was clearly demonstrated showing that the sensing process was rather physical rather photochemical. The demonstrated sensing element also showed high selectivity when exposed to various gases (see Fig. 5.1F–G). Wu et al. [15] exhibited a $CH_3NH_3PbBr_3$ humidity sensing element with excellent performance: (1) high sensitivity (down to 0.04% humidity could be sensed) (see Fig. 5.1A–D); (2) very fast response and recovery times (250 s and 30–70 s, respectively); (3) very good reversibility (after 10 cycles, from 7% to 98% humidity, 93.4% of the fluorescent intensity in the dry state could be recovered, see Fig. 5.1D). The sensing was recorded through (1) the PL quenching observed under the exposure to various moisture concentration (starting from 7% and level up to 98%), and (2) the emission wavelength redshift due to crystal strain [when the perovskite was coupled with 5,10,15,20-tetrakis (pentafluorephenyl) porphyrin] as a function of the humidity concentration.

The sensing mechanism was attributed to the establishment of hydrogen bonds between water molecules and iodines (see Fig. 5.1E). The authors declared that the adsorption of the water molecules was reversible only if the interaction was of a very short time duration. Instabilities were observed when the sensing elements exposed to humidity for 24 h and for concentrations higher than 50%.

Another example how to reverse a disadvantage to an advantage was the work done by Ren et al [17]: the authors leveraged the moisture sensitivity of $CH_3NH_3PbI_{3\text{-}x}Cl_x$ vertically aligned nanosheet arrays (thickness of 105 nm and width of 2 μm) to fabricate ultrasensitivity, fast, with strong stability against the decomposition by water molecules (due to the chlorine anion) and with impressive specificity. The deformation of perovskite crystal due to the formation of hydrogen bonds between halide ions and polar water molecules was utilized by the authors to show the promising potential of perovskites as efficient humidity sensors. Moreover, the noticed crystal deformation, known as hydrated crystal phase, was reversible after the drying of the material, empathizing even more their potential as efficient and low-cost humidity sensing elements. The latter demonstrated strong response of their conductivity when exposed to humidity: the calculated sheet resistance dropped from 1.28E8 Ω to 7.39E4 Ω with the RH tuned from 30% to 90%. This was attributed to the electron transfer, from the water molecules have infiltrated into the bulk of the perovskite, inducing changes in the density of perovskite's energy states. UPS measurements showed the lift up of the Fermi Level of the water absorbed samples an indication of n-doping the adsorbed water molecules induced into the perovskite semiconductor. To highlight that the responsivity of the demonstrated devices was faster than commercially available psychrometer: to reach the first RH peak it took 51 s for the psychrometer while took 21 s for the exhibited perovskite-based element. The fast response was attributed to the large surface over bulk area of the nanosheet array and of their vertically alignment. The latter, as XRD and XPS measurements showed, allowed the exposure of the sensing elements without showing any sign of decomposition, even after the exposure to humidity conditions of 90%.

5.3 Gas sensing

Organometallic perovskite films exhibit excellent gas sensing properties. Optical bleaching of $MAPbI_3$ films has been reported after the films exposed in ammonia gas molecules by Zhao et al. [18] for a short period. XRD measurements demonstrated a phase transformation of $MAPbI_3$ films either because NH_3 molecules intercalated into or formed new structures with the perovskite template–network. These structural changes were reversible (see Fig. 5.2A and B) only in case the interaction with the ammonia molecules was of short duration. UV–vis measurements demonstrated the color changes across the whole visible spectrum (400–800 nm) during the interaction with the ammonia gas; very fast the films changed color from dark brown to become fully transparent.

Cesium-based perovskite nanocrystals ($CsPbBr_3$) were demonstrated as very sensitive, low cost, and easy to use as optical spectrochemical probes for hydrochloric acid (HCl). The latter

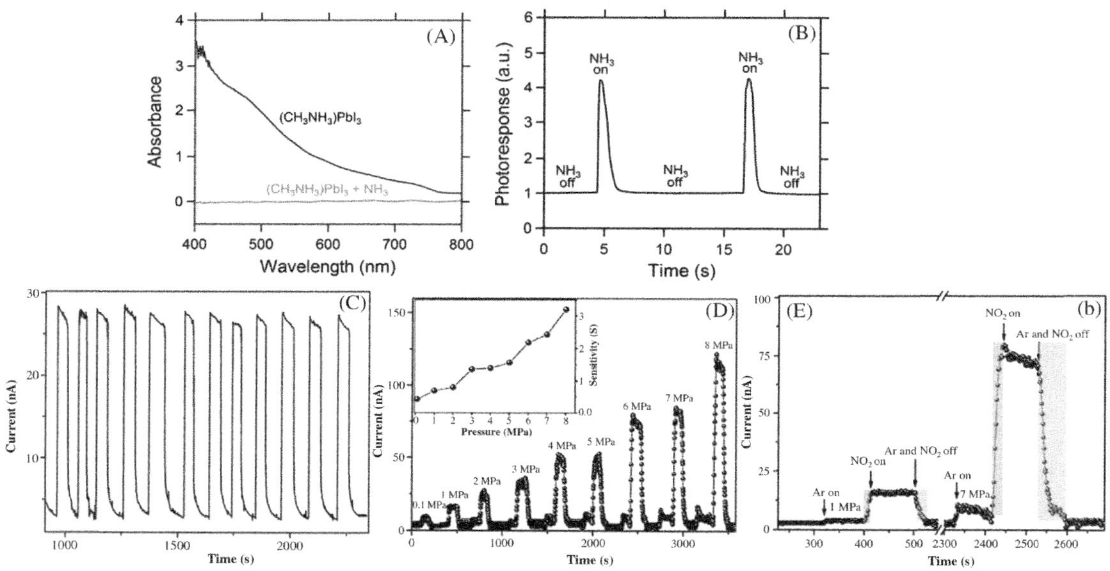

Figure 5.2 (A) The impact of the gas intrusion (in this case ammonia—NH_3) into the $MAPbI_3$ crystal that clearly affects its physical properties (absorbance) and (B) its electrical conductivity in a reversible way. (C) The response transients exposed to 30 ppm NO_2 after more than 12 cycles; (D) dynamic response curve of the $MAPbI_3$ based sensor, at room temperature, exposed in the same NO_2 concentration, under high pressures, while the inset displays the sensitivity of the $MAPbI_3$ sensor under the same conditions; (E) and transient curve upon the incorporation of high-pressure argon gas and NO_2 within the testing chamber at pressures ranging from 1 to 7 MPa. *Source*: (B) Reproduced from Y. Zhao, K. Zhu, Optical bleaching of perovskite $(CH_3NH_3)PbI_3$ through room temperature phase transformation induced by ammonia, Chem. Commun. 50 (2014) 1605–1607 with permission from the Royal Society of Chemistry; (E) Reproduced from X. Fu, S. Jiao, N. Dong, G. Lian, T. Zhao, S. Lv, et al., A CH3NH3PbI3 film for a room-temperature NO2 gas sensor with quick response and high selectivity, RSC Adv. 8 (2018) 390–395 with permission from the Royal Society of Chemistry.

is toxic and corrosive gas and is of high importance to be detected to low concentrations as low as 5 ppm. Chen et al. [19] showed that $CsPbBr_3$ due to anion exchange with HCl (Br ions were substituted by Cl anions) modulated their PL and UV−vis spectra as a function of the HCl concentration. The observed blueshift in the PL and absorption spectra was attributed to the interaction with the HCl that resulted in the formation of CsPb $(Br/Cl)_3$ without phase and morphology changes compared to the pristine reference $CsPbBr_3$ nanocrystal respective characteristics. The reaction times and the observed translation in PL and absorption wavelengths were shorter and larger respectively with the HCl concentration. The interactions with the HCl were reversible since the reverse anion exchange (Cl -> Br) could be realized through the reaction of $CsPb(Br/Cl)_3$ perovskite nanocrystals with HBr vapor. The measured PL and UV−vis spectra were similar to these ones of $CsPbBr_3$ reference nanocrystals. The detector showed good selectivity as well; no modulations of its optical properties (PL and UV−vis) were observed after its interaction with other volatile acids and bases such as HNO_3, CH_3COOH, and NH_4OH.

The sensing mechanism of organometallic perovskite nanostructures was mainly based on the variation of the chemical composition of their environment. Stoeckel et al. [20] demonstrated that the selected deposition technique of the solution-processed ($MAPbI_3$) films, one step (1S) or two step (2S), had an impact on the oxygen (O_2) sensing performance of the constructed films. Their findings showed very clearly that the morphology of the film is another important parameter beyond the well-known parameters such as (1) targeted gas concentration and (2) composition of the perovskite nanostructure. This works launched the possibility to tune sensor's sensitivity through controlling its morphology via different processing methods. The uniform morphology of the 2S $MAPbI_3$ films builds sensing elements with a great sensitivity to oxygen molecules leading to current enhancements of more than 3000 folds when the oxygen concentration varied from 0% to 100%. However, the 1 S $MAPbI_3$ films nonuniform morphology induced less sensitivity leading to a corresponding enhancement of 100 less than in the case of 2S films. Beyond the electrical measurements, the authors noticed an increase in the measured PL signal after the exposure of the perovskite films in oxygen environment. The PL signal enhanced with the oxygen concentration and this was attributed to the healing of the iodine vacancies that operated as the reaction centers of the sensing elements by the oxygen atoms intercalated within the bulk of the perovskite films. The interactions with the oxygen molecules were reversible; the electrical and optical status of the films returned back to its pristine statue after the removal of the oxygen gas making the nanostructured organohalide perovskites ideal materials for oxygen sensing.

The importance of the morphological characteristics (the higher the roughness the better the sensing performance) and of the surface to bulk ratio as a consequence of the deposition substrate were also demonstrated by Zhuang et al. [21]. The morphology of the films played an important role and offered a number of diffusion channels and reactive sites for the targeted gas molecules to interact with the sensing platform. Moreover, the authors exhibited that the doping of a $MAPbI_3$ film with thiocyanate ions (SCN^-) resulted in $CH_3NH_3PbI_{3-x}(SCN)_x$ chemiresistor sensing films with enhanced lifetime (1) against ambient (exposure to humidity) and (2) operational conditions [exposure to the targeted gases (acetone and NO_2 gases)]. The improved stability compared to the $MAPbI_3$ reference films was attributed to the stronger bonding between the SCN ions with the Pb atoms within the bulk of the constructed films that did not allow the decomposition of the films into their pristine elements. In all cases, the conductivity of films was increased due to charge transfer between the perovskite and the targeted gas. This allows the interaction with oxidative and reductive gases with the simultaneous enhancement of the film's conductivity. The sensing films were extremely sensitive to acetone and NO_2 vapors, with sensitivities of the order of 2 or 3.9 respectively. The selectivity of the $CH_3NH_3PbI_{3-x}(SCN)_x$ towards the reported gases compared to other targeted gases was also reported. The preference for acetone and NO_2 gas molecules was based on the strong sensitivity towards these gases (of the order of two and three) compared to low sensitivities demonstrated towards ethanol (0.64), methylbenzene (0.26), and water (0.14).

However, the long reaction and recovery times were indicators of good interaction between the sensing element and the targeted gas; there is an endeavor to shorten these times as a milestone to demonstrate better sensing performance and elements. This was the main highlight of the work done by Fu et al. [22] that exhibited response and recovery times of 5 s and 25 s respectively of $MAPbI_3$ films under NO_2 gas interaction (see Fig. 5.2C). Moreover, the authors explained how the sensitivity (down to one ppm) of the demonstrated sensing films was improved with the pressure of the targeted gas (NO_2) due to the increase of the adsorption energy. The adsorption of the NO_2 molecules with the perovskite reaction sites (mainly due to the establishment of hydrogen bonding between CH_3 and NO_2) resulted in electron transfer from the perovskite toward the adsorbed NO_2 molecules (see Fig. 5.2D–E).

For the first time, Kakavelakis et al. [23] demonstrated a self-powered, low cost, ultrafast, and sensitive organometallic hybrid-based solution-processed films of mixed halide $CH_3NH_3PbI_{3-x}Cl_x$ as ozone sensing element (see Fig. 5.3A). Its

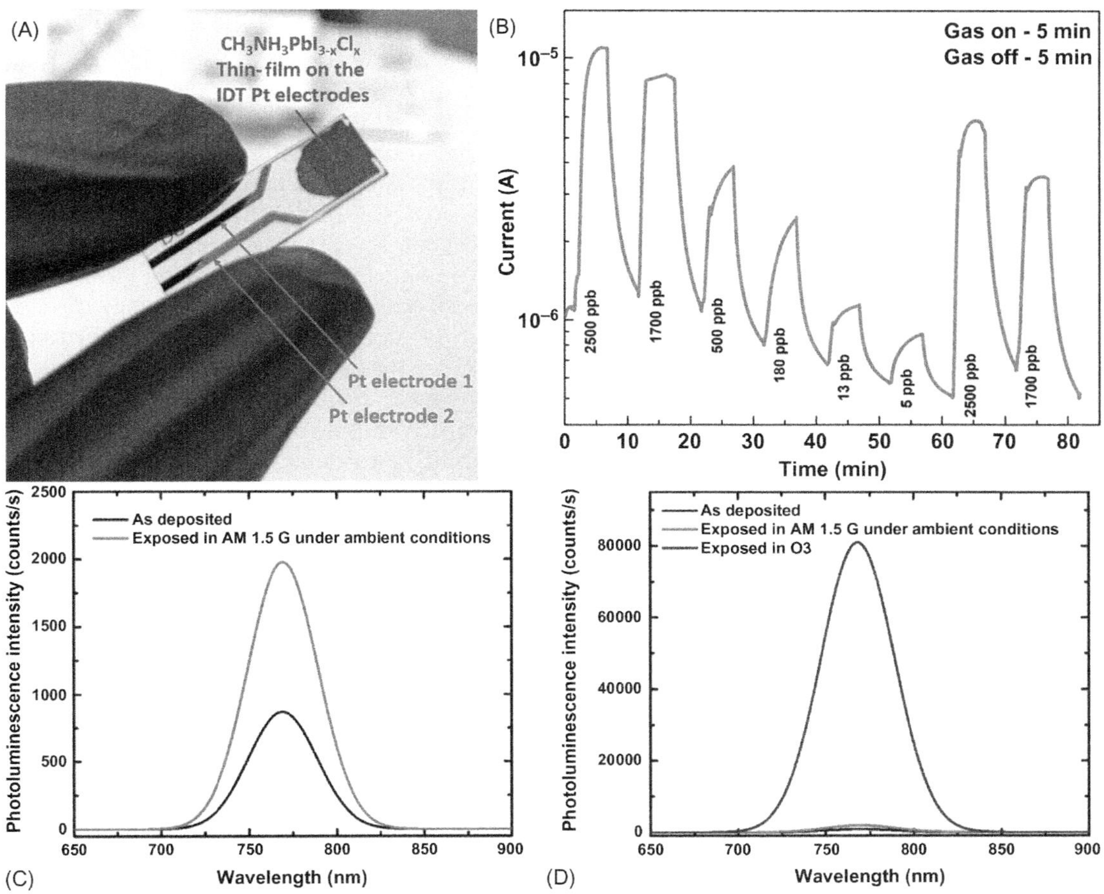

Figure 5.3 (A) The hybrid lead mixed halide sensor fabricated on the prepatterned interdigitated Pt electrodes glass substrate; (B) the electrical response of the hybrid lead mixed halide ozone sensing element exposed to various ozone concentrations; photoluminescence spectra of (C) the pristine unexposed (*black line - bottom curve*) and exposed (*red line - top curve*) hybrid lead mixed halide perovskite thin films; and (D) ozone exposed (*blue line - top curve*) hybrid lead mixed halide perovskite thin films. *Source*: Reproduced from G. Kakavelakis, E. Gagaoudakis, K. Petridis, V. Petromichelaki, V. Binas, G. Kiriakidis, et al., Solution processed CH3NH3PbI3-xClx perovskite based self – powered ozone sensing element operated at room temperature, ACS Sens. 3 (1) (2018) 135–142 with permission from the American Chemical Society.

operation was based on changes of its film resistance when exposed to different concentrations of ozone molecules. A decrease in sheet resistance was observed even in the exposure to 5 ppb of ozone molecules (see Fig. 5.3B). The authors explained the increase of their films' conductivity after exposure to ozone to the passivation of surface traps (due to unpaired Pb^{2+} ions) by electrons transferred from the ozone molecules. This hypothesis was supported by the enhancement of the received PL spectra after the prolonged exposure of the films in ozone treatment (see Fig. 5.3C–D).

One of the top priorities in the perovskite-related technologies is the improvement of these materials' stability under ambient conditions operation. The replacement of the organic part with the cesium (Cs) has generated some encouraging results to tackle this challenge. In the field of sensing, Chen et al. [24] recently proposed the use of $CsPbBr_3$ porous and interconnected network of nanocrystals to detect O_2 (oxidative gas) and volatile gases such as acetone and ethanol (reducing gases) with high sensitivity (detection of less than 1% of oxygen within N_2 environment), selectivity and at the same time with improved stability under operational conditions compared to the inorganic–organic respective optoelectronic systems. The authors found out that the light activation of the sensing medium improved their sensing ability and selectivity toward reductive and oxidative gas molecules. Without the photoexcitation of the sample, the latter exhibited very poor chemical sensitivity towards the targeted gas molecules. The photo excitation of the sample assisted the increase of the measured current from 0.1 pA to 1.5 nA. The interaction (the reaction centers are vacancies of bromide with lead—the unpaired lead atoms interact with the oxygen or acetone or ethanol molecules) with the O_2 molecules lead, due to charge transfer from the perovskite towards the oxygen molecules, the increase of the measured current to 2.57 nA. The respective current values when the perovskite platform was interacted with ethanol and acetone were 1.98 and 3.95 nA, respectively. Sensitivities down to the detection of 1 ppm of ethanol was demonstrated. The current values returned to its initial value when the oxygen or the acetone and ethanol gases removed, showing the reversibility of the demonstrating sensing element. After 2 weeks of storage and then exposure to oxygen molecules, the inorganic nanocrystals demonstrated similar sensitivities to the fresh prepared films. Under operational conditions, the inorganic perovskite nanocrystals could last under exposure to targeted gases and under light excitation for more than 7 h of operation.

Another interesting approach regarding photoluminescence-quenching probe (PLQPs) elements for oxygen sensing was proposed by Lin et al. [25]. The authors of this work revealed the role of the doping and its concentration of inorganic $CsPbCl_3$ nanocrystals with magnesium atoms, Mn^{+2} ($Mn:CsPbCl_3$) on optical de-excitation of such perovskites in presence of oxygen molecules. More particularly it was observed 53% quenching, when the concentration of the oxygen molecules varied from 0 to 100%, of the PL signal emitted by the Mn^{2+} dopants (@ 530 nm) on the surface of the $CsPbCl_3$ perovskite nanocrystals. To highlight, the concentration of the Mn^{2+} dopants (over the Pb ions replaced) (1) enhanced the energy transfer of the PL signal generated from the exciton recombination in the $CsPbCl_2$ to excite the Mn^{2+} dopants; (2) increased the PL signal generated from Mn^{2+} de-excitation and resulted in its longer decay times; and (3) increased the interaction/sensing ability of the $Mn:CsPbCl_3$ NCs with the oxygen molecules: stronger PL quenching and shorter lifetime of the fast decay part of the PL signal. The influence of the lifetime of the fast decay part of the Mn^{2+} received signal was perceived as the reaction sites between the oxygen molecules and the $Mn:CsPbCl_3$ NCs were the dopants on the near surface of the NCs (the location of the reaction sites was also affected by the temperature incurred the doping process, with the higher the doping temperature to facilitate the reaction centers (Mn^{2+}) with the oxygen molecules to locate on the near surface of the NCs). The demonstrated PLQPs demonstrated the operational characteristics an ideal gas sensing element should contain such as reversibility, responsivity (within 5 s, 97% of the PL intensity was quenched), and sensitivity.

Another demonstration of the stability superiority of visible light-activated inorganic perovskite-based sensing elements ($CsPbBr_3$ nanocubes) was the work by Brintakis et al. [26]. The sensing elements demonstrated very fast response and recovery times of the order of few hundreds seconds, high sensitivities down to 4 ppb of ozone gases and reversibility. The highlight of this work is the stability of the sensing elements: even after their storage for 2 weeks under ambient conditions the sensing performance remained the same as with this one of fresh samples.

5.4 Conclusion and challenges

Inorganic and inorganic–organic hybrid metal halide perovskite semiconductors have emerged as the rising star materials in various optoelectronic applications starting from solar cells, lasers, light-emitting diodes up to sensing elements. The large number of applications of the chemically synthesized perovskite semiconductors was attributed to their very attractive physical and electrical properties: morphology characteristics, morphological and constitutional physical properties, high bipolar conductivity, direct bandgap, solution-processed materials, and high sensitivity and reactivity to their environmental conditions and synthesis. In this chapter we reviewed the sensing mechanism, the sensing performance, and the parameters that affect the performance of these materials (films, nanocrystals, and nanocubes) as humidity and gas sensors. Room temperature operation, self-powered chemicoresistive and optical sensing elements, high and fast sensitivity, and quick reversibility are among their advantages compared to the state-of-the-art metal-oxide sensing elements have been demonstrated.

Despite the great potential the inorganic and hybrid metal halide perovskite semiconductors, in any form they have, as humidity and gas sensing elements have, there is a number of challenges should be tackled in order to dominate this field.

First of all more effort should be made by the scientific community to deeply understand the fundamental physics, chemistry, and materials science of these materials from the perspective of synthesis, structure, and physical and optical properties; this will allow the understanding of all the mechanisms are involved during the adsorption and de-adsorption of the targeted gas molecules. Advances related the fundamental science around this fascinating class of materials will shed the progress and optimize the "design rules" of this type of sensing elements: higher selectivity, faster response times, and fabrication of intergraded sensing systems.

Improving their long-term stability, under storage and under operational conditions, are the primary priorities. Suggestions could be, the deposition of graphene or any other material made membranes that could operate as humidity absorbers and protectors against from water molecule attachments (have different size of porous diameters that stop humidity molecules but allow the interaction with the targeted gas molecules). Another suggestion is the fabrication of less porous films (using mCVT reaction [11]) without pinholes that do not allow the intrusion of water molecules within the bulk of perovskite. This of course comes with the cost of less sensitivity regarding the sensing part. The use of inorganic perovskite nanocrystals was a step also toward more tolerant to the external environment conditions sensing elements. The hybridizing of perovskites with other functional materials such as porous membranes can also be used to extend their lifetime under operational conditions: high concentration of gas agents and long-time exposure times have been proven detrimental parameters for the sensing elements lifetime.

The sensitivity of the perovskite-based sensing elements can also be improved with the tuning of their constitution and that will allow us to construct heterostructure platforms where different areas can interact with different gas molecules. The pressure of the latter can also affect their sensitivity.

The concentration and the type of defects onto the bulk and the surface of the sensing elements control mainly their sensitivity to various gas molecule targets. More studies should be implemented regarding the optimum synthesis of the perovskite-based sensor templates in order to demonstrate fast reaction times (related to the conductivity of the films and less defect sites) and at the same time reaction sites (defect and doping sites). The doping of the outer platform of the perovskite films, nanocrystals, and nanocubes can also be a technique to control the sensitivity and selectivity of these sensing elements. The selection of the deposition substrate plays a crucial role in the sensing ability of these materials, and controls the surface to bulk ratio: the higher this ratio is the higher its sensitivity. Another more innovative proposal could be the exploitation of the continuous wave (CW) stimulated Raman spectroscopy (CW-SRS) tool to support the selectivity of the perovskite-based sensing elements. Two laser diodes with high spectral purity will be used: one to excite the Raman peaks of the targeted gas and the other one to trigger the stimulation emission. This emission can potentially be linked with the electrical signal received as a result of the adsorbed targeted gas molecules.

The improvement of response times (reaction and recovery times) can be improved with the constructions of (1) composites with graphene-based materials (e.g., reduced graphene oxide) that will infiltrate between the grains of perovskite. This is expected to improve the conductivity of these areas with a direct consequence to the characteristic times; and (2) improve the design principles to build longer grain size crystals, action that enhances the nanostructure conductivity.

Chemically synthesized perovskite materials allow though the control of their morphology, their constitution, and their compatibility with other porous materials to improve and control their humidity and gas sensing performance and applicability. Their reactivity, combined with their bipolar conductivity and their photoluminescence efficiency, with the external environment, allow the tailoring of their ability to transduce any external stimulus to electrical and optical signal as a function of the targeted gas molecule or molecules. We strongly believed in their sensing abilities and we are confident that this will be the case in the near future.

References

[1] P. Docampo, T. Bein, A long-term view on perovskite optoelectronics, Acc. Chem. Res. 49 (2) (2016) 339–346.

[2] Z. Shi, A.H. Jayatissa, Perovskite based solar cells: a review of recent progress, materials and processing methods, Materials 11 (5) (2018) 729.

[3] C. Petridis, G. Kakavelakis, E. Kymakis, The renaissance of graphene-related materials in photovoltaics with the emergence of metal-halide perovskite solar cells, Energy Environ. Sci. 11 (2018) 1030–1061.

[4] G. Kakavelakis, E. Kymakis, K. Petridis, 2D materials beyond graphene for metal halide perovskite solar cells, Adv. Mat. Interfaces 5 (2018) 1800339.

[5] M.M. Stylianakis, T. Maksudov, A. Panagiotopoulos, G. Kakavelakis, K. Petridis, Inorganic and hybrid perovskite based laser devices: a review, Materials 12 (2019) 859.

[6] Q.V. Le, H.W. Jang, S.Y. Kim, Recent advances toward high – efficiency halide perovskite light – emitting diodes: review and perspective, Small Methods 2 (2018) 1700419.

[7] J. Miao, F. Zhang, Recent progress on highly sensitive perovskite photodetectors, J. Mater. Chem. C 7 (2019) 1741–1791.

[8] Z. Zhu, Q. Sun, Z. Zhang, J. Dai, G. Xing, S. Li, et al., Metal halide perovskites: stability and sensing – ability, J. Mater. Chem. C 6 (2018) 10121–10137.

[9] A. Kojima, K. Teshima, Y. Shirai, T. Miyasaka, Organometal halide perovskites as visible – light sensitizers for photovoltaic cells, J. Am. Chem. Soc. 131 (2009) 6050–6051.

[10] R. Wang, M. Mujahid, Y. Duan, Z.K. Wang, J. Xue, Y. Yang, A review of perovskites solar cell stability, Adv. Funct. Mater. (2019) 1808843.

[11] B. Wang, T. Chen, Exceptionally stable $CH_3NH_3PbI_3$ films in moderate humid environmental condition, Adv. Sci. 3 (2016) 1500262.

[12] P.T. Moseley, Progress in the development of semiconducting metal oxide gas sensors: a review, Meas. Sci. Technol. 28 (2017) 082001.

[13] G. Grancini, V. D' Innocenzo, E.R. Dohner, N. Martino, A.R. Srimath Kandada, E. Mosconi, et al., $CH_3NH_3PbI_3$ perovskite single crystals: surface photophysics and their interaction with the environment, Chem. Sci. 6 (2015) 7305–7310.

[14] L. Hu, G. Shao, T. Jiang, D. Li, X. Lv, H. Wang, et al., Investigation of the interaction between perovskite films with moisture via the in situ electrical resistance measurement, ACS Appl. Mater. Interfaces 7 (45) (2015) 25113–25120.

[15] X. Wu, F. Li, Z. Cai, Y. Wang, F. Luo, X. Chen, An ultrasensitive and reversible fluorescence sensor of humidity using perovskite $CH_3NH_3PbBr_3$, J. Mater. Chem. C 4 (2016) 9651–9655.

[16] H.-H. Fang, S. Adjokatse, H. Wei, J. Jang, G.R. Blake, J. Huang, et al., Ultrahigh sensitivity of methylammonium lead tribromide perovskite single crystals to environmental gases, Sci. Adv. 2 (7) (2016) e1600534.

[17] K. Ren, L. Huang, S. Yue, S. Lu, M. Azam, Z. Wang, et al., Turning a disadvantage into an advantage: synthesizing high-quality organometallic halide perovskite nanosheet arrays for humidity sensors, J. Mater. Chem. C 5 (2017) 2504–2508.
[18] Y. Zhao, K. Zhu, Optical bleaching of perovskite $(CH_3NH_3)PbI_3$ through room temperature phase transformation induced by ammonia, Chem. Commun. 50 (2014) 1605–1607.
[19] X. Chen, H. Hu, Z. Xia, W. Gao, W. Gou, Y. Qu, et al., $CsPbBr_3$ perovskite nanocrystals as highly selective and sensitive spectrochemical probes for gaseous HCl detection, J. Mater. Chem. C 5 (2017) 309–313.
[20] M.A. Stoeckel, M. Gobbi, S. Bonacchi, F. Liscio, L. Ferlauto, E. Orgiu, et al., Reversible, fast and wide range oxygen sensor based on nanostructured organometal halide perovskite, Adv. Mater. 29 (2017) 1702469.
[21] Y. Zhuang, W. Yuan, L. Qian, S. Chen, G. Shi, High performance gas sensor based on a thiocyanate ion – doped organometal halide perovskite, Phys. Chem. Chem. Phys. 19 (2017) 12876–12881.
[22] X. Fu, S. Jiao, N. Dong, G. Lian, T. Zhao, S. Lv, et al., A $CH_3NH_3PbI_3$ film for a room-temperature NO_2 gas sensor with quick response and high selectivity, RSC Adv. 8 (2018) 390–395.
[23] G. Kakavelakis, E. Gagaoudakis, K. Petridis, V. Petromichelaki, V. Binas, G. Kiriakidis, et al., Solution processed $CH_3NH_3PbI_{3-x}Cl_x$ perovskite based self – powered ozone sensing element operated at room temperature, ACS Sens. 3 (1) (2018) 135–142.
[24] H. Chen, M. Zhang, R. Bo, C. Barugkin, J. Zheng, Q. Ma, et al., Superior self powered room temperature chemical sensing with light activated inorganic halides perovskites, Small 14 (2018) 1702571.
[25] F. Lin, F. Li, Z. Lai, Z. Cai, Y. Wang, O.S. Wolfbeis, et al., MnII-doped cesium lead chloride perovskite nanocrystals: demonstration of oxygen sensing capability based on luminescent dopants and host – dopant energy transfer, ACS Appl. Mater. Interfaces 10 (27) (2018) 23335–23343.
[26] K. Brintakis, E. Gagaoudakis, A. Kostopoulou, V. Faka, A. Argyrou, V. Binas, et al., Ligand-free all inorganic metal halide nanocubes for fast, ultra – sensitive and self-powered ozone sensors, Nanoscale Adv. 1 (2019) 2699–2706.

6

Vacancy formation in 2D and 3D oxides

Kapil Dhaka and Maytal Caspary Toroker
Department of Materials Science and Engineering, Technion - Israel Institute of Technology, Haifa, Israel

6.1 Role of defects in 2D and 3D phases

There are various chemical, electrical, and optical properties of defects in materials that play a crucial role in material science. An excellent example is semiconductors, where the small amount of perturbation by creating defects and imposing impurities, even in very small concentrations, determines the electrical conductivity [1–9]. Defects generally play a dominant role also in surface and interface properties and in nanotechnology. The material properties can be beneficial to tune in a desired manner, when one can control the nature, amount, and characteristics of defects based on defect engineering. Defect formation is also helpful to generate new functionalized material properties. There are numerous areas of research where the importance of the specific defects has been discussed to improve the performance of a material or device and sometimes useful to explain the unusual nature of materials through bridging between theory and experiments. First principle calculations have a strong contribution in this area of research over the decades. We will discuss a few examples in this chapter to review the fundamentals of oxygen vacancies present in oxide materials and their effects in the specific properties. Oxygen vacancies are generally behaving as electron donor and as a result, defect state appears in the bandgap region in most cases as shown for Lu_2O_3 and $SrTiO_3$ in Fig. 6.1 [10]. This causes enhancement of electron concentration in the cell or on the surfaces.

There are several types of studies which suggest the importance and effect of defects, especially oxygen defects, in two-dimensional (2D) and three-dimensional (3D) materials; few selected ones are discussing below.

2D Nanomaterials for Energy Applications. DOI: https://doi.org/10.1016/B978-0-12-816723-6.00006-X

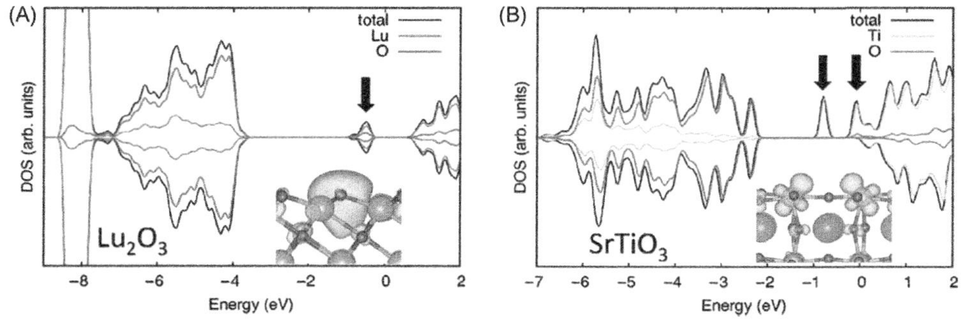

Figure 6.1 DOS for a surface oxygen vacancy at the Lu_2O_3 (A) and $SrTiO_3$ (B). The insets show the electron density of the defect state, indicated by the arrows in the DOS. *DOS*, density of states. *Source*: Reprinted with permission from T. Sarkar, S. Ghosh, M. Annamalai, A. Patra, K. Stoerzinger, Y. Lee, et al., The effect of oxygen vacancies on water wettability of transition metal based $SrTiO_3$ and rare-earth based Lu_2O_3, RSC Adv. 6 (2016) 109234–109240.

6.2 Effect and importance of oxygen vacancies on 2D and 3D materials

1. *2D electron gas*: Shen et al. [11] have studied the role of oxygen vacancies at the surface of $SrTiO_3$. They have used DFT + U level of theory [12]. From their calculations, the formation of a metallic state is only observed after introducing of oxygen vacancies. The charge carriers are strongly localized at the surface and deplete rapidly within a few layers from the surface, which is an indication of the formation of a two-dimensional electron gas (2DEG). Their work was motivated by previous studies on $SrTiO_3$ [13] angle-resolved photoemission spectroscopy observations of a highly metallic 2DEG at the (001) vacuum-cleaved surface of $SrTiO_3$ and the subsequent discussion on the possible role of oxygen vacancies for the appearance of such a state. 2DEG was earlier observed by Ohtomo et al. [14] on the $LaAlO_3/SrTiO_3$ heterointerface.

   In Fig. 6.2, Shen et al. [11] compared the carrier density of $SrTiO_3$ slabs in terms of number of carriers per unit cell area versus number of carriers per unit cell volume. Their 2D carrier density compares well with the experimental data ($n_{2D} = 2 \times 10^{14}$ cm^{-2}) [13,15]. They also observe that the number of carriers per area does not scale with the oxygen-vacancy concentration as it happens for the bulk electron doping n_{3D}. These results strongly indicate that the observed metallic state does not correspond to bulk electron doping or oxygen-vacancy concentration.

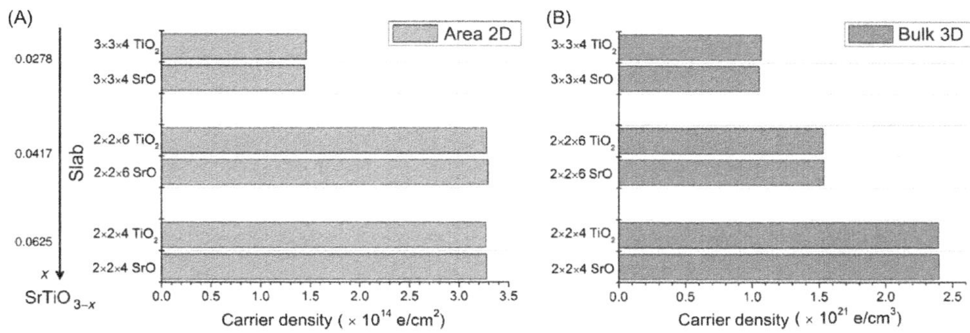

Figure 6.2 Calculated carrier density for the SrO and TiO_2 terminated slabs: (A) carrier density per surface area and (B) carrier density per volume. *Source*: Reprinted with permission from J. Shen, H. Lee, R. Valent, H.O. Jeschke, Ab initio study of the two-dimensional metallic state at the surface of $SrTiO_3$: importance of oxygen vacancies, Phys. Rev. B 86 (2012) 195119.

2. *Grain boundaries*: Hojo et al. [16] has studied the atomic and electronic structures of a (210)Σ5 grain boundary in CeO_2 using scanning transmission electron microscopy, electron energy loss spectroscopy (EELS), and theoretical calculations (Fig. 6.3) with LSDA + U formalism to account for the strong on-site coulomb repulsion [17]. Their results revealed that oxygen vacancies play a crucial role in the stable grain boundary structure of CeO_2. They have determined the Ce and oxygen sublattice structures and obtained evidence of oxygen nonstoichiometry at the grain boundary. The presence of oxygen vacancies was also confirmed by EELS measurements. The importance of considering nonstoichiometry in constructing grain boundary structural models is demonstrated, especially for systems where a high degree of nonstoichiometry is expected. This finding paves the way for comprehensive understanding of grain boundaries through atomic scale determination of atom and defect locations.
3. *Dissociation of H_2 molecules*: Oxides surfaces are sensitive to oxygen vacancies. Yang et al. [18] has studied the dissociation of H_2 molecule on the β-Ga_2O_3 (100)B surface by using PW91 [19] generalized gradient approximation (Fig. 6.4). They found that due to the oxygen vacancy, the neighboring surface Ga and O atoms get some extra electrons than on the stoichiometric Ga_2O_3 surface. The extra electrons around the surface O(I) atom are very near to the Fermi energy, and distribute on the p_z orbital, which are thus very active to hydrogen molecules. By calculating the 2D PES cuts for hydrogen molecules on the stoichiometric and oxygen vacancy included

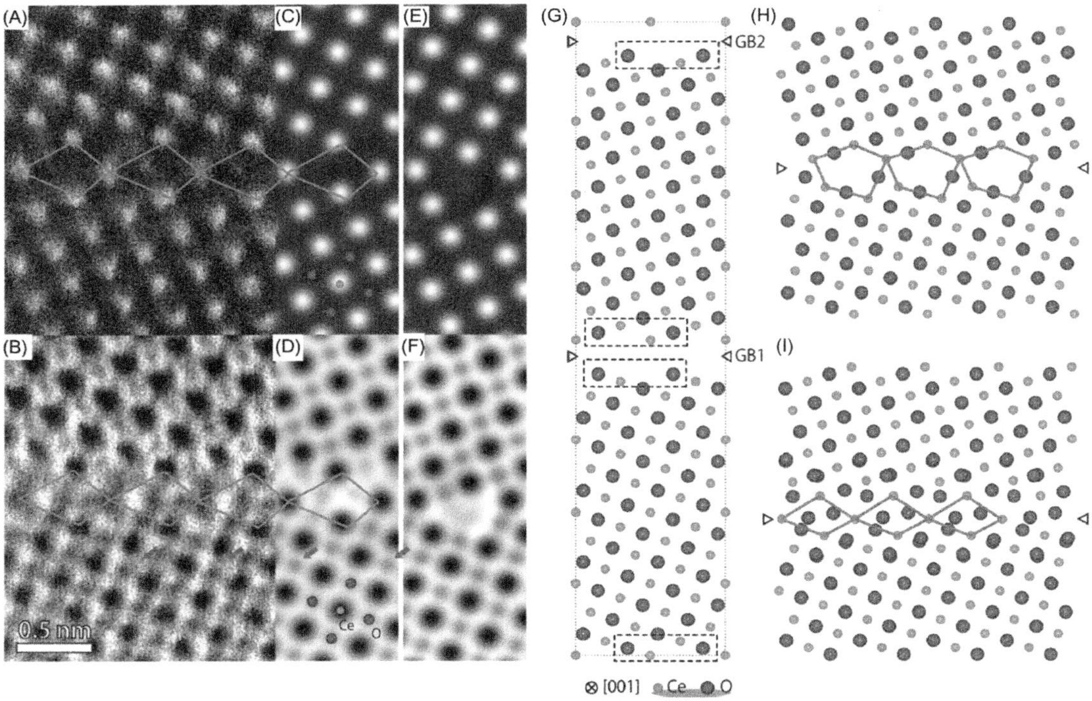

Figure 6.3 (A) Half-angle annular dark-field (HAADF) imaging and (B) annular bright-field (ABF) images of a [001] (210)Σ5 grain boundary in a CeO_2 thin film. (C) Simulated HAADF and (D) ABF images of the nonstoichiometric grain boundary model structure. (E) Simulated HAADF and (F) ABF images of the stoichiometric grain boundary model structure. The structural units of each boundary are indicated by polygons. (G) An initial supercell with two equivalent grain boundaries: GB1 and GB2. In the nonstoichiometric case, one of the four oxygen atoms marked with small dotted rectangles was systematically removed to introduce oxygen vacancies. (H) Stoichiometric and (I) nonstoichiometric stable grain boundary structures. The structural units of each boundary are indicated by polygons. From (A) to (F) are experimental measured, and (G) to (I) are the first principle results. *Source*: Reprinted with permission from H. Hojo, T. Mizoguchi, H. Ohta, S.D. Findlay, N. Shibata, T. Yamamoto, et al., Atomic structure of a CeO_2 grain boundary: the role of oxygen vacancies, Nano Lett. 10 (2010) 4668–4672.

β-Ga_2O_3 (100)B surfaces, they found that hydrogen molecules are very hard to get close to the stoichiometric Ga_2O_3 surface, but much easier to dissociate on the oxygen vacancy included Ga_2O_3 surface. Our results indicate the enhancement of surface activity to hydrogen molecules by introducing surface oxygen vacancies.

4. *Oxygen evolution reactions (OER) and oxygen reduction reactions (ORR)*: Similar to previous one, catalysis properties are sensitive with oxygen vacancies on oxide surfaces. Xu et al. [20] has demonstrated plasma-engraving strategy to produce

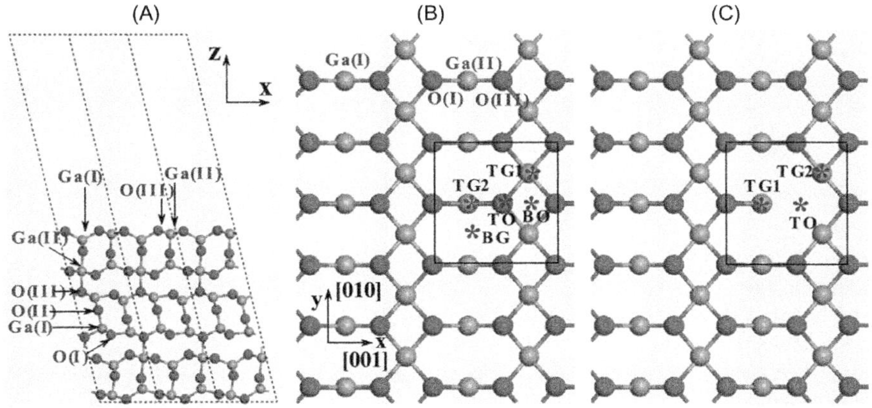

Figure 6.4 The β-Ga_2O_3 (100)B surface viewed along the (A) [010] and (B) [100] directions, and (C) the β-Ga_2O_3 (100)B surface with an O(III) vacancy viewed along the [100] direction. Gray and red balls respectively represent Ga and O atoms. The black squares in (B) and (C) represent the surface unit cells. *Source*: Reprinted with permission from Y. Yang, P. Zhang, Dissociation of H_2 molecule on the β-Ga_2O_3 (100) B surface: the critical role of oxygen vacancy, Phys. Lett. A. 374 (2010) 4169–4173.

Co_3O_4 nanosheets with oxygen vacancies and high surface area. The electrocatalytic performance of Co_3O_4 for oxygen evolution reaction (OER) is mainly affected by its surface area and the oxygen vacancies. In addition to the usual contribution of the high surface area, the specific activity results imply the significant role of the oxygen vacancies. The electrocatalytic activity for OER of metal oxides could be significantly enhanced through proper surface engraving by plasma. With the plasma engraving, although less electrocatalyst resided, more active sites and better catalytic activity were realized by obtaining the high surface area and oxygen vacancies. Zhuang et al. [21] has also observed the enhancement in OER activities on iron cobalt oxide in the presence of oxygen vacancies as mentioned in Fig. 6.5A. Xu et al. [22] also observed substantial contribution of oxygen vacancies, which exhibit remarkable OER performance. In the continuation of catalysis activities, Pei et al. [23] has also mentioned the enhancement of oxygen reduction reaction (ORR) activity on TiO_2 catalysis in the presence of engineered oxygen vacancies as mentioned in Fig. 6.5B and C.

5. *Ferromagnetism*: Presence of defects are generally affecting the spin orientation of the materials. Change et al. [24] has observed induced ferromagnetism in SnO_2 thin films due to the presence of oxygen vacancies. This behavior should be

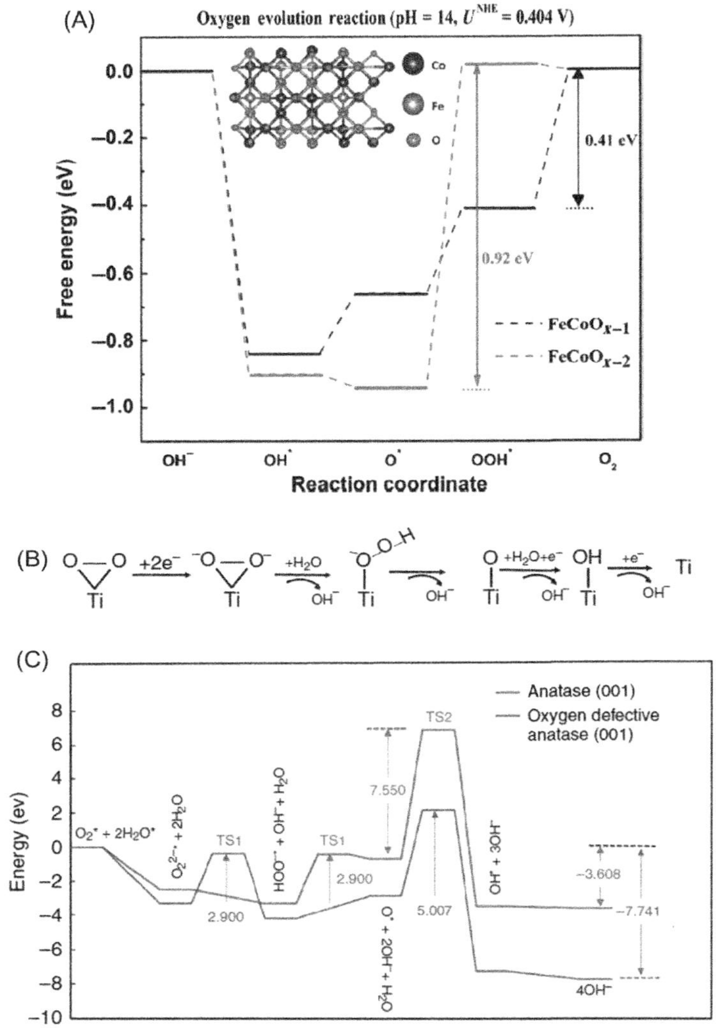

Figure 6.5 (A) OER mechanism and energy profiles: OER pathway in alkaline media showing the superior catalytic performance at the equilibrium potential ($U^{NHE} = 0.404$ V) of $Fe_1Co_1O_x$ with one oxygen vacancy. (B and C) ORR mechanism and energy profiles: (B) DFT-calculated ORR mechanism on the anatase TiO_2 surface, and (C) the energy profiles of ORR on the {001}-TiO_2 and {001}-TiO_{2_x}, where TS represents transition states. *DFT*, density functional theory. *Source*: Reprinted with permission from (A) L. Zhuang, Y. Jia, T. He, A. Du, Tuning oxygen vacancies in two-dimensional iron-cobalt oxide nanosheets through hydrogenation for enhanced, Nano Res. 11 (2018) 3509–3518; (B and C) D. Pei, L. Gong, X. Zhang, J. Chen, Y. Mu, A. Zhang, et al., Defective titanium dioxide single crystals exposed by high-energy {001} facets for efficient oxygen reduction, Nat. Commun. 6 (2015).

strongly weakened in thicker films, where the contribution of oxygen vacancies to the volume of sample is much smaller, and it completely disappears for post annealed SnO_2 films under oxygen atmosphere (Fig. 6.6).

6. *Vacancy diffusion*: Diffusion of atoms generally occurs in the presence of impurity such as dopants in the bulk. Diffusion can also play a role in properties of solids and device degradation. Dopants incorporated during growth may diffuse inside the growing material at the high temperatures used for high quality growth. Another reason of the diffusivity is

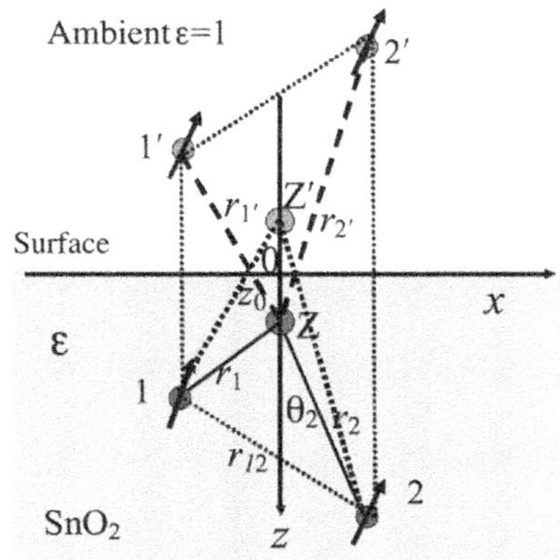

Figure 6.6 Defect at distance z_0 under the SnO_2 film surface. Two carriers (electrons or holes) 1 and 2 (shown in *green circles* with arrows) are localized near the defect with effective charge Z_e (shown in *red circle*). Carrier image charges are shown as 1′ and 2′ and defect image is Z′. *Source*: Reprinted with permission from G.S. Chang, E. Kurmaev, A.N. Morozovska, J.A. Mcleod, Oxygen-vacancy-induced ferromagnetism in undoped SnO_2 thin films, Phys. Rev. B 85 (2012) 165319.

sometimes due to the presence of heavy atoms in the cell. Ritzmann et al. [25] studied ternary oxide $LaFeO_3$ using DFT + U [26] level of theory and found that La vacancies are the reason for oxygen diffusion in the material (Fig. 6.7). The basic relationship for vacancy-mediated diffusion dictates that the oxygen diffusion coefficient (D_O) is the product of the vacancy diffusion coefficient (D_V) and the oxygen-vacancy concentration (C_V):

$$D_O = C_V D_V \tag{6.1}$$

$$D_V = \frac{1}{6} a^2 \nu e^{(Q/K_B T)} \tag{6.2}$$

Here ν is a pre-exponential factor, Q is the activation energy taken to be the enthalpic barrier, a is the jump length for oxygen diffusion [27].

7. *Conducting nature of oxide semiconductors*: Defect chemistry is majorly used to understand the conducting nature of semiconductors. There are various theoretical studies have been done in literature to suggest n- or p-type of conduction based on acceptor and donor vacancy formations in 2D surfaces as well as 3D bulk structures [3,5–7,28–32]. We will discuss this part in detail in the section 6.3 on how to calculate vacancy formation energy (VFE) by using ab initio methods.

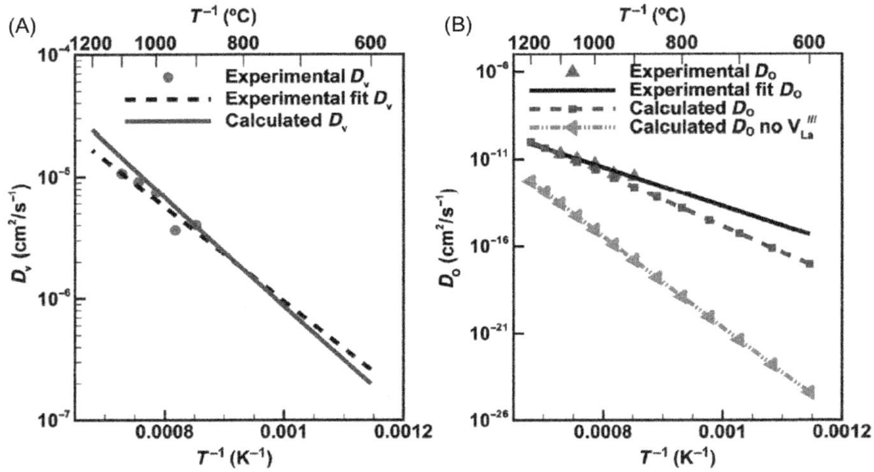

Figure 6.7 Computed and experimental diffusion coefficients: (A) vacancy diffusion coefficients (D_V) and (B) oxygen diffusion coefficients (D_O). Experimental data [25] (*red circles/triangles* and *black lines*), computed D_V (*solid blue line*), computed D_O considering V_{La}''' (*dashed blue line with squares*), and computed D_O neglecting V_{La}''' (*green line with triangles*). *Source*: Reprinted with permission from A.B. Munoz-garcia, M. Pavone, Ab initio evaluation of oxygen diffusivity in $LaFeO_3$: the role of lanthanum vacancies, MRS Commun. 3 (2013) 161–166.

6.3 Key quantities for the calculation of vacancy formations

Thermodynamic formalism is established to investigate the physical properties of point defects and relevant thermodynamic quantities, such as VFE, enthalpy, and entropy can be obtained from electronic-structure calculations. This allows calculation of defect and impurity thermodynamic equilibrium structures and charge carrier concentrations. In case of semiconductors, theoretical methods allow to calculate the relative stability of ionized states of a vacancy and provide the identification of the vacancy as either a deep and shallow defect. For guiding experimental fabrication, these theoretical models are developed to compliment experiments by adding the effects of reservoir conditions on defect formations. In the review chapter, we mainly focus on the calculations of VFE and thermodynamic quantities related to the changes that occur due to the generation of point defects in cells/supercells.

6.3.1 Calculation of vacancy formation energies

Ab initio calculations are a useful tool for understanding defect chemistry and effect of impurities. The Kohn–Sham (KS) density functional theory (DFT) based electronic structure calculation is a widely accepted and most successful method in this field. According to Kohn and Sham, DFT focuses on quantities in real, three-dimensional coordinate spaces, mainly on ground state electron density. The single particle KS equations, in principle, account for all ground state many body effects when used with exact exchange-correlation (XC) functional. The simplest approximation of XC functional is the local density approximation (LDA), where XC functional depends on the exchange-correlation energy per particle of a uniform electron gas of a given density. An important improvement over LDA is the generalized gradient approximation (GGA) of electron density where the XC functional depends on electron density and its spatial variation since these initial DFT approaches have been limited in their ability to predict properties associated with the electronic structure of materials such as "bandgap problem." Therefore more advanced functionals have been developed and used to overcome this deficiency, both by going beyond DFT and by implementing advanced functionals within DFT. In the sections 6.3.4 and 4.3.5, a more detailed explanations are discussed about these theoretical issues, including critical comparison of different approaches, such as DFT + U and hybrid functionals in the calculations of defect formations.

In this section, we present the equations for the atomic vacancy formation and the favorable conditions based on reservoir environment. In solid-state physics, the formation enthalpy is referred to as the energy difference between the resulting compound and the energy sum of individual elemental solids, and as a result, heat is generated or absorbed. In this work, we calculate the formation enthalpies from the ground state energies of compounds.

Vacancy formation energies of the neutral defect is defined by

$$H_f(\mathrm{D}) = E(\mathrm{D}) - E(\mathrm{P}) + \sum_i n_i(\mu_i^{\mathrm{ele}} + \Delta\mu_i) \tag{6.3}$$

where $E(\mathrm{D})$ is the total energy of the defective solid, $E(\mathrm{P})$ is the total energy of the perfect solid, and μ^{ele} are elemental reference energies of the constituent removed atoms from the host cells. These elemental energies are calculated by considering their standard states [(i.e., O_2(g), H_2(g), and TM(s)]. And $\mu_i = \mu_i^{\mathrm{ele}} + \Delta\mu_i$ is the chemical potential of a reservoir of atom i where $\Delta\mu_i$ is the correction in the standard elemental energies due to the reservoir conditions. We will discuss chemical potentials in detail in the section 6.3.2.

The defect formation enthalpy of a defect at a charge state q is given by

$$\Delta H_f(D,q) = E(D,q) - E(\mathrm{P}) + \sum_i n_i\left(\mu_i^{\mathrm{ele}} + \Delta\mu_i\right) + q\left(E_f + \varepsilon_{\mathrm{VBM}} + \Delta V\right) \tag{6.4}$$

where $E(D,q)$ is the total energy of charged defective solid, E_f is the Fermi energy level, $\varepsilon_{\mathrm{VBM}}$ is the valence band maxima for the perfect crystal. In case of charged vacancies, the correction to artificial coulomb interaction is introduced. It should be noted here that the valence band maximum (VBM) of host cell with a defect and with a perfect crystal are different reflected in the difference in average potential ΔV. The average potential difference ΔV is determined by the shift of the VBM energy from the perfect solid to the solid with vacancy far ($\sim$3Å) from the vacancy sites.

Atomic point defects and impurities are classified as shallow or deep acceptors or donors, where shallow defects or impurities are known for small ionization energies and easily release charge carriers in the host cell and enhance the conductivity of the system and render the material n- or p-type conductivity. However, deep donors or acceptors for those defects or impurities are not easily thermally ionized and would not contribute extra charge carrier's concentration in the conductivity. In the

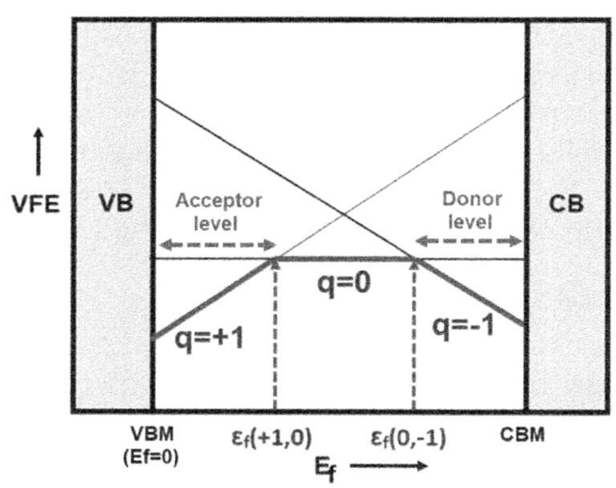

Figure 6.8 Schematic illustration of *VFE* with Fermi energy (E_f) for an example defect which has three standard charge states $q = +1$, 0, and -1. Solid black lines correspond to the formation energy as defined by Eq. (6.2). This defect exhibits two charge-state transition levels, $+1$ to 0 and 0 to -1 charge states which belong to acceptor and donor energy levels. The thick solid red lines indicate the energetically lowest and most favorable charge state for a given Fermi level. Donor and acceptor levels are defined as blue dashed lines. *VFE*, vacancy formation energies.

Fig. 6.8, we have shown schematic representation of VFE in the Fermi level region starting from VBM to CBM for an example defect with different charge states $q = +1$, 0, and -1. Due to ionization of the vacancies, varying charge state of the defect is shown in the form of transition levels. Therefore acceptor and donor level appear near to VBM and CBM, respectively. The basic concept is that when a vacancy is ionized then it contributes electrons or holes in the system and increases the charge concentration in the material. These extra charge carriers have a potential contribution in the conductivity. We have discussed this in section 6.3.5 in detail.

6.3.1.1 Elemental energy (μ^{ele})

Chemical potentials or elemental energies can be calculated by adding vibrational and thermal corrections in the calculated energy of the elemental atom. In case of atoms who has bulk standard states such as transition metals. For instance, bulk with a single atom such as Nickel atom is generally found in Ni-fcc structure and the energy is calculated from the fcc bulk structure. To find the ground state energy, following thermal, vibrational, and entropy corrections are introduced:

$$\mu^{\mathrm{ele}}(\mathrm{Ni}) = \frac{1}{4}\left[E_{\mathrm{Ni(fcc)}} + \frac{1}{2}\sum_{N=1}^{3(N-1)-3} h\nu_i + \sum_{N=1}^{3(N-1)-3} \frac{h\nu_i}{e^{(h\nu_i/kT)}-1} - T\Delta S_{\mathrm{Ni(fcc)}}\right] \tag{6.5}$$

where the first term is the total energy of the Ni-fcc unit cell which has 4 Ni atoms, second term denotes the zero-point energy corrections for the vibrational frequencies, third term is for the thermal energy correction due to vibrational frequencies, and last term belongs to the entropy of the system.

For instance, we have taken the example of O_2 (g) and H_2 (g) molecular free energy calculations, the thermal contributions are compressed into terms related to the enthalpy and entropy. For oxygen free energy can be calculated by these terms:

$$\mu^{\mathrm{ele}}(\mathrm{O}) = \frac{1}{2}\left[E_{\mathrm{O_2}} + \frac{1}{2}\sum_{N=1}^{3(N-1)-3} h\nu_i + \int_0^T C_{p,\mathrm{O_2}}TdT - T\Delta S_{\mathrm{O_2}}\right] \tag{6.6}$$

and similarly, for hydrogen free energy:

$$\mu^{\mathrm{ele}}(\mathrm{H}) = \frac{1}{2}\left[E_{\mathrm{H_2}} + \frac{1}{2}\sum_{N=1}^{3(N-1)-3} h\nu_i + \int_0^T C_{p,\mathrm{H_2}}TdT - T\Delta S_{\mathrm{H_2}}\right] \tag{6.7}$$

In both the equations, first and second terms are the energies and zero-point vibrational corrections of O_2 and H_2 molecules, respectively. Enthalpy is calculated by

$$H(T) - H(0\mathrm{K}) = \frac{1}{2}\left(\int_0^T C_{p,O_2} T dT\right) = K_B T\left[\frac{7}{2} + \frac{\theta_\nu}{T}\frac{1}{e^{(\theta_\nu/T)} - 1}\right] \quad (6.8)$$

Since O_2 and H_2, both have two atoms in the molecules, third term in Eqs. (6.3) and (6.4) is calculated similarly by Eq. (6.5). These thermal and vibrational corrections for a few selected molecules at available at NIST webbook [33].

In Eq. (6.5), H is the enthalpy, R is the gas constant, T is the absolute temperature, and Θ_ν is the vibrational temperature. The factor of 7/2 arises from the ideal gas and rigid-rotor approximations for a homonuclear diatomic molecule. Θ_ν is defined where ν is the vibrational frequency, h is Planck's constant, and k_B is Boltzmann's constant.

$$\Theta = \frac{h\upsilon}{k_B} \quad (6.9)$$

Forth terms are the entropy terms and are calculated by

$$s = s_e + s_t + s_r + s_\nu \quad (6.10)$$

$$= k_B\left[ln(q_e) + \left(ln(q_t) + \frac{5}{2}\right) + \left(ln(q_\nu) + \frac{\theta_\nu}{T}\frac{1}{e^{\frac{\theta_\nu}{T}} - 1}\right) + (ln(q_r) + 1)\right] \quad (6.11)$$

where

$$q_t = \left(\frac{2\pi M K_B T}{h^2}\right)^{\frac{3}{2}} V;\ q_r = \left(\frac{8\pi^2 I K_B T}{2h^2}\right);\ q_\upsilon = \left(\frac{1}{1 - e^{-(h\nu/kT)}}\right) \quad (6.12)$$

and $q_e = 3$ (triplet for O_2 molecule) and 1 (singlet for H_2 molecule), I = moment of inertia, M and V = mass and volume of the molecule.

Fig. 6.9 shows the calculated elemental energies of Ni(s), O(g), and H(g) atoms after adding all the thermal and vibrational corrections. For the demonstration, we have chosen three different DFT functional as PBE/PBE + U, and two hybrid functionals as PBE0 and HSE06. For the DFT calculations, we have used VASP quantum chemistry code [34,35]. Calculations of thermal corrections contributions are done using the PHONOPY software [36].

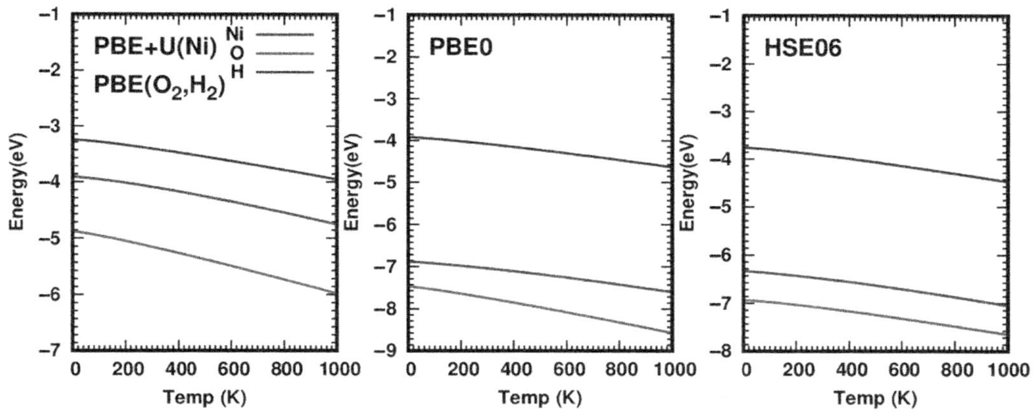

Figure 6.9 Variation in chemical potentials or elemental free energies with temperature after adding all the thermal and vibrational corrections of Ni(s), O(g), and H(g) atoms for PBE + U/PBE, PBE0, and HSE06 functional.

6.3.2 Chemical potentials and reservoir conditions ($\Delta\mu$)

Reservoir conditions always affect the formation of vacancies as mentioned in Eqs. (6.1) and (6.2). The chemical potentials of atoms are affected by the surrounding atoms in the cell and depend on reservoir conditions. Therefore it is required to set thermodynamic equations with the help of secondary phase materials. The main purpose to do this to avoid those conditions where the other secondary materials can be formed during the preparation of the main simple. In this process, we can define certain reservoir conditions which are suitable for our main sample material. There are several studies [2,5,37–40] have been done in past few decades where researcher have used this process to assign the reservoir conditions with the help ab initio calculations. In this chapter we are taking two examples in details to understand this as ternary oxides $PbTiO_3$ and $CuCrO_2$ [4,7].

In Fig. 6.10A, Scanlon et al. [7] have studied $CuCrO_2$ ternary oxide and considered Cu_2O, CuO, and Cr_2O_3 as secondary materials. Thermodynamic equations are used: $\mu_{Cu} + \mu_{Cr} + 2\mu_O = \Delta H$ ($CuCrO_2$), $\mu_{Cu} + 2\mu_O \le \Delta H(CuO_2)$, $2\mu_{Cu} + \mu_O \le \Delta H(Cu_2O)$, $2\mu_{Cr} + 3\mu_O \le \Delta H(Cr_2O_3)$, $\mu_{Cu} \le 0$, $\mu_{Cr} \le 0$, and $\mu_O \le 0$. Environment A in Fig. 6.10A corresponds to Cr-poor, Cu-poor, and O-rich conditions. Environment B is at the Cr-poor limit, with μ_{Cu} and μ_O having midrange values, with condition C corresponding to the Cu-rich limit of the stability region, with a midrange μ_{Cr} and a relatively low μ_O. Lastly, environment *D* represents Cr-rich,

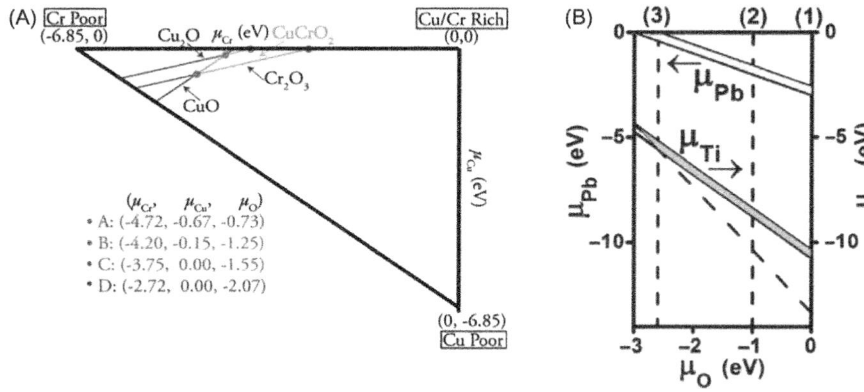

Figure 6.10 Illustration of the accessible range of (A) $CuCrO_2$ calculated using HSE06 in the form of chemical potentials (μ_{Cu}, μ_{Cr}) and different reservoir conditions (Cu/Cr/O-rich/poor). (B) $PbTiO_3$ allowed ranges for chemical potentials μ_{Pb}, μ_{Ti}, and μ_O, obtained from thermodynamic equations. *Source*: Reprinted with permission from (A) D.O. Scanlon, G.W. Watson, Understanding the p-type defect chemistry of $CuCrO_2$, J. Mater. Chem. 21 (2011) 3655; (B) Y. Yao, H. Fu, Charged vacancies in ferroelectric $PbTiO_3$: formation energies, optimal Fermi region, and influence on local polarization, Phys. Rev. B - Condens. Matter Mater. Phys. 84 (2011) 1–10.

O-poor, and Cu-rich limits, which should favor the formation of oxygen poor (or n-type) defects. Yao et al. [4] have derived similar equations but slightly different graphical representation shown in Fig. 6.10B. Thermodynamic equation at chemical equilibrium conditions, they have used following equations: $\mu_{Pb} + \mu_{Ti} + 3\mu_O = \Delta H$ ($PbTiO_3$), $\mu_{Pb} + \mu_O \leq \Delta H(PbO)$, $\mu_{Ti} + 2\mu_O \leq \Delta H(TiO_2)$, $\mu_{Cu} \leq 0$, $\mu_{Cr} \leq 0$, and $\mu_O \leq 0$. In this case the secondary materials are PbO and TiO_2.

6.3.3 Entropy corrections

In case of defect formation energies, the configurational entropy usually dominates and determines the temperature dependence of the defect concentration in thermodynamic equilibrium. These temperature-dependent defect formation energies include additional entropy contributions such as (1) vibrational energies (also known as zero-point energies due to phonons): this is due to the perturbation in the host cell by creating a defect. As a result, chemical bonds and electronic structures are redistributed and some energy is stored in the form of vibrations; (2) electronic contributions; and (3) magnetic excitations: generally, it belongs to the changes in spin orientations. All these entropy contributions in the formation energies are not always considered due to a few reasons. The first and main reason is the tarnished bandgap problem, especially for semiconductors or insulators, which arises from approximate DFT exchange-correlational functionals, which shows a bandgap error of a few electron volts and compare to this error, the entropy contributions are negligible. Another reason is that the calculations of vibrational and magnetic entropy are computationally costly by several orders of magnitude compared to nonthermal static ($T = 0K$) defect calculations and the contributions in the formation energies have very low effect or sometimes negligible in many materials. In case of metals, the bandgap problem is mostly absent due to efficient screening and therefore all these entropy contributions are essentially considered for an accurate portrayal of defect formation energies.

6.3.4 Charge transition levels

As shown in the schematic Fig. 6.8, the charge transition level $\varepsilon(q/q')$ is the Fermi-level position at which the formation energy of defect D at charge state q is equal to that at charge state q' namely, $\Delta H(D,q) = \Delta H(D,q')$ [41]. The transition energy level can be obtained by

$$\varepsilon(D, q/q') = \left[\frac{\Delta H_{D,q}(E_F = 0) - \Delta H_{D,q'}(E_F = 0)}{(q' - q)}\right] \tag{6.13}$$

Buckeridge et al. [2] studied binary oxides such as ZnO, SnO_2, and In_2O_3. They have calculated vacancy formation energies and charge transition states for these samples. Oxygen transition levels are found deep in donor in the bandgap region for ZnO and SnO_2 as shown in the Fig. 6.11. A defect is generally called deep when the energy required to remove electrons from the valence band or to add electrons to the conduction band is much larger than the thermal energy k_BT (which is 0.026 eV at 300K). From this definition, these transition levels are found deep from the CBM for ZnO and SnO_2 case.

6.3.5 Charge carrier and defect concentrations

Defect concentration at thermodynamic equilibrium and self-consistent Fermi energies are defined [32]. Defect concentrations are calculated by

$$[V_d^q] = N_{D_O} g \exp\left(\frac{E_f(V_d^q)}{kT}\right) \tag{6.14}$$

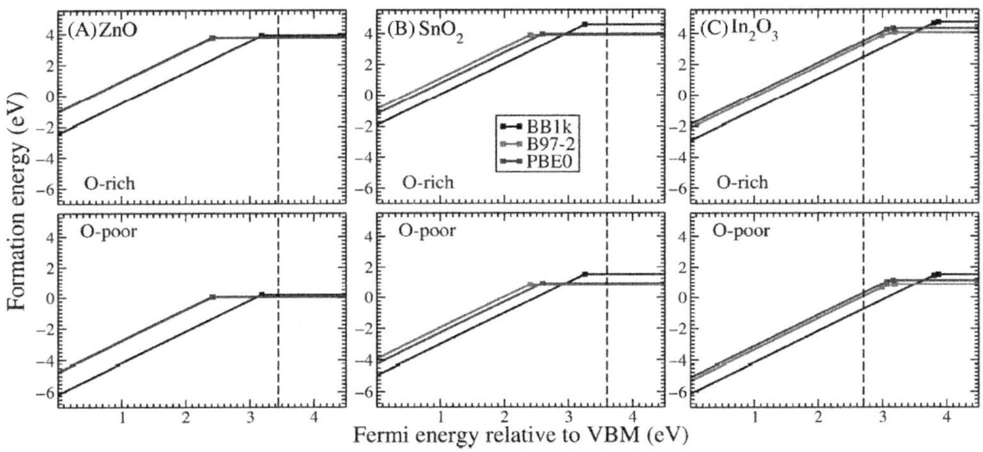

Figure 6.11 Formation energies as a function of Fermi energy relative to the VBM of oxygen vacancies in (A) ZnO, (B) SnO_2, and (C) In_2O_3, shown for O-rich and O-poor conditions. Results are given for three hybrid density functionals: BB1k (*black*), B97-2 (*red*), and PBE0 (*blue*). A vertical dashed line is placed at the position of the conduction band minimum for each system. The slopes indicate the charge states; a transition occurs where the slope of a line changes. For ZnO, the B97-2 result lies almost entirely below the PBE0 result. *VBM*, valence band maximum. *Source*: Reprinted with permission from J. Buckeridge, C.R.A. Catlow, M.R. Farrow, A.J. Logsdail, D.O. Scanlon, T.W. Keal, et al., Deep vs shallow nature of oxygen vacancies and consequent n-type carrier concentrations in transparent conducting oxides, Phys. Rev. Mater. 054604 (2018) 56–59.

where N_{D_O} is the density of sites of the host cell where vacancies are formed, g is the degeneracy of the defect state, k is Boltzmann's constant, and T is the temperature. Electron (n_0) and hole (p_0) concentrations are calculated as

$$n_0 = \int_{E_g}^{\infty} f_e(E)\rho(E)dE \tag{6.15}$$

$$p_0 = \int_{-\infty}^{0} f_h(E)\rho(E)dE \tag{6.16}$$

where E_g is the bandgap, $\rho(E)$ is the density of states of the host cell without vacancies, $f_e(E)$ is the Fermi–Dirac distribution function, where $f_h(E)$ defined as $f_h(E) = 1 - f_e(E)$. As each concentration is a function of Fermi energy (E_f), either via the defect formation energy or the Fermi–Dirac function, the equilibrium values can be calculated self-consistently at T, given the constraint of overall charge neutrality. Buckeridge et al. [2] have calculated charge carrier concentrations of binary oxides such as ZnO, SnO_2, and In_2O_3 by using hybrid functionals. There results suggested that all three oxides are showing mixed behavior in certain reservoir and thermodynamic conditions. In case of O-poor conditions (Fig. 6.12), ZnO and SnO_2 are shown n-type nature but at O-rich conditions, electron and hole concentrations are almost equal. On the other hand, In_2O_3 is found n-type in both conditions. In their study, they found that the calculations of vacancy and charge carrier concentrations are more helpful to understand the vacancy deep traps and give a clearer picture of conduction in these oxides.

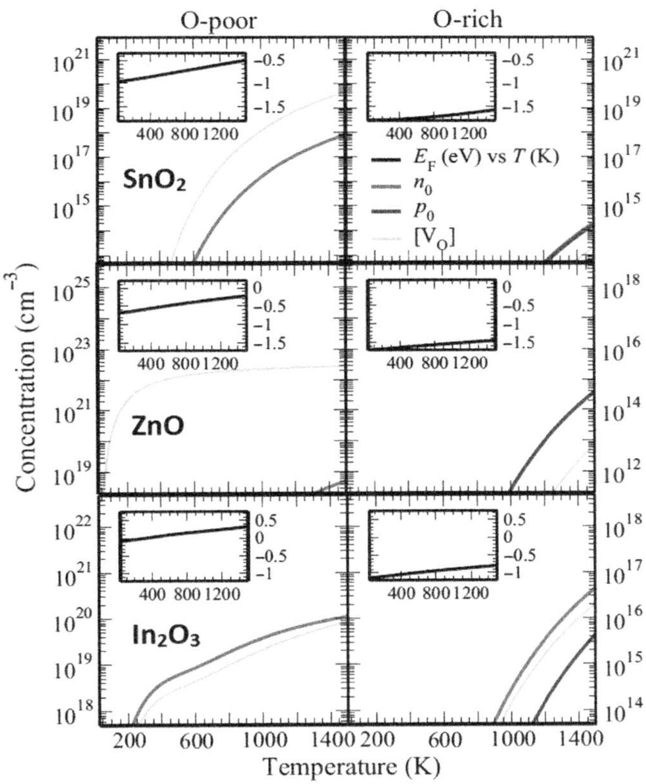

Figure 6.12 Calculated defect concentration ($[V_O]$, *green line*), electron (n_0, *red line*), and hole (p_0, *blue line*) carrier concentrations as a function of temperature *T* for SnO_2, ZnO, and In_2O_3 binary oxides, under O-poor and O-rich conditions. In this figure PBE0 hybrid functional is used but they have tested their calculations for B97-2 and BB1k functional too as mentioned in Fig. 6.11. The insets show the self-consistent Fermi energy E_f (*black line*) as a function of *T*, relative to the conduction band minimum. *Source*: Reprinted with permission from J. Buckeridge, C.R.A. Catlow, M.R. Farrow, A.J. Logsdail, D.O. Scanlon, T.W. Keal, et al., Deep vs shallow nature of oxygen vacancies and consequent n-type carrier concentrations in transparent conducting oxides, Phys. Rev. Mater. 054604 (2018) 56–59.

6.3.6 General comparable predictions for the point defect formation from theory and experiments

The prior importance is to compare the theory results with experimental predicted results. These comparisons are essential for validation of the computational approach. It should be noted here that experimental observations of defects in solids have their own limitations, which computational studies can aid in overcoming. Another importance for such comparisons is to interpret, validate, and explain experimental observations is a crucial asset of the first-principles calculations. Therefore the final goal is to reliably predicted structure and properties that can be experimentally implemented and observed. In this section we will discussion about such key quantities that can be experimentally predicted or measured.

6.3.6.1 Atomic structure

There are several studies where they mentioned about the symmetry break during atomic relaxation while creating vacancies or due to impurities [42–44]. Its normally depend on the size of the vacancy or impurity atom/s. These structural changes can be overserved from the experiments as well as theory. X-ray absorption fine structure (EXAFS) has been used for the measurement of atomic structures and bond lengths around an impurity [45]. This technique is ideally suitable for the measurement of local distortion around the defects and impurities, but it might fail when the impurity/defect concentration gets very high.

6.3.6.2 Scanning tunneling microscopy and spectroscopy

Scanning tunneling microscopy (STM) and scanning tunneling spectroscopy (STS) are the most powerful techniques for revealing the topographic structures of the surfaces [46,47]. These techniques are also can be used to measure defects on or slightly below surfaces. Insight into bulklike defects can be obtained from cross-sectional STM after cleavage, provided that the investigated cleavage surface is atomically flat, exhibits no states within the bulk bandgap, and has a low density of STM-observable surface. Feenstra et al. [47] had a detailed review article at STM and STS techniques.

6.3.6.3 *Defect concentrations*

As we discussed in the defect concentration case, calculated defect formation energy [in Eqs. (6.1) and (6.2)] is used to calculate defect concentrations. Experimentally, the impurities can be determined by secondary ion mass spectrometry or Rutherford backscattering spectrometry. Point defect concentrations determination is a bit tricky. A technique called electron paramagnetic resonance is one of the few techniques that can both identify the nature of a defect and accurately measure its concentration. Another less powerful technique is the positron annihilation spectroscopy [48] which can identify and measure point defects, but is typically limited to detection of vacancies. Less frequently used are electrical resistivity [49] and specific-heat measurements [50]. Resistivity measurements probe for the additional scattering due to defects. The specific heat associated with the creation of intrinsic defects (notably vacancies) can be separated from bulk contributions via its exponential increase with rising temperature or the characteristic time scale of defect formation. Another approach measures electrical noise and uses sophisticated theoretical tools to extract dynamical defect properties such as creation and annihilation rates or equilibrium concentrations [51].

6.3.6.4 *Nuclear magnetic resonance chemical shifts and Mössbauer spectroscopy*

Nuclear magnetic resonance (NMR) is used for molecules as well as solids to provide information about the chemical shift and structural information to study point defects [52]. When combined with first-principles calculations of chemical shifts, the approach allows an unambiguous determination of the microscopic structure. Whereas, Mossbauer spectroscopy probes changes in the nuclear energy levels and allows detection of interactions of point defects with neighboring atoms [53].

In addition of it, vibrational frequencies can be determined by the Raman spectroscopy or Fourier-transform infrared spectroscopy. These experimentally predicted results are directly compared with first principle calculations. These results generally give an idea of bonding formations and structural distribution in the material. Presence of defects often rise in local vibrational modes, which will reflect in the measurements.

6.3.6.5 Defect charge transition levels

Thermodynamic charge transition levels discussed in the sections 6.3.4 and 6.3.5 can be measured from experimental techniques such as deep-level transient spectroscopy or temperature-dependent Hall measurements [54], while optical levels can be observed in photoluminescence, absorption, or cathodoluminescence experiments. The identification of the underlying defect is greatly helped by comparison to theory, notably in complex cases [55].

Ritzmann et al. [25] have attempted to compare their theory results with experimental predicted Oxygen-vacancy concentration of $LaFeO_3$ ternary oxides. They demonstrated the enormous impact of La vacancies in raising the oxygen-vacancy concentration in $LaFeO_3$ as shown in Fig. 6.13. In this ternary oxide, La atoms are heavier compared to oxygen atoms and due to this, La vacancies are causing the increment in oxygen diffusivity. Comparisons of these diffusion coefficient results have already been discussed in Fig. 6.7.

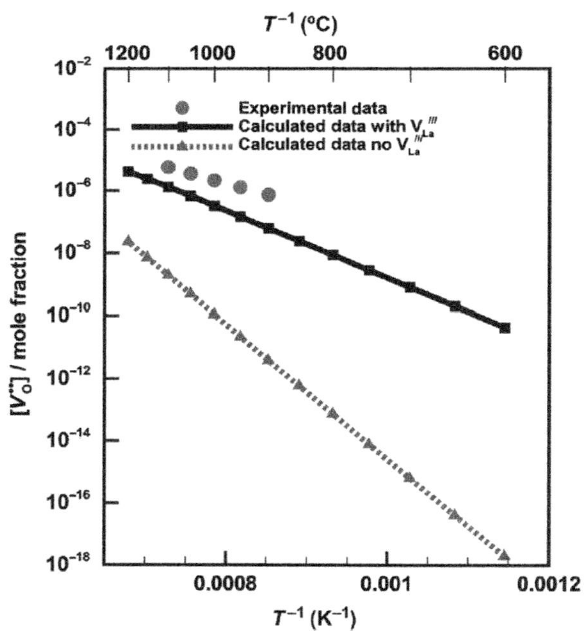

Figure 6.13 Theory [25] and experimental [56] predicted oxygen-vacancy concentrations (mole fraction) over the range of experimental temperatures. Experimental data were only tabulated between 900°C and 1100°C. Experimental data (*red circles*), computed data considering V_{La}''' (*black line with squares*), and computed data neglecting V_{La}''' (*dashed blue line with triangles*). *Source*: Reprinted with permission from A.B. Munoz-garcia, M. Pavone, Ab initio evaluation of oxygen diffusivity in $LaFeO_3$: the role of lanthanum vacancies, MRS Commun. 3 (2013) 161–166.

References

[1] C. Freysoldt, B. Grabowski, T. Hickel, J. Neugebauer, G. Kresse, A. Janotti, et al., First-principles calculations for point defects in solids, Rev. Mod. Phys. 86 (2014) 253–305. Available from: https://doi.org/10.1103/RevModPhys.86.253.

[2] J. Buckeridge, C.R.A. Catlow, M.R. Farrow, A.J. Logsdail, D.O. Scanlon, T.W. Keal, et al., Deep vs shallow nature of oxygen vacancies and consequent n-type carrier concentrations in transparent conducting oxides, Phys. Rev. Mater. 054604 (2018) 56–59. Available from: https://doi.org/10.1103/PhysRevMaterials.2.054604.

[3] W.B. Zhang, N. Yu, W.Y. Yu, B.Y. Tang, Stability and magnetism of vacancy in NiO: a GGA + U study, Eur. Phys. J. B 64 (2008) 153–158. Available from: https://doi.org/10.1140/epjb/e2008-00303-x.

[4] Y. Yao, H. Fu, Charged vacancies in ferroelectric $PbTiO_3$: formation energies, optimal Fermi region, and influence on local polarization, Phys. Rev. B - Condens. Matter Mater. Phys. 84 (2011) 1–10. Available from: https://doi.org/10.1103/PhysRevB.84.064112.

[5] S.B. Zhang, S.-H. Wei, A. Zunger, H. Katayama-Yoshida, Defect physics of the $CuInSe_2$ chalcopyrite semiconductor, J. Phys. Chem. Solids 60 (1999) 1865–1873. Available from: https://doi.org/10.1016/S0022-3697(99)00190-0.

[6] D.O. Scanlon, J. Buckeridge, C.R.A. Catlow, G.W. Watson, Understanding doping anomalies in degenerate p-type semiconductor LaCuOSe, J. Mater. Chem. C 2 (2014) 3429–3438. Available from: https://doi.org/10.1039/C4TC00096J.

[7] D.O. Scanlon, G.W. Watson, Understanding the p-type defect chemistry of $CuCrO_2$, J. Mater. Chem. 21 (2011) 3655. Available from: https://doi.org/10.1039/c0jm03852k.

[8] Y. Koyama, H. Arai, I. Tanaka, Y. Uchimoto, Z. Ogumi, First principles study of dopant solubility and defect chemistry in $LiCoO_2$, J. Mater. Chem. A 2 (2014) 11235–11245. Available from: https://doi.org/10.1039/C4TA01443J.

[9] F. Dionigi, P. Strasser, NiFe-based (Oxy)hydroxide catalysts for oxygen evolution reaction in non-acidic electrolytes, Adv. Energy Mater. 6 (2016). Available from: https://doi.org/10.1002/aenm.201600621.

[10] T. Sarkar, S. Ghosh, M. Annamalai, A. Patra, K. Stoerzinger, Y. Lee, et al., The effect of oxygen vacancies on water wettability of transition metal based $SrTiO_3$ and rare-earth based Lu_2O_3, RSC Adv. 6 (2016) 109234–109240. Available from: https://doi.org/10.1039/c6ra22391e.

[11] J. Shen, H. Lee, R. Valent, H.O. Jeschke, Ab initio study of the two-dimensional metallic state at the surface of $SrTiO_3$: importance of oxygen vacancies, Phys. Rev. B 86 (2012) 195119. Available from: https://doi.org/10.1103/PhysRevB.86.195119.

[12] J. Zaanen, A.I. Liechtenstein, V.I. Anisimov, Density-functional theory and strong interactions: orbital ordering in Mott–Hubbard insulators, Phys. Rev. B 52 (1995) R5467.

[13] R. Weht, X.G. Qiu, F. Bertran, A. Nicolaou, O. Copie, T. Kondo, et al., Two-dimensional electron gas with universal subbands at the surface of $SrTiO_3$, Nature 469 (2011) 189. Available from: https://doi.org/10.1038/nature09720.

[14] A. Ohtomo, H.Y. Hwang, A high-mobility electron gas at the LaAlO 3/$SrTiO_3$ heterointerface, Nature 427 (2004) 423–427.

[15] W. Meevasana, P.D.C. King, R.H. He, S. Mo, M. Hashimoto, A. Tamai, et al., Creation and control of a two-dimensional electron liquid at the bare

$SrTiO_3$ surface, Nat. Mater. 10 (2011). Available from: https://doi.org/10.1038/NMAT2943.
[16] H. Hojo, T. Mizoguchi, H. Ohta, S.D. Findlay, N. Shibata, T. Yamamoto, et al., Atomic structure of a CeO_2 grain boundary: the role of oxygen vacancies, Nano Lett. 10 (2010) 4668–4672. Available from: https://doi.org/10.1021/nl1029336.
[17] S.L. Dudarev, G.A. Botton, S.Y. Savrasov, C.J. Humphreys, A.P. Sutton, Electron-energy-loss spectra and the structural stability of nickel oxide: an LSDA + U study, Phys. Rev. B 57 (1998) 1505–1509. Available from: https://doi.org/10.1103/PhysRevB.57.1505.
[18] Y. Yang, P. Zhang, Dissociation of H_2 molecule on the β-Ga_2O_3 (100) B surface: the critical role of oxygen vacancy, Phys. Lett. A 374 (2010) 4169–4173. Available from: https://doi.org/10.1016/j.physleta.2010.08.031.
[19] J.P. Perdew, K. Burke, M. Ernzerhof, Generalized gradient approximation made simple [Phys. Rev. Lett. 77, 3865 (1996)], Phys. Rev. Lett. 78 (1996). Available from: https://doi.org/10.1103/PhysRevLett.78.1396. 1396–1396.
[20] L. Xu, Q. Jiang, Z. Xiao, X. Li, J. Huo, S. Wang, et al., Oxygen vacancies plasma-engraved Co 3 O 4 nanosheets with oxygen vacancies and high surface area for the oxygen evolution reaction, Angew. Chem. 55 (2016) 5277–5281. Available from: https://doi.org/10.1002/anie.201600687.
[21] L. Zhuang, Y. Jia, T. He, A. Du, Tuning oxygen vacancies in two-dimensional iron-cobalt oxide nanosheets through hydrogenation for enhanced, Nano Res. 11 (2018) 3509–3518. Available from: https://doi.org/10.1007/s12274-018-2050-8.
[22] Y. Yin, Wenjing Xu, Fenglei Lyu, Yaocai Bai, Aiqin Gao, Ji Feng, et al., Porous cobalt oxide nanoplates enriched with oxygen vacancies for oxygen evolution reaction, Nano Energy. 43 (2018) 110–116.
[23] D. Pei, L. Gong, X. Zhang, J. Chen, Y. Mu, A. Zhang, et al., Defective titanium dioxide single crystals exposed by high-energy {001} facets for efficient oxygen reduction, Nat. Commun. 6 (2015). Available from: https://doi.org/10.1038/ncomms9696.
[24] G.S. Chang, E. Kurmaev, A.N. Morozovska, J.A. Mcleod, Oxygen-vacancy-induced ferromagnetism in undoped SnO_2 thin films, Phys. Rev. B 85 (2012) 165319. Available from: https://doi.org/10.1103/PhysRevB.85.165319.
[25] A.B. Munoz-garcia, M. Pavone, Ab initio evaluation of oxygen diffusivity in LaFeO3: the role of lanthanum vacancies, MRS Commun. 3 (2013) 161–166. Available from: https://doi.org/10.1557/mrc.2013.28.
[26] O.K. Andersen, V.I. Anisimov, J. Zaanen, Band theory and Mott insulator: Hubbard U instead of Stoner I, Phys. Rev. B 44 (1991) 943.
[27] K.A. Marino, E.A. Carter, First-principles characterization of Ni diffusion kinetics in β-NiAl, Phys. Rev. B 78 (2008) 184105. Available from: https://doi.org/10.1103/PhysRevB.78.184105.
[28] A. Janotti, J.B. Varley, P. Rinke, N. Umezawa, G. Kresse, C.G. Van De Walle, Hybrid functional studies of the oxygen vacancy in TiO_2, Phys. Rev. B 81 (2010) 085212. Available from: https://doi.org/10.1103/PhysRevB.81.085212.
[29] C. Mitra, C. Lin, J. Robertson, A.A. Demkov, Electronic structure of oxygen vacancies in $SrTiO_3$ and $LaAlO_3$, Phys. Rev. B 86 (2012) 155105. Available from: https://doi.org/10.1103/PhysRevB.86.155105.
[30] K. Hoang, M.D. Johannes, Defect chemistry in layered transition-metal oxides from screened hybrid density functional calculations, J. Mater. Chem. A 2 (2014) 5224–5235. Available from: https://doi.org/10.1039/C4TA00673A.
[31] P.G. Sundell, M.E. Björketun, G. Wahnström, Thermodynamics of doping and vacancy formation in $BaZrO_3$ perovskite oxide from density functional

calculations, Phys. Rev. B - Condens. Matter Mater. Phys. 73 (2006) 1–10. Available from: https://doi.org/10.1103/PhysRevB.73.104112.
[32] D.B. Laks, C.G. Van De Walle, G.F. Neumark, P.E. Blöchl, S.T. Pantelides, Native defects and self-compensation in ZnSe, Phys. Rev. B 45 (1992) 10965–10978. Available from: https://doi.org/10.1103/PhysRevB.45.10965.
[33] J. Mizusaki, M. Yoshihiro, S. Yamauchi, K. Fueki, No title, J. Solid State Chem. 58 (1985) 257–266.
[34] G. Kresse, From ultrasoft pseudopotentials to the projector augmented-wave method, Phys. Rev. B 59 (1999) 1758–1775. Available from: https://doi.org/10.1103/PhysRevB.59.1758.
[35] G. Kresse, J. Furthmüller, Efficiency of ab-initio total energy calculations for metals and semiconductors using a plane-wave basis set, Comput. Mater. Sci. 6 (1996) 15–50. Available from: https://doi.org/10.1016/0927-0256(96)00008-0.
[36] A. Togo, F. Oba, I. Tanaka, First-principles calculations of the ferroelastic transition between rutile-type and $CaCl_2$-type SiO_2 at high pressures, Phys. Rev. B - Condens. Matter Mater. Phys. 78 (2008). Available from: https://doi.org/10.1103/PhysRevB.78.134106.
[37] A. Walsh, Defect physics of the kesterite thin-film solar cell absorber Cu2ZnSnS4, Appl. Phys. Lett. 96 (2010) 021902. Available from: https://doi.org/10.1063/1.3275796.
[38] T. Jing, Y. Dai, X. Ma, W. Wei, B. Huang, The photocatalytic properties of ultrathin bismuth oxychloride nanosheets: a first principles study, Phys. Chem. Chem. Phys. 18 (2016) 7261–7268. Available from: https://doi.org/10.1039/C5CP07437A.
[39] T. Hisatomi, J. Kubota, K. Domen, Recent advances in semiconductors for photocatalytic and photoelectrochemical water splitting, Chem. Soc. Rev. 43 (2014) 7520–7535. Available from: https://doi.org/10.1039/C3CS60378D.
[40] S.B. Zhang, S. Wei, A. Zunger, Stabilization of ternary compounds via ordered arrays of defect pairs, Phys. Rev. Lett. 78 (1997) 4059–4062.
[41] F.H. Taylor, J. Buckeridge, C.R.A. Catlow, Defects and oxide ion migration in the solid oxide fuel cell cathode material LaFeO 3, Chem. Mater. 28 (2016) 8210–8220. Available from: https://doi.org/10.1021/acs.chemmater.6b03048.
[42] M. Nolan, S.C. Parker, G.W. Watson, The electronic structure of oxygen vacancy defects at the low index surfaces of ceria, Surf. Sci. 595 (2005) 223–232. Available from: https://doi.org/10.1016/j.susc.2005.08.015.
[43] R. Leitsmann, F. Chicker, P. Pla, Boron-silicon complex defects in GaAs: an ab initio study, J. Appl. Phys. 109 (2011) 063533. Available from: https://doi.org/10.1063/1.3561373.
[44] A. Bhattacharya, C. Carbogno, B. Böhme, M. Baitinger, Y. Grin, M. Scheffler, Formation of vacancies in Si- and Ge-based clathrates: role of electron localization and symmetry breaking, Phys. Rev. Lett. 118 (2017) 236401. Available from: https://doi.org/10.1103/PhysRevLett.118.236401.
[45] B. Laboratories, Extended x-ray absorption fine structure-its strengths and limitations as a structural tool, Rev. Mod. Phys. 53 (1981) 769.
[46] Y. Zhang, T. Gao, Y. Gao, S. Xie, Q. Ji, K. Yan, et al., Defect-like structures of graphene on copper foils for strain relief investigated by high-resolution scanning tunneling microscopy, ACS Nano. 5 (2011) 4014–4022.
[47] R.M. Feenstra, Cross-sectional scanning tunnelling 1 microscopy of III-v semiconductor structures, Semicond. Sci. Technol. 9 (1994) 2157–2168.
[48] M.J. Puska, R.M. Nieminen, Theory of positrons in solids and on solid surfaces, Rev. Mod. Phys. 66 (1994) 841. Available from: https://doi.org/10.1103/revmodphys.66.841.

[49] G. Trumpy, J. Traff, R.M.J. Cotterill, K. Petersen, Positron lifetimes and trapping probabilities observed separately for vacancies and dislocations in aluminium, J. Phys. F Met. Phys. 2 (1972) 459.
[50] Y. Kraftmakher, Equilibrium vacancies and thermophysical properties of metals, Phys. Rep. 299 (1998) 79.
[51] J. Kindel, K. Lee, H.K. Sanatorium, E. Lindman, R. Stenzel, L. Angeles, Thermal-equilibrium properties of vacancies in metals through current-noise measurements, Phys. Rev. Lett. 36 (1975) 38.
[52] M. Minier, R. Andreani, C. Minier, Quadrupolar nuclear resonance in electron-irradiated aluminum: electric field gradients around vacancies and interstitials, Phys. Rev. B 18 (1978) 102.
[53] G. Czjzek, W.G. Berger, Hyperfine interactions at the final positions of 56 Fe (n, γ) 57 Fe recoil nuclei in bcc iron-aluminum alloys, Phys. Rev. B 1 (1970) 957.
[54] M. Lannoo, J. Bourgoin, Point Defects in Semiconductors I: Theoretical Aspects, Springer Series in Solid-State Sciences, Springer-Verlag, Berlin, 1981.
[55] B. Hourahine, R. Jones, A.N. Safonov, S. Öberg, P.R. Briddon, S.K. Estreicher, Identification of the hexavacancy in silicon with the B4 80 optical center, Phys. Rev. B 61 (2000) 12594.
[56] T. Ishigaki, S. Yamauchi, J. Mizusaki, K. Fukei, H. Naito, T. Adachi, Diffusion of oxide ions in LaFe03 single crystal, J. Solid State Chem. 55 (1984) 50–53.

7

2D materials for smart energochromic sunscreen devices

Valery A. Barachevsky
Photochemistry Centre RAS, FSRC "Crystallography and Photonics", Moscow, Russia

7.1 Introduction

A practically important problem of human life is to protect humans from the harmful effects of solar radiation, especially short-wave UV light. Meanwhile, radiation penetrating into the residential and industrial premises through the glazing plays a vital role. In addition to the direct protection from solar radiation, material costs arise to ensure a comfortable life. This is manifested in the need to use electricity to cool the premises in summer and to heat them in the cold season. At present, these problems are being solved with the use of constant light protective devices, namely blinds, curtains, translucent reflecting mirrors, antireflection treatments, or specifically designed glazing.

Currently, there is a need for smart devices that dynamically change their light transmission in the visible region of the spectrum in accordance with changes in the intensity of solar radiation. These devices meet modern requirements and are economically viable. Smart glazing will save 40% of energy costs compared to modern glazing [1].

These devices can be created using the so-called energochromic materials, which reversibly change color and other physical and chemical properties under external stimuli (light, heat, electrical voltage, etc.). These include photo-, thermo-, and electrochromic materials discussed in this review. Dynamic change of color (light transmission) of photochromic materials is based directly on the change in the intensity of solar radiation and, as a rule, does not require additional energy costs for regulating the light transmission. For the operation of

2D Nanomaterials for Energy Applications. DOI: https://doi.org/10.1016/B978-0-12-816723-6.00007-1

thermo- and electrochromic materials, electrical devices are necessary to ensure the controlled heating of thermochromic materials and the change in electrical voltage in the case of electrochromic materials.

In this chapter, we analyze the achievements and shortcomings of the development and use of two-dimensional (2D) thin-film photo-, thermal, and electrochromic materials in sun-protection devices.

7.2 2D energochromic materials

7.2.1 Photochromic materials

The simplest case for sunscreen applications are photochromic materials, which reversibly change their color under activating radiation, which is absorbed by the initial A (λ_1) and photoinduced B (λ_2) forms. The color intensity adaptively varies with the intensity of the activating radiation and depends on temperature (Scheme 7.1).

7.2.1.1 Inorganic systems

The most well-known inorganic photochromic materials are photochromic glasses containing silver halides. The beginning of the development of sunscreen photochromic materials of this type dates back to 1966, when Corning created ophthalmic photochromic borosilicate and aluminoborosilicate glasses with doped silver halides (chlorine or bromine) [2]. The photochromism of these glasses is based on the reversible photodissociation of silver halides to form atomic silver, determining the photoinduced neutral color (Scheme 7.2, Fig. 7.1) [3].

$$A \underset{\lambda_2,\,\Delta T}{\overset{\lambda_1}{\rightleftharpoons}} B$$

Scheme 7.1

$$\mathrm{AgHal} \underset{h\nu_2,\,\Delta T}{\overset{h\nu}{\rightleftharpoons}} \mathrm{Ag^0} + \mathrm{Hal^0}$$

Scheme 7.2

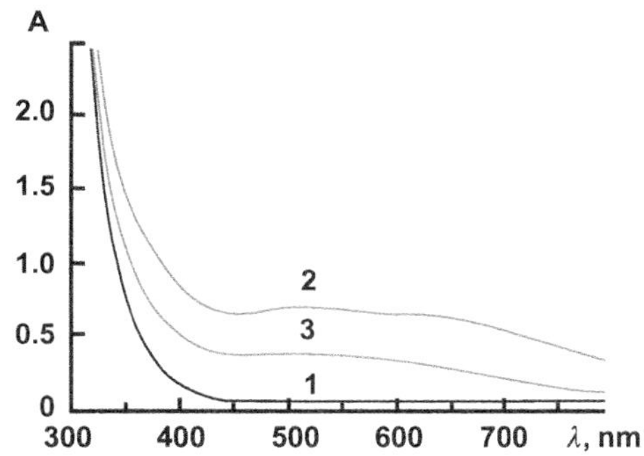

Figure 7.1 Absorption spectra of photochromic silicate glass based on AgBr before (1) and after sun irradiation (2) and in the process of spontaneous dark relaxation (3).

The advantages of photochromic silicate materials include the practically unlimited operating life under the sunlight and neutral photoinduced coloring due to formation of atomic silver as in photographic emulsions. The implementation of such a photochemical reaction in a glass matrix ensures the reversibility of the process with the formation of the initial silver halide molecules. The main disadvantages of photochromic silicate glasses are low molar extinction coefficients of silver halides and their photoproducts, which lead to the need to manufacture thick 3D materials, as well as technological and cost problems. In addition, the high-temperature glassmaking technology and the weight of the photochromic glass material turned out to be economically unprofitable for obtaining glazing of large sizes.

Attempts have been made to use this mechanism of photochromism in 2D photochromic sol–gel [4–6] and polymer matrices [7]. Photochromic sol–gel films containing AgCl particles were prepared by dissolving silver nitrate and trichloroacetic acid in the precursor sol [4]. Methyltrimethoxysilane and 3-glycidoxypropyltrimethoxysilane were also used as the ormosil matrices to produce sol–gel photochromic materials [5]. 3-Chloropropyltrimethoxysilane and bromophenyltrimethoxysilane were added as halogen sources and silver colloidal dispersion was introduced into the precursor sol. The coating became transparent and photochromic after precipitation of $Ag(Cl_1 - xBr_x)$ microcrystals at temperature above 300°C. Photochromic performances were improved by insertion of a SiO_2 buffer layer between the substrate and photochromic layer, the substitution of Cl with Br and the incorporation of a minute amount of Cu [6].

The size of the metallic Ag precipitate determines the color of the resulting films, showing a shift to the red in the absorption spectra as the size of the Ag particles increases from 8 nm (clear yellow) to 30 nm (purple) [6].

Polymer layers obtained by thermal polymerization of hybrid compounds of AgCl nanoparticles with methyl methacrylate showed weak photochromism [7]. More efficient photochromism was found for AgCl nanoparticle layers in a polyurethane matrix [8]. In this case, as in the case of photochromic glasses, the introduction of copper salts ($CuCl_2$) into the composition increased the speed of photocoloration of such a composite material. Unfortunately, the resulting composite photochromic materials based on sol–gel and polymer matrices are inferior in their characteristics to photochromic silicate glasses and cannot yet be of practical importance.

Thin films of a number of metal oxides (MoO_3, WO_3, V_2O_5, Nb_2O_5, TiO_2, and ZnO) exhibit photochromic properties due to electron transfer involving protons, leading to the formation of reduced colored oxides [9–11].

Of particular interest are photochromic composite films based on tungsten oxide WO_3 [12]. Films fabricated from peroxoisopoly-tungstic acid and a transparent urethane resin have effective photochromic properties under sunlight at any time of the year [12]. The photochromic properties of WO_3-based materials result from the reduction of W^{6+} to W^{5+} in the WO_3 host upon UV irradiation [13–16].

Tungsten oxide-based photochromic films were prepared by using methylcellulose as a film matrix and polyols such as ethylene glycol, propylene glycol, and glycerin as dispersing agents [17]. The use of glycerin as well as an increase in its concentration and water improves the properties of photochromic films, which is manifested in an increase in their photosensitivity and the rate of dark relaxation to the initial state, respectively.

The most practically significant results were obtained when creating photochromic films by the sol–gel technology [18–20]. Highly transparent tungsten oxide/3-(triethoxysilyl)propyl methacrylate/tetraethoxysilane hybrid xerogels prepared by the traditional sol–gel process manifest photochromic transformation with controlled behavior by cation diffusion [18]. A multilayered photochromic film containing WO_3, TiO_2, and SiO_2 sols and a solid electrolyte (an iodine/iodide redox couple) fabricated using sol–gel chemistry and dip-coating shows homogeneous coloring ($T_{vis} = 76\%$ down to 35%) within 15 min under 0.75 sun illumination (750 W m^{-2}) (Fig. 7.2) [19].

The same method was used to prepare porous orthorhombic tungsten oxide (o-WO_3) thin films stabilized by nanocrystalline anatase TiO_2, with a ruthenium-based dye as sensitizer [20].

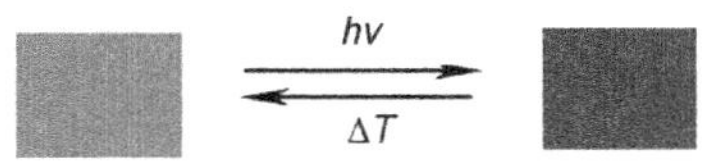

Figure 7.2 Photochromic transformations of the photochromic film containing WO_3, TiO_2, and SiO_2 sols and a solid electrolyte.

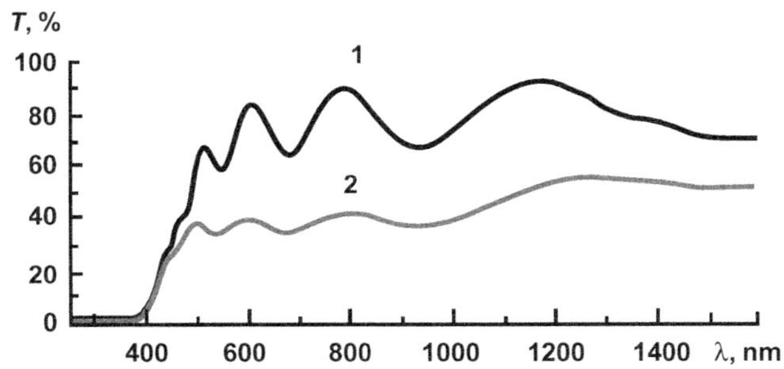

Figure 7.3 Transmission spectra of the deposited film of oxygen-containing yttrium hydride before (1) and after (2) sun irradiation.

Photochromism is shown by deposited films of oxygen-containing yttrium hydride (YO_xH_y) [21]; the photoinduced absorption which is observed not only in the visible but also in the infrared (IR) spectral region (Fig. 7.3).

The switchable optical properties can be tailored by changing the chemical composition of YHO films [22]. Increasing oxygen concentration in the film was accompanied with a decrease in the photochromic response. The resulting yellow transparent material with smallest bandgap showed best photochromic performance.

Similar photochromic properties are inherent in lanthanide oxy-hydride thin films (270–350 nm) based on Gd, Dy, and Er [23].

The photochromism of polyoxometalate layers is based on reversible photoinduced electron transfer from oxygen to the metal, which leads to the restoration of metal centers and to the appearance of color [24]. Photochromic films of polyoxolanthanate [25], polyoxomolybdate (Fig. 7.4) [26], and a number of sulfonic polyoxometalates [27] were obtained.

It has been shown that the photocoloration efficiency of films based on polyoxotungstate $H_3PW_{12}O_{40}$ can be increased by a factor of 5 by coating their surface with a YSO: Pr^{3+}

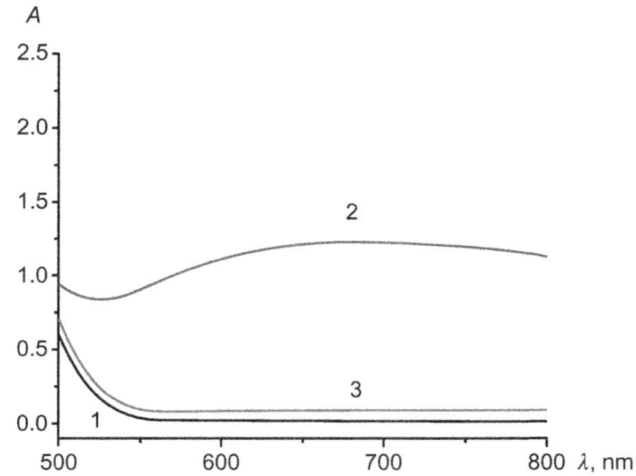

Figure 7.4 Absorption spectra of molybdenum polyoxometalate before (1) and after UV irradiation (2) and after subsequent exposure in the dark (3).

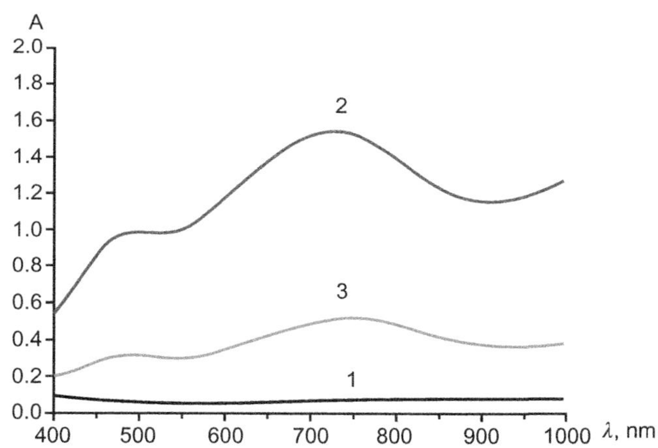

Figure 7.5 Absorption spectra of the urethanesil film containing phosphotungstic acid before (1) and after UV irradiation (2) and in the process of spontaneous dark relaxation (3).

upconverting thin layer, which converts the visible component into UV activating irradiation under sunlight [28].

Layers of urethanesils (hybrid ormosils with a polydimethylsiloxane backbone linked by urethane bonds) containing phosphotungstic acid have not only photochromic properties (Fig. 7.5) but also high adhesion to silicate glass [29].

7.2.1.2 Organic systems

2D photochromic sunscreen materials are preferably fabricated using organic photochromic compounds, which differ from inorganic photochromic substances by high molar extinction coefficients. In turn, they usually have selective absorption

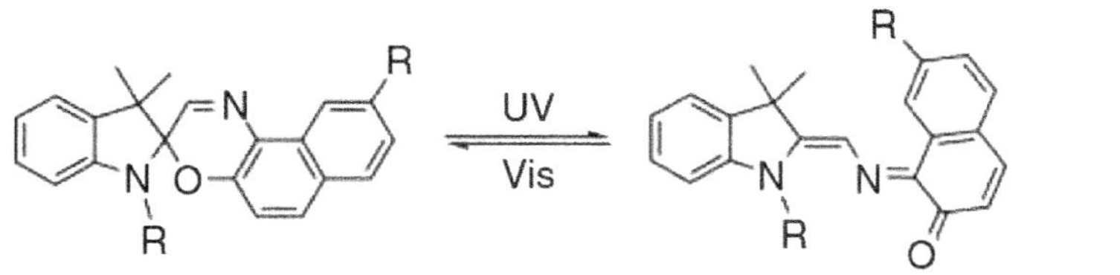

Scheme 7.3

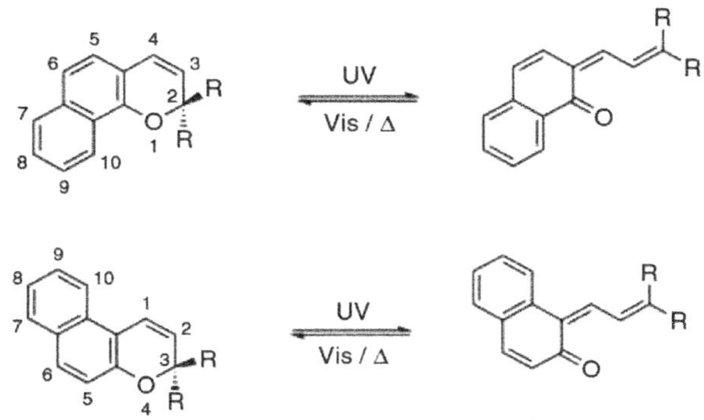

Scheme 7.4

bands of photoinduced forms and low durability under sunlight as compared with photochromic silicate glasses. Intensive synthetic research is needed to eliminate these shortcomings. Nevertheless, at present, various companies have created a range of photochromic ophthalmic polymer lenses based on organic photochromic compounds using the most photodegradation-resistant spirooxazines (Scheme 7.3) and chromenes (Scheme 7.4) [30].

The photochromic ophthalmic lenses are manufactured using several technologies, namely, production of organic glass containing spirooxazines or chromenes by thermal or photopolymerization, the introduction of a photochromic compound into the polymeric glass surface by diffusion, and the deposition of a polymer layer on the lens surface by spin-coating and triplexing of the photochromic film between two polymeric lenses [31].

Photochromic polymer ophthalmic lenses have many years of commercial success. Unfortunately, due to the degradation of photochromic compounds [32], their service life under sunlight does not exceed 3 years [31]. Apparently, for this reason, photochromic polymeric materials have not yet been created for glazing of buildings and vehicles of various types.

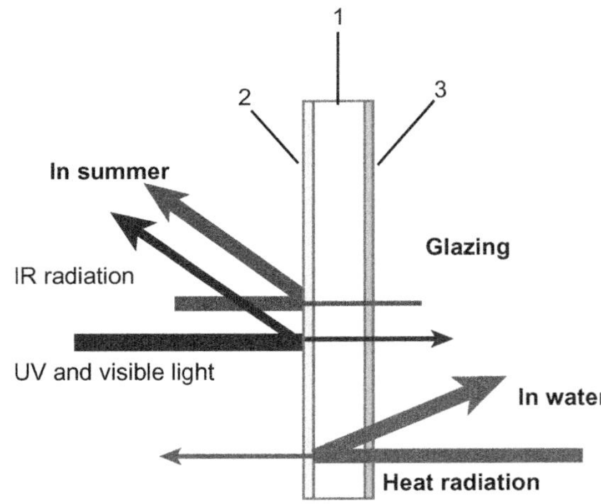

Figure 7.6 Structure of a multilayer photochromic sun-protective and thermo-saving polymer film: (1) poly(ethylene terephthalate) substrate; (2) IR reflecting layer; and (3) photochromic polymer layer.

In this regard, the development of multilayer photochromic energy-saving polymer films (Fig. 7.6) was undertaken using photochromic spirooxazines [33,34] and naphthopyrans [35] as the least photodegradable and the most temperature-stable photochromic compounds.

In addition to the poly(ethylene terephthalate) substrate, the design of a multilayer photochromic polymer film (Fig. 7.6, 1) includes an IR reflecting metallic (Fig. 7.6, 2) and photochromic polymer (Fig. 7.6, 3) layers. The use of a copper-containing metal coating deposited in vacuum on the front side of the polymer substrate (Fig. 7.6, 1) provides not only a reflection of the thermal component of solar radiation in summer but also removes the thermal load from the photochromic layer cast on the reverse side of the substrate (Fig. 7.6, 3). As a result, the photocoloration efficiency of the photochromic compound increases (Fig. 7.7).

In the cold season, the IR reflecting layer prevents the heat radiation from escaping the room through the glazing. Thus such a multilayer film used for glazing not only provides the comfort of a person in the room but also saves electricity used for air conditioning in summer and heating in winter.

These multilayer films are characterized by constant reflection of IR radiation of up to 95%; reversible decrease in the visible light intensity of up to 50%; constant decrease in the UV radiation intensity of up to 99%; self-gluing film can be used to supplement existing glazing; and the film provides shock and blast resistance for glazing. The application of these films opens up new functional possibilities for glazing used in the building

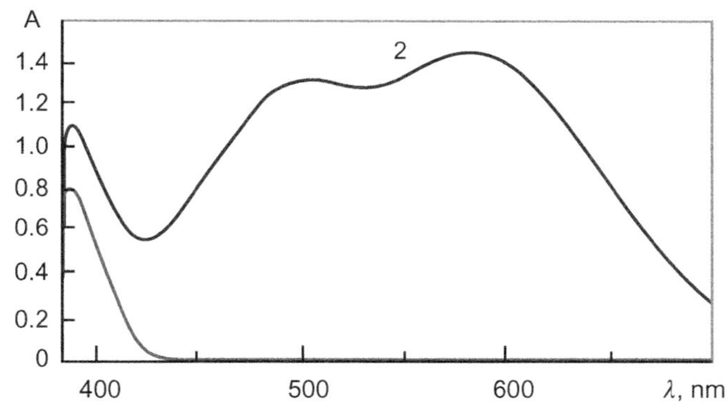

Figure 7.7 Absorption spectra of photochromic polymer layer before (1) and after UV irradiation (2).

industry and transport vehicles. Unfortunately, these films, like ophthalmic lenses, have a limited service life under sunlight (not more than 3 years). In connection with this, research of new photochromic 2D materials among both organic and inorganic systems is continued.

7.2.1.3 Organic–inorganic systems

Organic–inorganic photochromic materials have inherent advantages of both organic (effective photocolorability, low weight, flexibility, etc.) and inorganic (high thermal and mechanical strength) materials [6]. These materials are fabricated by introducing photochromic organic molecules into sol–gel silicate and aluminosilicate matrices. The sol–gel process [36] is used most commonly for the preparation of photochromic hybrid materials. Unlike inorganic materials, these photochromic materials are prepared at low temperatures (<150°C). The materials have nanostructured porosity, which ensures high efficiency of transformation of photochromic molecules. At the same time, they have a high transparency.

For the first time, a sol–gel photochromic material was obtained using one of the photochromic fulgides (Aberchrome 670) [37]. Subsequently, samples of such materials with photochromic spiropyrans [38–40] and spirooxazines [40–42] were designed. However, the most interesting materials of this type are those containing photochromic naphthopyrans, which differ from spiro compounds by a higher resistance to irreversible phototransformations and a wide variety of colors of the photoinduced merocyanine form [6,43–45].

A photochromic ormosilic coating was obtained on the basis of naphthopyran [44].

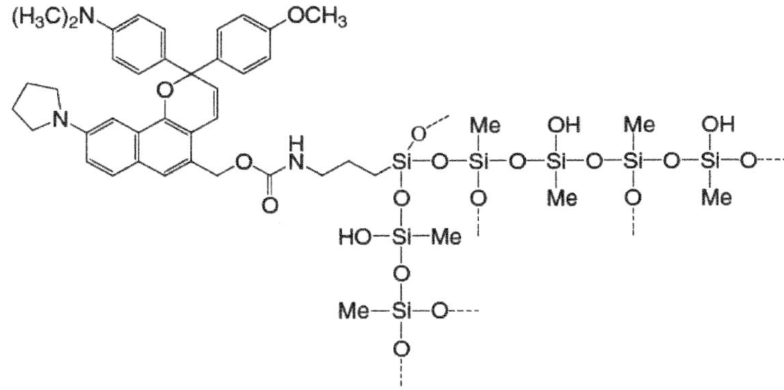

The photoinduced spectral changes of such coating are shown in Fig. 7.8.

The coating material based on sol–gel mesoporous coating matrix with embedded naphthopyran was used to produce a photochromic window, which reduced visible light transmission by 50% in the tinted state and bleached back to the clear state within 30 min in the dark [45].

Analysis of the presented results on the development of 2D photochromic sunscreen materials shows that, despite the success in the development of 3D photochromic silver halide materials, 2D sunscreen materials have not yet been introduced into production. The most promising inorganic oxide coatings developed so far, as well as silver halide films, do not meet the application requirements due to low photosensitivity and slow relaxation characteristics. For practical use, developed multilayer photochromic sunscreens and energy-saving films, as well as sol–gel coatings using naphthopyrans, which are successfully employed in the manufacture of polymeric ophthalmic lenses, are of interest. Meanwhile, one should bear in mind that the durability of such materials under sunlight is limited to 3 years.

7.2.2 2D thermochromic materials

By 2D thermochromic materials are meant materials of two types:

- passive thermochromic coatings and
- electrically controlled thermochromic materials.

7.2.2.1 Passive thermochromic coatings

Initially, layers based on vanadium oxides and polymers that turned into a light-scattering state when heated were considered

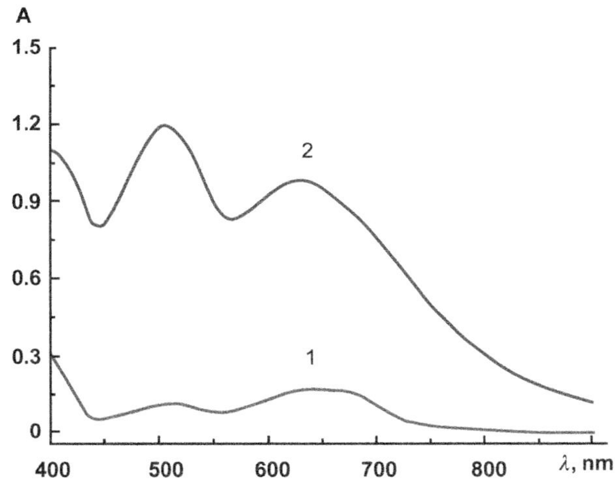

Figure 7.8 Absorption spectra of photochromic ormosilic coating based on naphthopyran before (1) and after (2) UV irradiation.

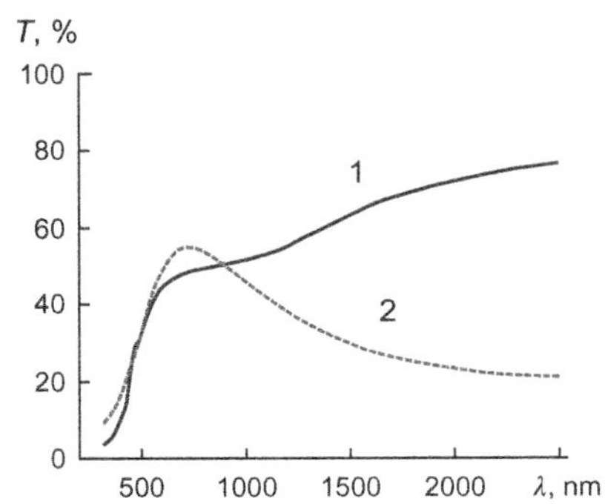

Figure 7.9 Transmission spectra of the thermochromic VO_2 film (50 nm) before (1) and after (2) heating up to 70°C.

to be the main thermochromic systems [46], but the key prospects are associated with the creation of optically transparent coatings based on these oxides.

A feature of coatings based on vanadium oxides is that a reversible thermally induced change in the light transmission is observed in the region of 750–2500 nm and is absent in the visible spectral region (Fig. 7.9) [47]. At the same time, these coatings are characterized by a constant low initial light transmission (about 40%) and a high transition temperature to the colored state (68°C), which requires the use of heaters.

The thermochromism of these layers is based on the reversibly thermoinduced phase transformation of the initial monoclinic semiconductor structure into a tetragonal metal-like structure (rutile) at $T > 68°C$ [48]. For practical use, these layers must be improved toward a lower phase transformation temperature, higher initial light transmission in the visible spectral region, and higher efficiency of photocoloration [49], and also the initial yellow color must be eliminated [50]. The first drawback can be corrected by introducing additives, in particular, tungsten [51,52]. By varying the tungsten content, the color transformation temperature can be controlled in the range of 15°C–68°C [53]. Several methods have been proposed for increasing the initial light transmission of these films in the visible region by introducing Mg^{2+} ions into the oxides [54–57]. As a result, it was possible to increase the initial light transmission to 51% with a concentration of magnesium ions of 7.2% [54]. SnO_2 and ZnO layers were used as buffer layers when applying Mg-doped VO_2 coatings [58]. At the same time, the service life of films under solar light increased [59]. The use of V_2O_5 as a buffer layer made it possible to maintain a high crystalline state of the VO_2 film [60].

An increase in light transmission was observed upon the introduction of Zn [61] or F [62–64] and with the use of substances with a high refractive index in a multilayer structure, for example, $TiO_2/VO_2/TiO_2$ [65]. An effect of Fe^{3+}/Mg^{2+} [66] and Al^{3+} [67] ions on the magnitude of the initial light transmission and the temperature of phase transformations was revealed. The deposition of a SiO_2 layer on top of the VO_2 layer made it possible to increase the initial light transmission to 56.3%, and the change in photoinduced absorption increased to 10.3% [68]. When using a SiO_2 coating with a refractive index of $n = 1.299$, the initial light transmission of the binary layer reached 80%, and the photoinduced change in absorption was 10.2% [69].

The initial light transmission of VO_2 in the visible spectral region sharply increases when VO_2 nanoparticles with a size of 20 nm are used instead of deposited films (Fig. 7.10) [70,71].

Simultaneously, the phase transformation temperature of the thermochromic coating decreased [72]. The degree of thermal modulation of light transmission depended on the nanoparticle size [73].

A high initial light transmission with acceptable thermal modulation of light transmission was achieved for the VO_2-based nanogrid perforated films [74] and for VO_2 thin films prepared by acid etching and self-patterning [75].

By means of nanolithography, it was possible to fabricate periodic 2D VO_2 nanonet, nanodome, and nanoparticle arrays with highly efficient light transmission modulation [76]. PVP-VO_2 composite films with the use of mesoporous VO_2 nanocrystals with pore size of about 2–10 nm synthesized by hydrothermal method manifested acceptable thermochromic properties [77].

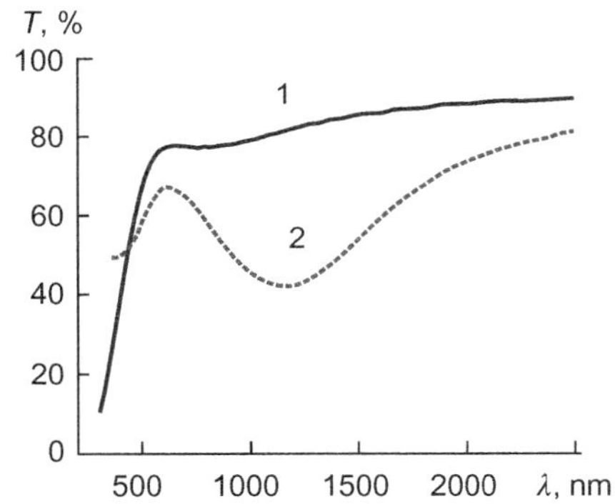

Figure 7.10 Transmission spectra of a thermochromic coating based on VO_2 nanoparticles of 50 nm size before (1) and after heating at 70°C (2).

Polymer coatings comprising composite VO_2 nanoparticles with SiO_2 and F shells (VO_2@SiO_2-F) seem to be most acceptable [78]. VO_2/TiO_2 bilayer films exhibit an increase in visible light transmittance and a higher transmittance modulation in the near-IR region [79]. The TiO_2 top layer not only enhances visible light transmittance but also protects the VO_2 layer from oxidation. Eu/W- and Tb/W-codoped VO_2 nanoparticles were found to be efficient in improving properties of single W doping samples [80].

A high-performance process has been developed for producing VO_2 nanoparticles from amorphous V_2O_5; the nanoparticles served to form thermochromic coatings on the surface of large-sized glasses [81].

Commercially attractive VO_2 elastomer colloidal films were manufactured using a low-temperature scalable synthesis and an overall simple fabrication method [82]. The application of nanoinks based on VO_2 nanoparticles was used to fabricate films using spin-coating and intense pulsed light sintering on a quartz substrate at applied voltages and at low cost [83]. A method has been developed for fabrication of mechanically flexible VO_2-based thermochromic films using graphene as the thin 2D support for the growth of high-quality VO_2 and as a shuttle to transfer the VO_2 layer to a flexible substrate [84].

The preparation method of sol–gel derived VO_2 thin films provides a facile synthesis technique with high levels of control over the film morphology and optical properties [85]. Thermosensitive microgels synthesized directly from hydroxypropyl cellulose and acrylic acid in pure water with the use of W-VO_2 nanoparticles possess a high initial transmission of 80%, thermoinduced transformations at 60°C, and modulation of IR transmission of 36% [86]. The VO_2@SiO_2/poly(*N*-isopropylacrylamide) hybrid nanothermochromic microgels synthesized through suspension polymerization demonstrated a 164% increase in the solar modulating ability at a relatively high average luminous transmittance (38.4%) [87].

Recently, the possibility of using sunscreens based on halide perovskites has been considered [88]. The thermochromism of these substances is based on their unusual crystallization behavior, namely, inverse temperature crystallization [89]. The thermochromic transformation of the biconstituent $MAPbBr_{3-x}I_x$ perovskite is due to a spectral shift of the absorption band (Fig. 7.11) [90].

The reversible thermochromic behavior was found for the thin film of dihydrated methylammonium lead iodide [91] and a halide perovskite, cesium lead iodide/bromide [92]. The 2D organic–inorganic hybrid complex $[(PyCH_2NH_3)_6][Pb_5I_{22}] \cdot 3H_2O$ obtained by the reaction of 2-aminomethyl-pyridine with PbI_2 in concentrated HI aqueous solution exhibited a reversible color change from orange to red at 80°C in the solid state in the layered perovskite inorganic sheets [93].

7.2.2.2 *Electrically controlled thermochromic materials*

Among organic thermochromic systems, some specific new generation thermochromic polymer materials exhibit promising properties from an applicative point of view [94]. However, only a few of them can be used to create sunscreens [95]. These include gel-type thermotpopic polymers that switch from a transparent to a scattering state due to a reversible thermochemical dissolution in water and a thermally induced modification in the length of the polymer molecules [46]. The switching temperature can be regulated to within 1.5°C in the 9°C–90°C range. Phase-separating polymer blends consisted of

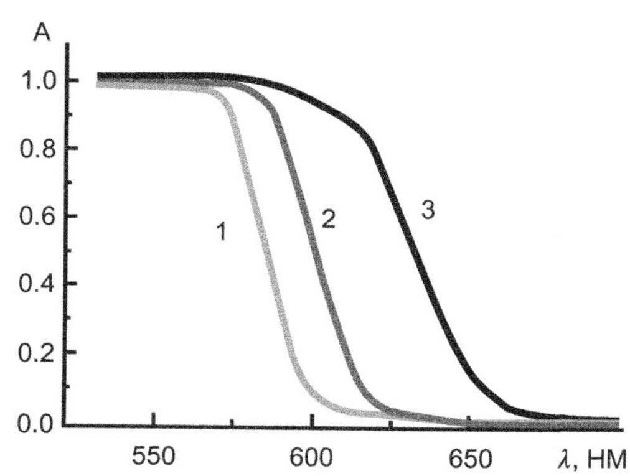

Figure 7.11 Absorption spectra of perovskite crystals obtained from inks at 60°C (1), 90°C (2), and 120°C (3).

a thermoplastic component embedded in a crosslinked matrix, for example, a poly(propylene oxide) and a styrene-hydroxyethyl methacrylate copolymer thermally crosslinked with a trifunctional isocyanate, and changed transmittance from 92% to 6% at 25°C and 50°C, respectively [96]. Due to their biodegradability and commercial availability at low costs, biopolymers are promising in the field of thermotropic materials [97]. A smart window assembly equipped with additional heating by indium tin oxide electrodes (ITO) had been reported [98].

The thermotropism of liquid crystals in polymer systems is achieved by tuning the orientation in response to temperature stimulus [99]. In a sandwich configuration smart window, films based on orientation transition of liquid crystals are good candidates for energy-efficient smart windows [100,101].

Thermochromic polymer systems obtained by doping the polymer matrix with thermochromic additives are of special interest for sun protection [95]. Such materials, in contrast to those discussed above, retain transparency when darkened, which is important from a practical point of view.

Ligand-exchange thermochromic and leuco dye-developer-solvent systems, which reversibly change the transmittance between 61% and 34% under sun irradiation and in the dark, respectively, were used as thermochromic additives. Unfortunately, leuco dye-developer-solvent systems require protection from the environment and are used mainly in the microencapsulated form, which prevents the creation of transparent thermochromic coatings [102].

Analysis of the results of development of thermochromic coatings shows that polymers that reversibly switch from the transparent to scattering state under voltage have some commercial success. However, the most practically attractive thermochromic systems are vanadium oxide-based passive thermochromic optically transparent coatings being currently developed; these materials, however, require improved optical properties in the visible spectral region. The advantage of these coatings is the effective thermally induced reversible change in the IR absorbance and the virtual absence of thermally induced modulation of the light flux in the visible range. In this connection, new thermochromic systems are being sought, in particular, based on perovskite crystals.

7.2.3 2D electrochromic materials

Electrochromic windows have attracted the most attention because they offer dynamic modulation in a broad spectral range ($0.3 < \lambda < 3\ \mu m$) with electric control [103]. However, compared with photochromic and thermochromic windows, they require full device configuration and electric supply, which complicates their design and increases the cost.

Electrochromic systems are complex multilayer structures that include substrates, usually made of silicate glass with applied conductive coatings, an electrochromic layer, and an electrolyte. For a number of applications in the manufacture of glazing, electrochromic materials must be flexible, which requires the use of polymer substrates. In addition, all components of materials must be transparent in the visible range and colorless, electrochemically and photochemically stable, and electrolytes must be solid or gel [104].

7.2.3.1 Inorganic systems

Among electrochromic materials, deposited tungsten oxide WO_3 layers occupy a leading position [105]. Although the physical mechanism for coloration is not fully understood, it likely originates from a combination of polaronic absorption at W^{5+} (or W^{4+}) sites and intraband transitions involving the electrons injected into the previously empty WO_3 *d* band states (Scheme 7.5) [106].

The electroinduced change in the light transmission of an electrochromic device from a clear, transparent WO_3 state to a dark-blue Li_xWO_3 state is shown in Fig. 7.12 [104].

It is shown that amorphous films using high-porous tungsten oxide a-WO_3 have electrochromic performance parameters, including an optical modulation of 70% at 700 nm, colored and bleached interchange times on the order of seconds, and a coloration efficiency $>130\ cm^2\ C^{-1}$ [107].

The electrochromic performance of the WO_3 oxides can be improved through nanostructuring [108,109]. Nanostructured films are characterized by enhanced coloration efficiencies, high charge capacities and optical contrast, and short switching times and long durability compared with crystalline thin films. Mixed nanostructured WO_3/TiO_2 oxide electrodes constitute a good candidate for smart window applications because of the excellent coloration and stability properties [110]. Enhanced electrochromic properties were found for the electrochromic

$$WO_3 + xe^- + xM^+ \rightarrow M_xWO_3 \left(M = H^+, Li^+, Na^+\right)$$

Scheme 7.5

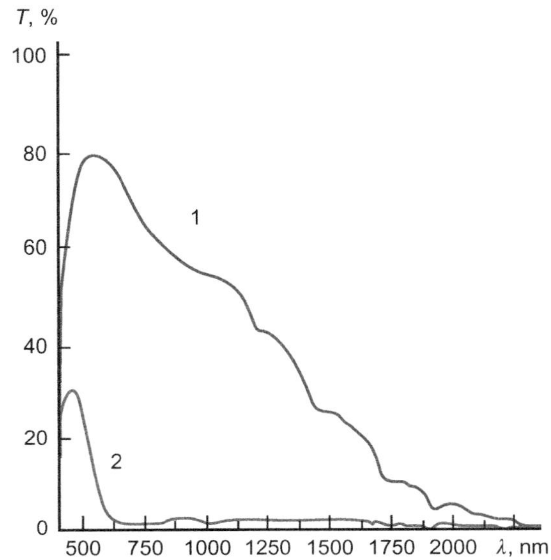

Figure 7.12 Transmission spectra of a full WO_3-based electrochromic device without (1) and with voltage application (2).

film based on WO_3 nanotree-like structures [111]. It is characterized by large active surface area, which provides a large optical modulation of 74.7% at 630 nm at a low potential of −0.2 V, shorter response times of 2.64 s for bleaching and 7.28 s for coloration, and a high coloration efficiency of 75.35 $cm^2\ C^{-1}$.

The films of Ag/WO_3 nanocomposite show enhanced coloration efficiency [112]. These films are characterized by high transmission modulation in the visible region (about 35%) (Fig. 7.13) and fast switching speed (about 6 s).

Besides WO_3, there are other cathodically coloring transition metal oxides, namely TiO_2, Nb_2O_5, and MoO_3 but they show low coloration efficiency and poor durability [113].

There are also anodically coloring metal oxides, notably NiO, IrO_2, and V_2O_5 [113–115].

Electrochromic layers using cathodically (WO_3) and anodically (NiO) coloring metal oxides with acceptable properties were developed [110].

Electrochromic evacuated glazing prototypes with dimensions up to 40 cm × 40 cm were fabricated with the use of WO_3 oxides [116]. These prototypes exhibit excellent optical and thermal performance with an approximately 60% reversible dynamic change of transmittance in the visible range at coloration efficiency up to 92 $cm^2\ C^{-1}$ and mid-pane U-values as low as 0.86 $W\ m^2\ K^{-1}$.

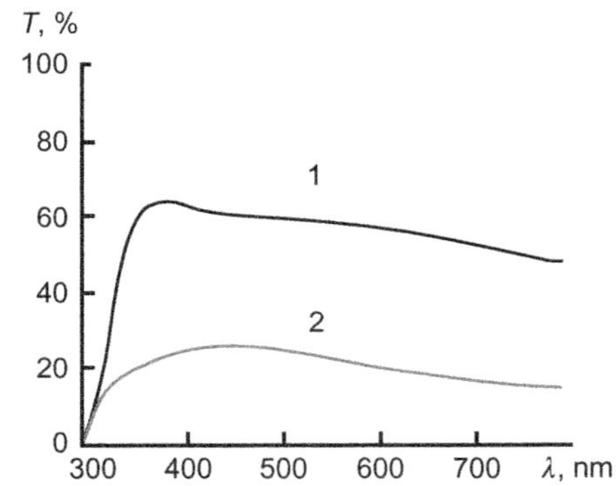

Figure 7.13 Transmission spectra of Ag/WO_3 nanocomposite films for bleached (1) and colored (2) states at 1.0 V switched voltage.

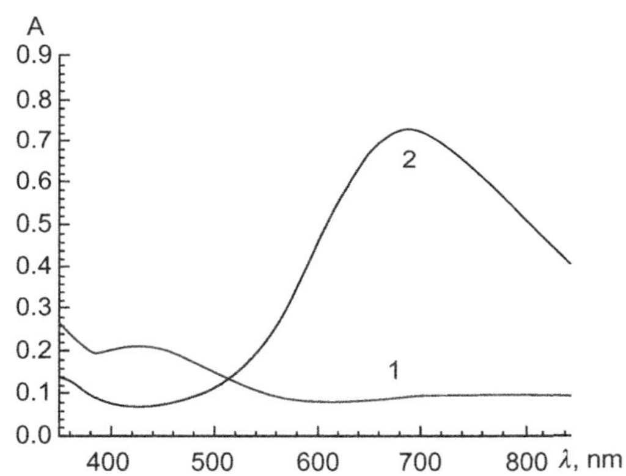

Figure 7.14 Adsorption spectra of the multilayer layer system including Prussian blue nanoparticles and linear poly (ethylene imine) at 1.5 V (1) and 0.6 V (2).

Other widely used electrochromic inorganic materials that manifest an electroinduced intense blue color arising from intervalence charge transfer between mixed iron oxidation states is Prussian blue (iron hexacyanoferrate, $[Fe^{III}Fe^{II}(CN)_6]$) [117]. A multilayer system including Prussian blue nanoparticles and linear poly(ethylene imine) providing electrochromic coloration in the visible region (Fig. 7.14) has been proposed for electrochromic coatings for glazing [118]. A multilayer film based on the poly(aniline) polycation and a negatively ionized Prussian blue nanoparticle dispersion turns green over the potential range from −0.2 to 0.6 V [119]. A window with a size exceeding 100 cm^2 with coloration/decoloration switching with a change in absorbance by more than 80% within approximately 2 min was prepared using a film containing Prussion Blue and anthraquinone-2,6-disulfonate [120].

The use of molybdate hexacyanoferrate with brownish red electroinduced color and Prussian blue in a thin sol–gel film provides a 40% transmittance modulation of solar irradiation in visible wavelengths (300–900 nm) by controlling the potential between −0.5 V and +1.0 V [121].

Aloe vera gel was used as an electrolyte in smart windows based on Prussian blue, due to low cost, ease of availability, and nontoxicity [122].

Besides the metal oxide materials, electrochromism was found for other materials, including organic systems and conjugated polymers.

7.2.3.2 Organic systems

Electrochromic systems based on viologens (4,4-bipyridine compounds) have been well known since long ago [123]. They exhibit a reversible redox transformation between a colorless dication form (V^{2+}) and a deep-blue [123] or green [114,124] radical cation (V^{+-}) form (Scheme 7.6).

The neutral species MV are characterized by yellow-brown color.

These systems continue to improve [125]. Electrochromic devices that change color depending on the voltage (Fig. 7.15) have been developed [126–128].

The possibility of designing electrochromic devices that reversibly show an electroinduced neutral color has been reported [129]. The devices containing benzene viologen, naphthalene viologen, and benzothiadiazole viologen revealed vivid color changes, high optical contrast, fast response time, and intense fluorescence emission [130]. The electrochromic systems with diheptyl viologen displayed various coloration behaviors depending on the concentration of viologens [131]. A series of novel chalcogen viologens (S, Se, and Te) with the redshift of absorption to the visible range were successfully synthesized [132].

Viologen-functionalized silica core–shell nanocomposites displayed reversible redox processes and electrochromic properties [133].

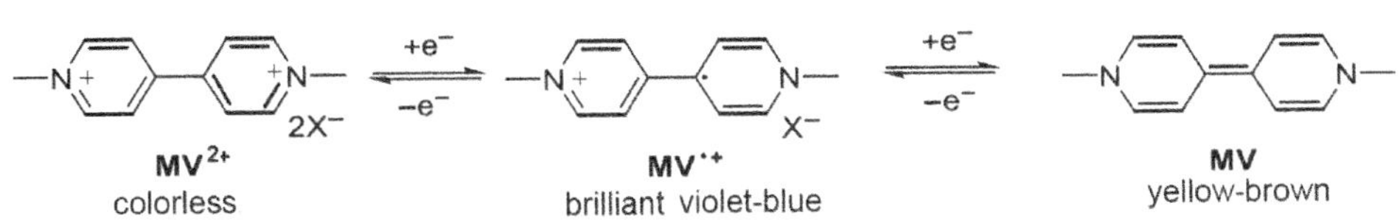

Scheme 7.6

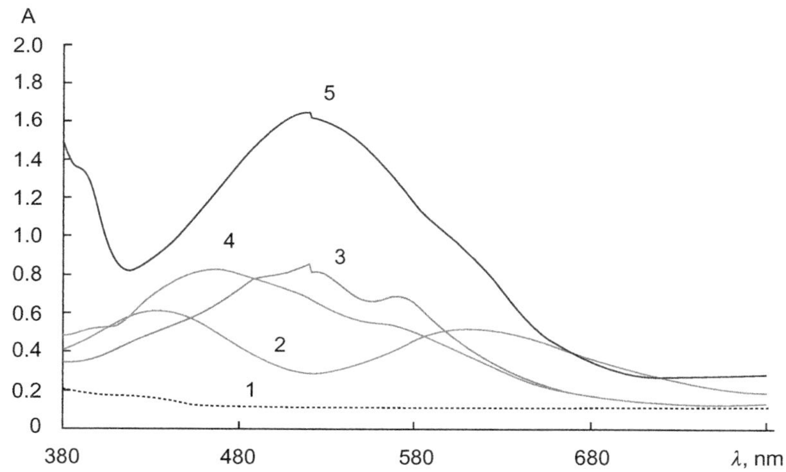

Figure 7.15 Absorption spectra of an electrochromic device containing a gel with viologens at 0 V (1), −0.7 V (2), −0.9 V (3), −1.1 V (4), and −1.7 V (5).

A simplified single-layer solid flexible electrochromic material based on a self-sticking polymer gel and novel hydroxyl viologen derivatives has been prepared by incorporating eletrochromic materials into a gel, resulting in the ITO/EC gel/ITO configuration [134]. A higher optical contrast (up to 82%) and coloration efficiency, together with the absence of any performance degradation provide its applications in smart windows.

Considerable attention is paid to the development of electrochromic devices using triphenylamine-based molecules [135]. The triphenylamine derivative is anchored to a mesoporous TiO_2 support as the electrochrome [136].

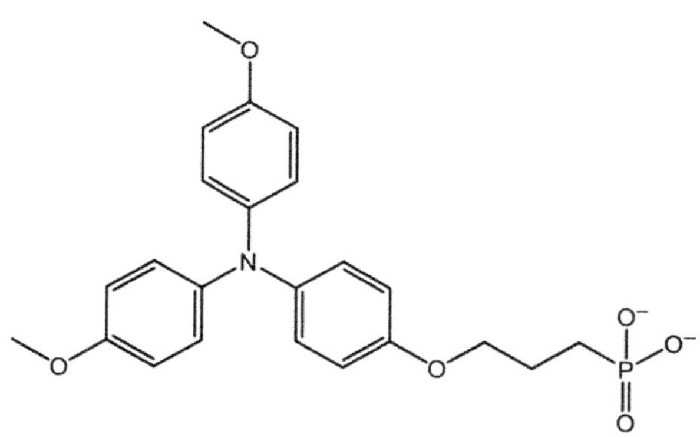

With the use of triphenylamine-based organic dyes, a new energy-harvesting electrochromic window based on the fusion of two technologies, namely organic electrochromic windows and dye-sensitized solar cells, has been developed [137]. Such photoelectrochromic window allows active tuning of the transmittance by varying the applied potential via a photovoltaic cell.

An electrochromic device containing triphenylamine derivatives (as an anodic material) and heptyl viologen (as a catodic material) is characterized by panchromatic spectral properties, namely, the reversible change between a colorless and transparent black states, enhanced coloring contrast, switching time, and long-term stability [138].

7.2.3.3 Polymeric systems

Recent years have witnessed significant advances in the context of the design and synthesis of D–A type conjugated polymers with specific D units containing A units in the backbone or side chain [139] as well as π-conjugated polymers employing dibenzo pentacycles (fluorenes, carbazoles, dibenzothiophenes, and dibenzofuran) as the backbones [140].

A series of polymers exhibiting electrochromic properties was synthesized using viologens, namely, carbazole-substituted viologen conducting polymers [141] and viologen-containing poly(2-isopropyl-2-oxazoline) [142].

Considerable attention is paid to the development of electrochromic materials based on polyaniline (Fig. 7.16) [143].

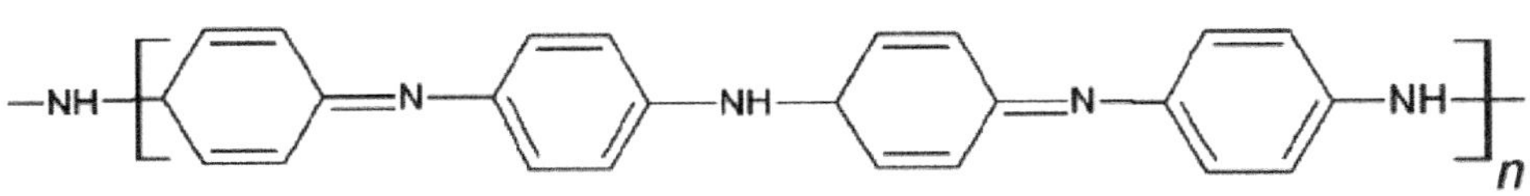

Among the conducting polymers, polyaniline exhibits excellent environmental stability, easy deposition, stability in aqueous solutions, and relatively high level of electrical conductivity [144]. Water-soluble polyaniline covalently bonded to modified acetylferrocene gives a significant enhancement effect on the electrochemical activities [143]. The flexible multicolor polyaniline thin films with modified nanostructure are prepared by combination of galvanostatic and cyclic voltammetric electrodeposition techniques [145].

The development of polyaniline-based electrochromic devices using biodegradable components in electrolytes attracts attention [146,147].

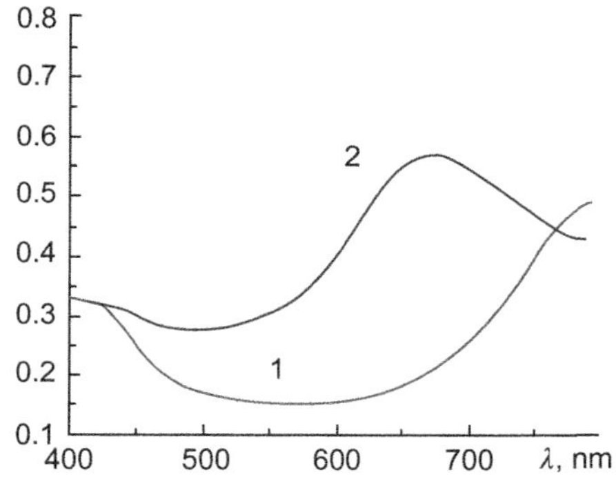

Figure 7.16 Absorption spectra of the electrochromic devices with a polyaniline layer at potentials of −2.0 V (1) and + 2.0 V (2).

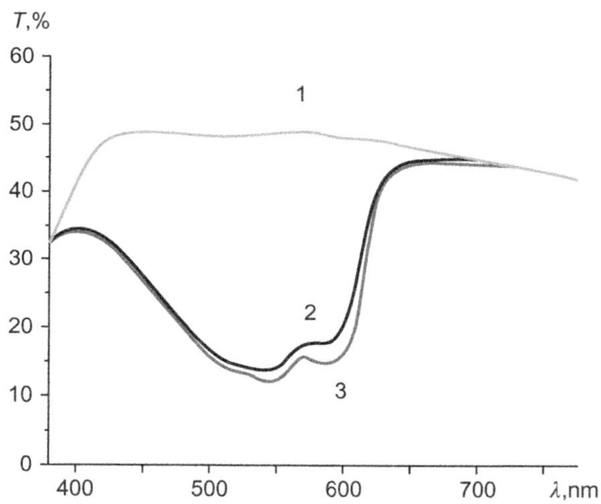

Figure 7.17 Absorption spectra of the electrochromic devices with a poly(3,4-ethylenedioxythiophene): poly (styrene sulfonate) layer at potentials of + 0.8 V (1), 0 V (2), and −0.5 V (3).

Polythiophene and its derivatives constitute a group of sulfur-containing heterocyclic conductive polymers which is characterized by a high conductivity ($\leq$500 S cm^{-1}) [148]. These polymers reversibly change color from clear to magenta (Fig. 7.17) [149].

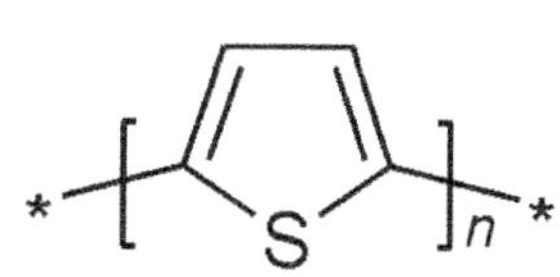

The electrochromic properties depend not only on the electron transfer of polymers but also on the presence of ionic conduction in the electrolyte. In this connection, a new structure including two thiophene monomers linked by poly(ethylene oxide) providing a high-quality film, the better coloration efficiency, and shorter response time has been developed [150].

Polymeric electrochromic films based on a star-shaped thiophene derivative consisting of one central core of phenyl and three arms of bithiophene exhibited excellent electrochromic properties with a multicolor change between orange-yellow, green, and blue colors and the fast switching speed [151].

The use of five vibrant colored-to-colorless dioxythiophene-based polymers gave rise to new electrochromic polymer blends that reversibly switched between highly desirable black and transmission states for achromatic optical transitions [152].

In synchronization with electrochromic processes, the sprayed polythiophene polymer films containing *trans*-stilbene and fumaronitrile groups can repeatedly switch to nonfluorescent state from fluorescent state [153].

Triphenylamine-containing electrochromic polymers with great potential applications in smart windows have experienced an exponential growth of research interests [135]. These include polyimides, polyamides, poly(triphenylamine)s, triphenylamine-based conjugated polymers, other polymers derived from triphenylamine, namely epoxy, polyurethanes, polyazomethines, polybenzoxazines, polysiloxanes, polymethacrylates, and poly (pyridinium salts).

The polymer films based on 4-[bis(4-thiophen-2-yl-phenyl) amino]benzaldehyde showed reversible color changes from yellow to black (Scheme 7.7, Fig. 7.18) [154].

Electrochemically synthesized triarylamine-xanthone-cored conjugated polymers were found to exhibit good stability and good coloration efficiency in different spectral ranges depending on the compound structure [155].

These polymeric materials with promising optical and electrical properties are of great potential for practical applications.

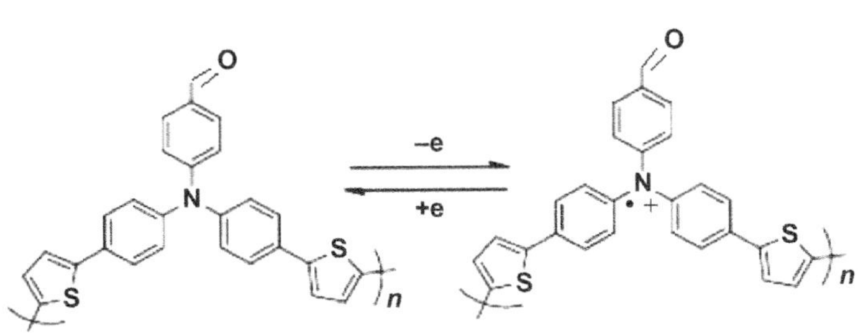

Scheme 7.7

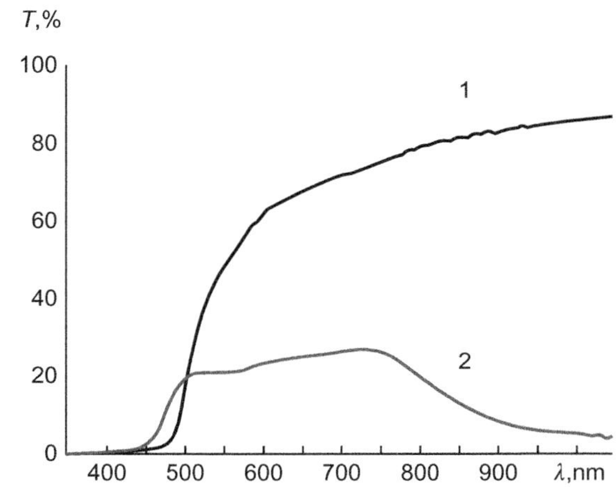

Figure 7.18 Transmission spectra of the polymer films based on 4-[bis(4-thiophen-2-yl-phenyl)amino]benzaldehyde at voltages of 0 V(1) and 2.1 V (2).

Pyrrole-based electrochromic polymers with exceptional optical contrasts and fast response time in the visible and near-infrared regions are very attractive for making smart windows [156,157].

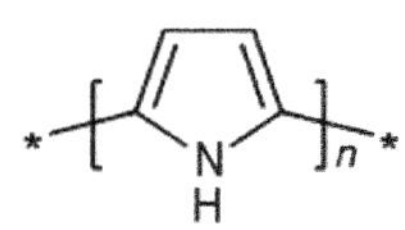

Steady progress in the field of the development of electrochromic polymers has been achieved. A wide range of building blocks exist today, and continuous efforts and known structure–property relationships provide the synthesis of electrochromic materials with required properties [158].

Further upgrading of electrochromic polymers are associated with improved design and component composition of electrochromic layers, developing these polymers with fine color tuning, technology transfer from laboratory conditions to practical use, and development of environmentally friendly processes in synthesis and manufacturing [159].

Other important electrochromic substances are transition metal–ligand complexes manifesting rich redox activities of metal cations and their ligand complexes [105,114], especially polypyridyl complexes [160], cyclometalated ruthenium complex with a redox-active amine substituent and three carboxylic acid groups [161], metal porphyrins [117], and metal phthalocyanines [162]. Metallopolymeric films based on a bis-cyclometalated ruthenium complex bridged by 1,3,6,8-tetra(2-pyridyl)pyrene

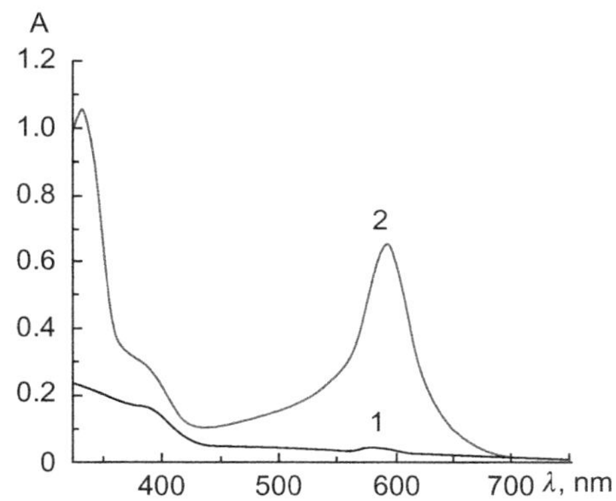

Figure 7.19 Absorption spectra of the the film based on iron (II)-acetate and 1,4-bis (2,2′:6′,2″-terpyridin-4′-yl) benzene in bleached at 1.5 V (1) and colored at 0 V (2) forms.

exhibited electrochromic behavior in the near-IR region (1200–2500 nm) upon switching of the two well-separated $Ru^{II}/^{III}$ processes at low potentials [163]. Unfortunately, most electrochromic metal-organic complexes exhibit electroinduced color change, which prevents their use in sun-protective glasses. In this respect, layer-by-layer self-assembling electrochromic thin films based on iron(II) acetate and 1,4-bis(2,2′:6′,2″-terpyridin-4′-yl) benzene are of interest [164]. They are reversibly colored blue when voltage is applied (Fig. 7.19).

Comparative analysis of the presented results on the development of electrochromic systems of various types based on inorganic and organic substances, as well as polymers exhibiting electrochromism, shows that systems based on WO_3 oxides are most suitable for glazing in their electrochromic and optical properties and the durability under sunlight. Electrochromic polymers that undergo transformations between a colorless state and a state with a neutral color also have good prospects.

7.3 Conclusion

Design of smart sunscreen glazing for various objects is a topical issue, as it not only provides comfortable conditions for life but also reduces energy costs during operation. At the same time, glazing with additional coatings should reduce the UV radiation of the sun to a safe level, ensure effective controlled modulation of visible radiation in accordance with changes in illumination, and significantly reduce the intensity of the IR component of solar radiation. Unlike building glazing, vehicle glazing should provide color reproduction. Reducing the intensity of UV radiation is easily attained by incorporating UV absorbers into sunscreen materials. The effect of IR radiation is decreased by using thin-film 2D metallized IR reflective coatings applied to the glazing or by controlled modulation of the IR absorption of smart materials. The controlled dynamic color change and light transmission in the visible spectral region is possible only with the use of 2D energochromic (photo-, thermo-, and electrochromic) materials. The absence of color distortion is possible only in the case of materials with a neutral color in the activated state.

Thermochromic and electrochromic materials can be used only as parts of special electrical devices, so they are quite expensive. In contrast, multilayer photochromic polymer films, which are characterized by simplicity of design and manufacture, automatic change of light transmission depending on the illumination in the sunlight, are of the greatest interest for widespread use.

The recent advances in photochromic, thermochromic, and electrochromic systems as well as 2D materials described in this short review indicate the possibility of widespread use of energochromic materials in sun-protection devices for various purposes. A number of energochromic materials are already in use. However, further prospects for their application depend on the improvement of their properties, namely, color neutrality, switching rate, bleached/colored state contrast, and durability. This is possible under the condition of further in-depth theoretical and experimental research of materials and improvement of the design of devices based on them.

For sunscreen photochromic organic materials, most important is to increase the service life under sunlight. Thermochromic materials must have transition temperature of 25°C and a switching range as wide as 5°C for maximizing energy savings under all different climates. Successful use of electrochromic materials depends on the creation of efficient electrochromic devices, reducing the cost and energy expenditure.

7.4 Acknowledgments

This work was supported by the Ministry of Science and Higher Education within the State assignment FSRC "Crystallography and Photonics," RAS.

References

[1] N. DeForest, A. Shehabi, J. O'Donnell, G. Garcia, J. Greenblatt, United States energy and CO_2 savings potential from deployment of near-infrared electrochromic window glazing, Build. Environ. 89 (2015) 107–117.

[2] W.H. Armistead, S.D. Stookey, Photochromic silicate glasses sensitized by silver halides, Science. 144 (1964) 150–154.

[3] A.V. Dotsenko, L.V. Glebov, V.A. Tsekhomsky, Physics and Chemistry of Photochromic Glasses, CRC Press, Boca Raton, New York, 1998.

[4] M. Zayat, D. Einot, R. Reisfeld, Reversible photochromism of Sol-Gel SiO_2 : AgCl films, J. Sol-Gel Sci. Technol. 10 (1997) 203–211.

[5] H. Tomonaga, T. Morimoto, Photochromic coatings containing Ag ($Cl_1 - xBrx$) microcrystals, J. Sol-Gel Sci. Technol. 19 (2000) 681–685.

[6] R. Pardo, M. Zayat, D. Levy, Photochromic organic–inorganic hybrid materials, Chem. Soc. Rev. 40 (2011) 672–687.

[7] X. Dong, J. Wang, X. Feng, et al., Fabrication and characterization of nanometer-sized AgCl/PMMA hybrid materials, Modern Appl. Sci. 2 (2008) 49–54.

[8] H. Miyazaki, H. Shimoguchi, H. Suzuki, T. Ota, Synthesis of photochromic AgCl-urethane resin composite films, Adv. Mater. Sci. Eng. 2012 (2012). Article ID 784202.

[9] J.N. Schrauben, R. Hayoun, C.N. Valdez, M. Braten, L. Fridley, J.M. Mayer, Titanium and zinc oxide nanoparticles are proton-coupled electron transfer agents, Science. 336 (2012) 1298–1301.

[10] A.I. Gavrilyuk, Photochromism in WO_3 thin films, Electrochim. Acta. 44 (1999) 3027–3037.

[11] Y. Wang, L. Pan, Y. Li, A.I. Gavrilyuk, Hydrogen photochromism in V_2O_5 layers prepared by the sol-gel technology, Appl. Surf. Sci. 314 (2014) 384–391.

[12] H. Miyazaki, T. Ishigaki, T. Ota, Photochromic smart windows employing WO_3-based composite films, J. Mater. Sci. Res. 6 (2017) 62–66.

[13] H. Miyazaki, Y. Baba, M. Inada, A. Nose, H. Suzuki, T. Ota, Fabrication of photochromic tungsten oxide based composite film using peroxoisopolytungstic acid, Bull. Chem. Soc. Jpn. 84 (2011) 1390–1392.

[14] H. Miyazaki, M. Inada, H. Suzuki, T. Ota, Fabrication of WO_3-based composite films and improvement its photochromic properties by copper doping, Bull. Chem. Soc. Jpn. 85 (2012) 1053–1056.

[15] H. Miyazaki, T. Ishigaki, H. Suzuki, T. Ota, Improvement of the photochromic properties of WO_3-based composite films by phosphorus addition, Bull. Chem. Soc. Jpn. 87 (2014) 838–841.

[16] H. Miyazaki, T. Ishigaki, H. Suzuki, T. Ota, Effect of film thickness and air atmosphere on photochromic properties of WO_3-based composite films, Bull. Chem. Soc. Jpn. 89 (2015) 20–23.

[17] S. Yamazaki, H. Ishida, D. Shimizu, K. Adachi, Photochromic properties of tungsten oxide/methylcellulose composite film containing dispersing agents, ACS Appl. Mater. Interf. 7 (2015) 26326–26332.

[18] K. Adachi, M. Tokushige, K. Ovata, S. Yamazaki, Y. Iwadate, Kinetics of coloration in photochromic tungsten(VI) oxide/silicon oxycarbide/silica hybrid xerogel: an insight into cation self-diffusion mechanisms, ACS Appl. Mater. Interf. 8 (2016) 14019–14028.

[19] M. Hočevar, U.O. Krašovec, A photochromic single glass pane, Sol. Ener. Mater. Sol. Cells. 186 (2018) 111–114.

[20] S. Balaji, Y. Djaoued, A.-S. Albert, R. Bruning, N. Beaudoin, J. Robichaud, Porous orthorhombic tungsten oxide thin films: synthesis, characterization, and application in electrochromic and photochromic devices, J. Mater. Chem. 21 (2011) 3940–3948.
[21] T. Mongstad, C. Platzer-Bjorkman, J.P. Maehlen, L.P. Mooij, Y. Pivak, B. Dam, et al., A new thin film photochromic material: oxygen-containing yttrium hydride, Sol. Ener. Mater. Sol. Cells. 95 (2011) 3596–3599.
[22] C.C. You, D. Moldarev, T. Mongstad, D. Primetzhofer, M. Wolff, E.S. Marstein, et al., Enhanced photochromic response in oxygen-containing yttrium hydride thin films transformed by an oxidation process, Sol. Ener. Mater. Sol. Cells 166 (2017) 185–189.
[23] F. Nafezarefi, H. Schreuders, B. Dam, S. Cornelius, Photochromism of rare-earth metal-oxy-hydrides, Appl. Phys. Lett. 111 (2017). 103903-1–103903-5.
[24] M.-S. Wang, G. Xu, Z.-J. Zhang, G.-C. Guo, Inorganic–organic hybrid photochromic materials, Chem. Commun. 46 (2010) 361–376.
[25] H. Zhang, L. Duan, Y. Lan, E. Wang, C. Hu, Synthesis, crystal structure, and photochromism of novel two-dimensional supramolecular networks based on Keggin-type polyoxoanion and lanthanide coordination cations, Inorg. Chem. 42 (2003) 8053–8058.
[26] Y. Wang, H. Li, C. Wu, Y. Yang, L. Shi, L. Wu, Chiral heteropoly blues and controllable switching of achiral polyoxometalate clusters, Angew. Chem. Int. Ed. 52 (2013) 4577–4581.
[27] K. Hakouk, O. Oms, A. Dolbecq, H.E. Moll, J. Marrot, M. Evain, et al., Sulfonium polyoxometalates: a new class of solid-state photochromic hybrid organic–inorganic materials, Inorg. Chem. 52 (2013) 555–557.
[28] M.J. Kang, E.ю.G. Santoro, Y.S. Kang, Enhanced efficiency of functional smart window with solar wavelength conversion phosphor – photochromic hybrid film, ACS Omega 3 (2018) 9505–9512.
[29] K.M.F. Rossi de Aguiar, E.P. Ferreira Neto, S. Blunk, J.F. Schneider, C.A. Picon, C.M. Lepienski, et al., Hybrid urethanesil coatings for inorganic surfaces produced by isocyanate free and sol-gel routes: synthesis and characterizations, RSC Adv. 6 (2016) 19160–19172.
[30] H. Tian, J. Zhang, Photochromic Materials: Preparation, Properties and Applications, Wiley-VCH Verlag GmbH & Co. KGaA, Weinheim, Germany, 2016.
[31] J.C. Crano, W.S. Kwak, C.N. Welch, Applied Photochromic Polymer Systems, Blackie, Glasgow, 1992.
[32] V. Malatesta, Photodegradation of organic photochromes, in: J.C. Crano, R.J. Guglielmetti (Eds.), Organic Photochromic and Thermochromic Compounds, Kluwer Academic/Plenum Publishers, New York, 1999.
[33] A.A. Dunaev, A.O. Ayt, V.A. Barachevsky, Polyfunctional multilayer photochromic, Nanotechnique 1 (2006) 75–81 (Rus.).
[34] V.A. Barachevsky, Photonics of organic pho-tochromic systems: modern trends, J. Photochem. Photobiol. A 196 (2008) 180–189.
[35] V.A. Barachevsky, Advances in photonics of organic photochromism, J. Photochem Photobiol. A. 354 (2018) 61–69.
[36] H. Frenkel-Mullerad, D. Avnir, The chemical reactivity of sol – gel materials: hydrobromination of ormosils, Chem. Mater. 12 (2000) 3754–3759.
[37] V.R. Kaufman, D. Ltvy, D. Avnir, A photophysical study the sol/gel transition in silica: structural dynamics and oscillations, room-temperature phosphorescence and photochromic gel glasses, J. Non-Cryst. Solids. 82 (1986) 103–109.

[38] D. Levy, S. Einhorn, D. Avnir, Application of the sol-gel process for the preparation of photochromic information-recording materials: synthesis, properties, mechanisms, J. Non-Cryst. Solids. 113 (1989) 137–145.
[39] D. Levy, D. Avnir, Effects of the changes in the properties of silics cage along the gel/xerogel transition on the photochromic behavior of trapped spiropyrans, J. Phys. Chem. 92 (1988) 4734–4738.
[40] B. Schaudel, C. Guermeur, C. Sanchez, K. Nakatanib, J.A. Delaire, Spirooxazine- and spiropyran-doped hybrid organic–inorganic matrices with very fast photochromic responses, J. Mater. Chem. 7 (1997) 61–65.
[41] C.W. Kim, S.W. Oh, Y.H. Kim, H.G. Cha, Y.S. Kang, Characterization spironaphthooxazine doped photochromic glass: the effect of matrix polarity and pore size, J. Phys. Chem. C. 112 (2008) 1140–1145.
[42] J. Biteau, F. Chaput, J.-P. Boilot, Photochromism of spirooxazine-doped gels, J. Phys. Chem. 100 (1996) 9024–9031.
[43] P. Coelho, L.M. Carvalho, L.F.F.F. Goncalves, C.J.R. Silva, A.M.F. Campos, G.M.J.M. Oliveira, Photochromic hybrid sol-gel films containing naphthopyrans, J. Sol-Gel Sci. Technol. 56 (2010) 203–211.
[44] Q. Meng, G. Wang, H. Jiang, Y. Wang, S. Xie, Preparation of a fast photochromic ormosil matrix coating for smart windows, J. Mater. Sci. 40 (2013) 5862–5870.
[45] L.Y.L. Wu, Q. Zhao, H. Huang, R.J. Lim, Sol-gel based photochromic coating for solar responsive smart window, Surf. Coating Technol. 320 (2017) 601–607.
[46] C.G. Granqvist, Chromogenic materials for transmittance control of large-area windows, Solid State Mater. Sci. 16 (1990) 291–308.
[47] C.G. Granqvist, Recent progress in thermochromics and electrochromics: a brief survey, Thin Sol. Films. 614 (2016) 90–96.
[48] G.B. Smith, C.G. Granqvist, Green Nanotechnology: Solutions for Sustainability and Energy in the Built Environment, CRC Press, Boca Raton, FL, 2010.
[49] C.G. Granqvist, S.-Y. Li, I.B. Pehlivan, G.A. Niklasson, Progress in chromogenic materials and devices: new data on electrochromics and thermochromics, Mater. Res. Soc. Symp. Proc. 1492 (2013) 99–110.
[50] F. Xu, X. Cao, H. Luo, P. Jin, Recent advances in VO_2-based thermochromic composites for smart windows, J. Mater. Chem. C. 6 (2018) 1903–1919.
[51] C.G. Granqvist, Transparent conductors as solar energy materials: a panoramic review, Sol. Energy Mater. Sol. Cells. 91 (2007) 1529–1598.
[52] S.-Y. Li, G.A. Niklasson, C.G. Granqvist, Thermochromic fenestration with VO_2-based materials: three challenges and how they can be met, Thin Sol. Films. 520 (2012) 3823–3828.
[53] M.K. Dietrich, F. Kuhl, A. Polity, P.J. Klar, Optimizing thermochromic VO_2 by co-doping with Wand Sr for smart window applications, Appl. Phys. Lett., 110, 2017, p. 141907.
[54] N. Mlyuka, G. Niklasson, C.-G. Granqvist, Mg doping of thermochromic VO_2 films enhancesthe optical transmittance and decreases the metal-insulator transition temperature, Appl. Phys. Lett. 95 (2009) 171909.
[55] M.K. Dietrich, B.G. Kramm, M. Becker, B.K. Meyer, A. Polity, P.J. Klar, Influence of doping with alkaline earth metals on the optical properties of thermochromic VO_2, J. Appl. Phys., 117, 2015, p. 185301.
[56] S.-Y. Li, N.R. Mlyuka, D. Primetzhofer, A. Hallén, G. Possnert, G.A. Niklasson, et al., Bandgap widening in thermochromic Mg-doped VO_2 thin films: quantitative data based on optical absorption, Appl. Phys. Lett. 103 (2013) 161907.

[57] S. Hu, S.-Y. Li, R. Ahuja, C.G. Granqvist, K. Hermansson, G.A. Niklasson, et al., Optical properties of Mg-doped VO_2: absorption measurements and hybrid functional calculations, Appl. Phys. Lett. 101 (2012) 201902.
[58] M. Panagopoulou, E. Gagaoudakis, N. Boukos, E. Aperathitis, G. Kiriakidis, D. Tsoukalas, et al., Thermochromic performance of Mg-doped VO_2 thin films on functional substrates for glazing applications, Sol. Ener. Mater. Sol. Cells. 157 (2016) 1004–1010.
[59] G.-T. Pan, Y.-L. Yang, S. Chong, N. Arjun, T.C.-K. Yang, Y.-C. Lai, The durability study of thermochromic vanadium dioxide films with the addition of barrier coatings, Vacuum. 145 (2017) 158–168.
[60] S. Long, X. Cao, G. Sun, N. Li, T. Chang, Z. Shao, et al., Effects of V_2O_3 buffer layers on sputtered VO_2 smart windows: improved thermochromic properties, tunable width of hysteresis loops and enhanced durability, Appl. Surf. Sci. 441 (2018) 764–772.
[61] M. Jiang, S. Bao, X. Cao, Y. Li, S. Li, H. Zhou, et al., Improved luminous transmittance and diminished yellow color in VO_2 energy efficient smart thin films by Zn doping, Ceram. Int. 40 (2014) 6331–6334.
[62] K.A. Khan, C.G. Granqvist, Thermochromic sputter-deposited vanadium oxyfluoride coatings with low luminous absorbance, Appl. Phys. Lett. 55 (1989) 4–6.
[63] W. Burkhardt, T. Christmann, S. Franke, W. Kriegseis, D. Meister, B.K. Meyer, et al., Tungsten and fluorine co-doping of VO_2 films, Thin Sol. Films. 402 (2002) 226–231.
[64] P. Kiri, M.E.A. Warwick, I. Ridley, R. Binions, Fluorine doped vanadium dioxide thin films for smart windows, Thin Sol. Films. 520 (2011) 1363–1366.
[65] C.G. Granqvist, P.C. Lansaker, N.R. Mlyuka, G.A. Niklasson, E. Avendano, Progress in chromogenics: new results for electrochromic and thermochromic materials and devices, Sol. Energy Mater. Sol. Cells. 93 (2009) 2032–2039.
[66] C. Ji, Z. Wu, L. Lu, X. Wu, J. Wang, X. Liu, et al., High thermochromic performance of Fe/Mg co-doped VO_2 thin films for smart window applications, J. Mater. Chem. C 6 (2018) 6502–6509.
[67] C. Jia, Z. Wu, X. Wu, J. Wang, J. Gou, Z. Huang, et al., Al-doped VO_2 films as smart window coatings: reduced phase transition temperature and improved thermochromic performance, Sol. Ener. Mater. Sol. Cells 176 (2018) 174–180.
[68] D. Kolenaty, J. Houska, J. Vlcek, Improved performance of thermochromic VO_2/SiO_2 coatings prepared by low-temperature pulsed reactive magnetron sputtering: prediction and experimental verification, J. Alloys Comp. 767 (2018) 46–51.
[69] J. Zhang, J. Wang, C. Yang, H. Jiad, X. Cuia, S. Zhao, et al., Mesoporous SiO_2/VO_2 double-layer thermochromic coating with improved visible transmittance for smart window, Sol. Ener. Mater. Sol. Cells. 162 (2017) 134–141.
[70] S.-Y. Li, G.A. Niklasson, C.G. Granqvist, Nanothermochromics: calculations for VO_2 nanoparticles in dielectric hosts show much improved luminous transmittance and solar energy transmittance modulation, J. Appl. Phys. 108 (2010) 063525.
[71] S.-Y. Li, G.A. Niklasson, C.G. Granqvist, Nanothermochromics with VO_2-based coreshell structures: calculated luminous and solar optical properties, J. Appl. Phys. 109 (2011) 113515.

[72] Y. Choi, D.M. Sim, Y.H. Hur, H.J. Han, Y.S. Jung, Synthesis of colloidal VO_2 nanoparticles for thermochromic applications, Sol. Ener. Mater. Sol. Cells. 176 (2018) 266–272.
[73] L. Miao, R. Chen, J. Zhou, C. Liu, Y. Peng, J. Gao, et al., Depressed haze and enhanced solar modulation capability for VO_2-based composite films with distinct size effects, RSC Adv. 6 (2016) 90813–90823.
[74] C. Liu, I. Balin, S. Magdassi, I. Abdulhalim, Y. Long, Vanadium dioxide nanogrid films for high transparency smart architectural window applications, Opt. Express 23 (2015) 124–132.
[75] N. Wang, Y.K. Peh, S. Magdassi, Y. Long, Surface engineering on continuous VO_2 thin films to improve thermochromic properties: top-down acid etching and bottom-up selfpatterning, J. Coll. Interf. Sci. 512 (2018) 529–535.
[76] Y. Ke, X. Wen, D. Zhao, R. Che, Q. Xiong, Y. Long, Controllable fabrication of two-dimensional patterned VO_2 nanoparticle, nanodome, and nanonet arrays with tunable temperature-dependent localized surface plasmon resonance, ACS Nano. 11 (2017) 7542–7551.
[77] S. Wu, S. Tian, B. Liu, H. Tao, X. Zhao, R.G. Palgrave, et al., Facile synthesis of mesoporous VO_2 nanocrystals by a cotton-template method and their enhanced thermochromic properties, Sol. Ener. Mater. Sol. Cells. 176 (2018) 427–434.
[78] I. Top, R. Binions, M.E.A. Warwick, C.W. Dunnill, M. Holdynski, I. Abrahams, VO_2/TiO_2 bilayer films for energy efficient windows with multifunctional properties, J. Mater. Chem. C. 6 (2018) 4485–4493.
[79] N.A. Fleer, K.E. Pelcher, J. Zou, K. Nieto, L.D. Douglas, D.G. Sellers, et al., Hybrid nanocomposite films comprising dispersed VO_2 nanocrystals: a scalable aqueous-phase route to thermochromic fenestration, ACS Appl. Mater. Interf. 9 (2017) 38887–38900.
[80] N. Wang, Q.S. Goh, P.L. Lee, S. Magdassi, Y. Long, One-step hydrothermal synthesis of rare earth/W-codoped VO_2 nanoparticles: reduced phase transition temperature and improved thermochromic properties, J. Alloys Comp. 711 (2017) 222–228.
[81] Y. Kim, S. Yu, J. Park, D. Yoon, A.M. Dayaghi, K.J. Kim, et al., High-throughput roll-to-roll fabrication of flexible thermochromic coatings for smart windows with VO_2 nanoparticles, J. Mater. Chem. C 6 (2018) 3451–3458.
[82] T. Moot, C. Palin, S. Mitran, J.F. Cahoon, R. Lopez, Designing plasmon-enhanced thermochromic films using a vanadium dioxide nanoparticle elastomeric composite, Adv. Opt. Mater. 4 (2016) 578–583.
[83] J.W. Youn, S.-J. Lee, K.-S. Kim, D.U. Kim, Adhesion characteristics of VO_2 ink film sintered by intense pulsed light for smart window, Appl. Surf. Sci. 441 (2018) 508–514.
[84] H. Kim, Y. Kim, K.S. Kim, H.Y. Jeong, A.-R. Jang, S.H. Han, et al., Flexible thermochromic window based on hybridizeVO_2/grapheme, ACS Nano 7 (2013) 5769–5776.
[85] M.M. Seyfouri, R. Binions, Sol-gel approaches to thermochromic vanadium dioxide coating for smart glazing application, Sol. Ener. Mater. Sol. Cells 159 (2017) 52–65.
[86] Y.-S. Yang, Y. Zhou, F.B.Y. Chiang, Y. Long, Tungsten doped VO_2/ microgels hybrid thermochromic material and its smart window application, RSC Adv. 7 (2017) 7758–7762.
[87] Y. Wang, F. Zhao, J. Wang, A.R. Khan, Y. Shi, Z. Chen, et al., VO_2@SiO_2/Poly (*N*-isopropylacrylamide) hybrid nanothermochromic microgels, Ind. Eng. Chem. Res. 57 (2018) 12801–12808.

[88] Y. Ke, C. Zhou, Y. Zhou, S. Wang, S.H. Chan, Y. Long, Emerging thermal-responsive materials and integrated techniques targeting the energy-efficient smart window application, Adv. Funct. Mater. (2018) 1–18. 1800113.
[89] N. Ahn, S.M. Kang, J.-W. Lee, M. Choi, N.-G. Park, Thermodynamic regulation of $CH_3NH_3PbI_3$ crystal growth and its effect on photovoltaic performance of perovskite solar cells, J. Mater. Chem. A 3 (2015) 19901–19906.
[90] M. De Bastiani, M.I. Saidaminov, I. Dursun, L. Sinatra, W. Peng, U. Buttner, et al., Thermochromic perovskite inks for reversible smart window applications, Chem. Mater. 29 (2017) 3367–3370.
[91] A. Halder, D. Choudhury, S. Ghosh, A.S. Subbiah, S.K. Sarkar, Exploring thermochromic behavior of hydrated hybrid perovskites in solar cells, J. Phys. Chem. Lett. 6 (2015) 3180–3184.
[92] J. Lin, M. Lai, L. Dou, C.S. Kley, H. Chen, F. Peng, et al., Thermochromic halide perovskite solar cells, Nat. Mater. 17 (2018) 261–267.
[93] H. Yu, Z. Wei, Y. Hao, Z. Liang, Z. Fu, H. Cai, Reversible solid-state thermochromism of a 2D organic-inorganic hybrid perovskite structure based on iodoplumbate and 2-aminomethyl-pyridine, New J. Chem. 41 (2017) 9586–9589.
[94] A. Seeboth, D. Lötzsch, R. Ruhmann, O. Muehling, Thermochromic polymers—function by design, Chem. Rev. 114 (2014) 3037–3068.
[95] A. Seeboth, R. Ruhmann, O. Mühling, Thermotropic and thermochromic polymer based materials for adaptive solar control, Mater. 3 (2010) 5143–5168.
[96] A. Beck, W. Körner, H. Scheller, J. Fricke, W. Platzer, V. Wittwer, Control of solar insolation via thermochromic light-switching gels, Sol. Energ. Mater. Sol. Cells 36 (1995) 339–347.
[97] H. Watanabe, Intelligent window using a hydrogel layer for energy efficiency, Sol. Ener. Mater. Sol. Cells 54 (1998) 203–211.
[98] A. Szilagyi, T. Gyenes, G. Filipcsei, M. Zrinyi, Thermotropic polymer gels: smart gel glass, Macromol. Symp. 227 (2005) 357–366.
[99] H. Khandelwal, A.P.H.J. Schenning, M.G. Debije, Infrared regulating smart window based on organic materials, Adv. Ener. Mater. 7 (2017) 1602209-1–1602209-18.
[100] H. Kakiuchida, M. Tazawa, K. Yoshimura, A. Ogiwara, Thermal control of transmittance/diffraction states of holographic structures composed of polymer and liquid crystal phases, Sol. Ener. Mater. Sol. Cells 94 (2017) 1747–1752.
[101] J. Sun, H. Wang, L. Wang, H. Cao, H. Xie, X. Luo, et al., Preparation and thermo-optical characteristics of a smart polymer-stabilized liquid crystal thin film based on smectic A–chiral nematic phase transition, Smart Mater. Struct. 23 (2014) 125038-1–125038-10.
[102] R. Kulčar, M. Friškovec, N. Hauptman, A. Vesel, G.M. Klanjšek, Colorimetric properties of reversible thermochromic printing inks, Dyes Pigm. 86 (2010) 271–277.
[103] C.G. Granqvist, Electrochromics for energy efficiency and indoor comfort, Pure Appl. Chem 80 (2008) 2489–2498.
[104] Y. Wang, E.L. Runnerstom, D.J. Milliron, Switchable materials for smart windows, Annu. Rev. Chem. Biomol. Eng. 7 (2016) 11.1–11.22.
[105] R.J. Mortimer, D.R. Rosseinsky, P.M. Monk, Electrochromism and Electrochromic Devices, Wiley-VCH, Weinheim, Germany, 2015.
[106] C.G. Granqvist, Electrochromics for smart windows: oxide-based thin films and devices, Thin Sol. Films. 564 (2014) 1–38.

[107] W. Cheng, J. He, K.E. Dettelbach, N.J.J. Johnson, R.S. Sherbo, C.P. Berlinguette, Photodeposited amorphous oxide films for electrochromic windows, Chem 4 (2018) 1–12.
[108] J.M. Wang, X.W. Sun, Z. Jiao, Application of nanostructures in electrochromic materials and devices: recent progress, Materials 3 (2010) 5029–5053.
[109] C.-P. Li, C. Engtrakul, R.C. Tenent, C.A. Wolden, Scalable synthesis of improved nanocrystalline, mesoporous tungsten oxide films with exceptional electrochromic performance, Sol. Ener. Mater. Sol. Cells 132 (2015) 6–14.
[110] N.N. Dinh, D.H. Ninh, T.T. Thao, T. Vo-Van, Mixed nanostructured Ti-W oxides films for efficient electrochromic windows, J. Nanomater. (2012) 1–7. ID 781236.
[111] Y. Li, W.A. McMaster, H. Wei, D. Chen, R.A. Caruso, Enhanced electrochromic properties of WO_3 nanotree-like structures synthesized via a two-step solvothermal process showing promise for electrochromic window application, ACS Appl. Nano Mater. 1 (2018) 2552–2558.
[112] H. Najafi-Ashtiani, S. Gholipour, A. Rahdar, Surface plasmon resonance effect for a new structure of Ag/WO_3 nanorod-shell nanocomposits and application in smart window, J. Mol. Struct. 1169 (2018) 25–30.
[113] C.G. Granqvist, Oxide electrochromics: an introduction to devices and materials, Sol. Ener. Mater. Sol. Cells 99 (2012) 1–13.
[114] R.J. Mortimer, Electrochromic materials, Annu. Rev. Mater. Res. 41 (2011) 241–268.
[115] H.B. Shim, W.Y. Kim, I.-K. Kang, H.C. Lee, Microstructural effects on nickel oxide film properties in an infrared electrochromic window for shutter-less infrared sensor application, IEEE Sens. J. 17 (2017) 6522–6528.
[116] S. Papaefthimiou, G. Leftheriotis, P. Yianoulis, T.J. Hyde, P.C. Eames, Y. Fang, et al., Development of electrochromic evacuated advanced glazing, Energy Buildings 38 (2006) 1455–1467.
[117] R.J. Mortimer, Organic electrochromic materials, Electrochim. Acta 44 (1999) 2971–2981.
[118] D.M. DeLongchamp, P.T. Hammond, High contrast electrochromism and controllable dissolution of assembled Prussion Blue/polymer nanocomposites, Adv. Funct. Mater. 14 (2004) 224–232.
[119] D.M. DeLongchamp, P.T. Hammond, Multiple-color electrochromism from layer-by-layer-assembled polyaniline/Prussian blue nanocomposite thin films, Chem. Mater. 16 (2004) 4799–4805.
[120] R. Yan, L. Liu, H. Zhao, Y.G. Zhu, C. Jia, M. Han, et al., A TCO-free Prussian blue-based redox-flow electrochromic window, J. Mater. Chem. C 4 (2016) 8997–9002.
[121] H.-Y. Liao, T.-C. Liao, W.-H. Chen, C.-H. Chang, L.-C. Chen, Molybdatehexacyanoferrate (MoOHCF) thin film: a brownish red Prussian blue analog for electrochromic window application, Sol. Ener. Mater. Sol. Cells 145 (2016) 8–15.
[122] O. Nanda, N. Gupta, R. Grover, K. Saxena, Self-powered electrochromic window using green electrolyte, AIP Adv. 8 (2018). 095117-1–095117-6.
[123] C. Bird, A. Kuhn, Electrochemistry of the viologens, Chem. Soc. Rev. 10 (1981) 49–82.
[124] H. Oh, D.G. Seo, T.Y. Yun, S.B. Lee, H.C. Moon, Novel viologen derivatives for electrochromic ion gels showing a green colored state with improved stability, Org. Electr. 51 (2017) 490–495.
[125] L. Striepe, T. Baumgartner, Viologens and their application as functional materials, Chem. Eur. J 23 (2017) 16924–16940.

[126] Y. Alesanco, A. Viñuales, J. Palenzuela, I. Odriozola, G. Cabeñero, J. Rodriguez, et al., Multicolor electrochromics: rainbow-like devices, ACS Appl. Mater. Interf 8 (2016) 14795–14801.
[127] H.-C. Lu, M.-H. Lee, S.-Y. Kao, K.-C. Ho, T.-H. Chang, Multicolor electrochromic devices based on phenyl and heptyl viologens immobilized with UV-cured polymer electrolyte, Sol. Ener. Mater. Sol. Cells 177 (2018) 75–81.
[128] A. Chaudhary, D.K. Pathak, S. Mishra, P. Yogi, P.R. Sagdeo, R. Kumar, Polythiophene-viologen bilayer for electro-trichromic device, Sol. Ener. Mater. Sol. Cells 188 (2018) 249–254.
[129] Y. Alesanco, A.I. Vinuales, G. Cabañero, J. Rodriguez, R. Tena-Zaera, Colorless to neutral-color electrochromic devices based on asymmetric viologens, ACS Appl. Mater. Interf 8 (2016) 29619–29627.
[130] Y. Shi, J. Liu, M. Li, J. Zheng, C. Xu, Novel electrochromic- uorescent bi-functional devices based on aromatic viologen derivates, Electrochim. Acta 285 (2018) 415–423.
[131] T.Y. Yun, H.C. Moon, Highly stable ion gel-based electrochromic devices: effects of molecular structure and concentration of electrochromic chromophores, Org. Electr. 56 (2018) 178–185.
[132] G. Li, L. Xu, W. Zhang, K. Zhou, Y. Ding, F. Liu, et al., Narrow-bandgap chalcogenoviologens for electrochromism and visible-light-driven hydrogen evolution, Angew. Chem. Int. Ed. 57 (2018) 4897–4901.
[133] J. Wang, J.-F. Wang, M. Chen, D.-J. Qian, M. Liu, Fabrication, characterization, electrochemistry, and redox-induced electrochromism of viologen-functionalized silica core-shell nano-composites, Electrochim. Acta 251 (2017) 562–572.
[134] M. Pan, Y. Ke, L. Ma, S. Zhao, N. Wu, D. Xiao, Single-layer electrochromic device based on hydroxyalkyl viologens with large contrast and high coloration efficiency, Electrochim. Acta 266 (2018) 395–403.
[135] H. Yen, G. Liou, Recent Advances in triphenylamine-based electrochromic derivatives and polymers, Polym. Chem. 9 (2018) 3001–30018.
[136] J.R. Jennings, W.Y. Lim, S.M. Zakeeruddin, M. Gratzel, Q. Wang, A flow electrochromic window, ACS Appl. Mater. Interf. 7 (2015) 2827–2832.
[137] E. Amasawa, N. Sasagawa, M. Kimura, M. Taya, Design of a new energy-harvesting electrochromic window based on an organic polymeric dye, a cobalt couple, and PProDOT-Me_2, Adv. Energy Mater. (2014). 1400379-1–1400379-9.
[138] J.-T. Wu, G.-S. Liou, Novel panchromatic shutter based on ambipolar electrochromic system without supporting electrolyte, Chem. Commun. 54 (2018) 2619–2622.
[139] X. Lv, W. Li, M. Ouyang, Y. Zhang, D.S. Wright, C. Zhang, Polymeric electrochromic materials with donor–acceptor structures, J. Mater. Chem. C 5 (2017) 12–28.
[140] K. Lin, S. Chen, B. Lu, J. Xu, Hybrid π-conjugated polymers from dibenzo pentacyclic centers: precursor design, electrosynthesis and electrochromics, Sci. China Chem. 60 (2017) 38–53.
[141] R. Kirankumar, W.-C. Huang, H.-F. Chen, P.-Y. Chen, Electropolymerization and characterization of carbazole substituted viologen conducting polymers: the effects of electrolytes and potential applications of the polymers, J. Electroanal. Chem. 826 (2018) 198–206.
[142] J. Nam, Y. Jung, J.H. Joe, W.-D. Jang, Dual stimuli-responsive viologen-containing poly(2-isopropyl-2-oxazoline) and its multi-modal electrochromic phase transition, Polym. Chem. 9 (2018) 3662–3666.

[143] S. Xiong, R. Wang, S. Li, B. Wu, J. Chu, X. Wang, et al., Electrochromic behaviors of water-soluble polyaniline with covalently bonded acetyl ferrocene, J. Electronic Mater 47 (2018) 3974–3982.
[144] Y. Li, Z. Li, F. Zheng, Polyaniline/silver/cerium nitrate ternary composite: synthesis, characterization and enhanced electrochemical properties, J. Appl. Polym. Sci. 132 (2015) 42785-1–42785-7.
[145] K. Zhou, H. Wang, J.T. Jiu, J.B. Liu, H. Yan, K. Suganuma, Polyaniline films with modified nanostructure for bifunctional flexible multicolor electrochromic and supercapacitor applications, Chem. Eng. J. 345 (2018) 290–299.
[146] A. Danine, L. Manceriu, A. Fargues, A. Rougier, Eco-friendly redox mediator gelatin-electrolyte for simplified TiO_2-viologen based electrochromic devices, Electrochim. Acta 258 (2017) 200–207.
[147] M. Pan, S. Zhao, L. Ma, N. Wu, D. Xiao, All-in-one electrochromic devices with biological tissues used as electronic components, Sol. Energ. Mater. Sol. Cells 189 (2019) 27–32.
[148] L. Nyholm, G. Nyström, A. Mihranyan, M. Strømme, Toward flexible polymer and paper-based energy storage devices, Adv. Mater. 23 (2011) 3751–3769.
[149] A.W. Lang, Y. Li, M. De Keersmaecker, D.E. Shen, A.M. Osterholm, L. Berglund, et al., Transparent wood smart windows: polymer electrochromic devices based on poly(3,4-ethylenedioxythiophene): poly (styrene sulfonate) electrodes, ChemSusChem. 11 (2018) 854–863.
[150] X. Liu, Y. Hu, L. Shen, G. Zhang, T. Cao, J. Xu, et al., Novel copolymers based on PEO bridged thiophenes and 3,4-ethylenedioxythiophene: electrochemical, optical, and electrochromic properties, Electrochim. Acta 288 (2018) 52–60.
[151] Y. Dong, F. Luo, L. Chen, F. Yuan, Y. Hou, W. Li, et al., Multi-color electrochromism containing green color based on electrochemically polymerized star-shaped phenyl bithiophene, Phys. Chem. Chem. Phys. 20 (2018) 12923–12928.
[152] L.R. Savagian, A.M. Österholm, D.E. Shen, D.T. Christiansen, M. Kuepfert, J.R. Reynolds, Conjugated polymer blends for high contrast black-to-transmissive electrochromism, Adv. Opt. Mater. (2018) 1800594-1–1800594-9.
[153] J. Wu, Y. Han, J. Liu, Y. Shi, J. Zheng, C. Xu, Electrofluorochromic and electrochromic bifunctional polymers with dual-state emission via introducing multiple C—H $\cdots$ π bonds, Org. Electr. 62 (2018) 481–490.
[154] J. Zeng, Z. Wan, H. Lia, P. Liua, W. Deng, Visible and near-infrared electrochromic properties of polymers based on triphenylamine derivatives with acceptor groups, Sol. Ener. Mater. Sol. Cells 178 (2018) 223–233.
[155] H.F. Higginbotham, M. Czichy, B.K. Sharma, A.M. Shaikh, R.M. Kamble, P. Data, Electrochemically synthesized xanthone-cored conjugated polymers as materials for electrochromic windows, Electrochim. Acta 273 (2018) 264–272.
[156] X. Chen, W. Qiao, Z.Y. Wang, Visible and near-infrared electrochromic thiophene–diketopyrrolopyrrole polymers, RSC Adv. 7 (2017) 15521–15526.
[157] W.T. Neo, Q. Ye, S.-J. Chua, J. Xu, Conjugated polymer-based electrochromics: materials, device fabrication and application prospects, J. Mater. Chem. C 4 (2016) 7364–7376.
[158] J.Y. Lee, S.-Y. Han, I. Cho, B. Lim, Y.-C. Nah, Electrochemical and electrochromic properties of diketopyrrolopyrrole-based conjugated polymer, Electrochem. Commun. 83 (2017) 102–105.

[159] T. Abidin, Q. Zhang, K.-L. Wang, D.-J. Liaw, Recent advances in electrochromic polymers, Polymer. 55 (2014) 5293–5304.
[160] S. Shankar, M. Lahav, M.E. van der Boom, Coordination-based molecular assemblies as electrochromic materials: ultra-high switching stability and coloration efficiencies, J. Am. Chem. Soc. 137 (2015) 4050–4053.
[161] B.-B. Cui, Y.-W. Zhong, J. Yao, Three-state near-infrared electrochromism at the molecular scale, J. Am. Chem. Soc. 137 (2015) 4058–4061.
[162] E.B. Orman, Z. Odabas, A.R. Ozkaya, High electrochemical versatility and applicability with metal phthalocyanines carrying peripheral 2,3-dihydro-1H-inden-5-yloxy substituents: rich redox behavior, oxygen electrocatalysis and electrochromism, J. Electrochem. Soc. 165 (2018) H530–H548.
[163] C.-J. Yao, J. Yao, Y.-W. Zhong, Metallopolymeric films based on a biscyclometalated ruthenium complex bridged by 1,3,6,8-tetra(2-pyridyl) pyrene: applications in near-infrared electrochromic windows, Inorg. Chem. 51 (2012) 6259–6263.
[164] M. Schott, M. Beck, F. Winkler, H. Lorrmann, D.G. Kurth, Fabricating electrochromic thin films based on metallo-polymers using layer-by-layer self-assembly: an attractive laboratory experiment, J. Chem. Educ. 92 (2014) 364–367.

8

2D thermoelectrics

Yu-Qiao Zhang[1,2] and Hiromichi Ohta[1,2]
[1]*Graduate School of Information Science and Technology, Hokkaido University, Kita, Sapporo, Japan* [2]*Research Institute for Electronic Science, Hokkaido University, Kita, Sapporo, Japan*

8.1 Introduction

8.1.1 Thermoelectrics

Thermoelectric technology is usually referred to the direct conversion from temperature difference into electricity based on Seebeck effect, which is promising to address the recent energy problems arising from waste heat discharges [1]. In 1821 T. J. Seebeck discovered that a voltage is generated between two ends of a metal bar by introducing a temperature difference in the bar. Thus when electric loads are connected at both ends of the metal bar, electric current can be obtained. When a temperature difference (ΔT) is introduced to both ends of a thermoelectric material, the carriers will diffuse from the hot side to the cold side due to the difference of the chemical potential (ΔV) of both ends when the outside current circuit is closed. The ΔV is proportional to ΔT, where the ratio is called thermopower ($\equiv$Seebeck coefficient, S).

However, in 1834, J.C.A. Peltier discovered that heating or cooling of the junctions occurs during electric current application to the heterogeneous metal circuit. This phenomenon is so called Peltier effect, which can convert electricity into temperature difference and has been commercially applied such as in electronic refrigerators.

Seebeck effect, which realizes the power generation from waste heat, is the main focus in thermoelectric field. Generally, the maximum power generation efficiency ($\eta_{\max}$) of thermoelectric materials could be calculated by:

$$\eta_{\max} = \frac{T_h - T_c}{T_h} \cdot \frac{\sqrt{1+\mathrm{ZT}}-1}{\sqrt{1+\mathrm{ZT}}+T_c/T_h}$$

2D Nanomaterials for Energy Applications. DOI: https://doi.org/10.1016/B978-0-12-816723-6.00008-3

where ZT is a dimensionless figure of merit, T_h is the temperature of hot side, and T_c is the temperature of cold side. Therefore ZT indicates the overall performance of a thermoelectric material:

$$ZT = \frac{S^2 \cdot \sigma \cdot T}{\kappa}$$

where Z is a figure of merit, T is the absolute temperature, S is the thermopower (≡ Seebeck coefficient), σ is electrical conductivity, and κ is thermal conductivity. Thus a good thermoelectric material should have large S, which is required to obtain high voltage, high σ, which is required to reduce internal resistance of the material, and low κ, which is required to introduce large temperature difference into both ends of the material [2–5].

8.1.2 Benefit of 2D thermoelectrics

Basically, the S of a thermoelectric material can be expressed by the following Mott equation [6]:

$$S = \frac{\pi^2}{3}\frac{k_B{}^2 T}{e}\left\{\frac{d[\ln(\sigma(E))]}{dE}\right\}_{E=E_F} = \frac{\pi^2}{3}\frac{k_B{}^2 T}{e}\left\{\frac{1}{n}\cdot\frac{dn(E)}{dE} + \frac{1}{\mu}\cdot\frac{d\mu(E)}{dE}\right\}_{E=E_F}$$

where k_B, e, n, and μ are the Boltzmann constant, electron charge, carrier concentration, and carrier mobility, respectively. Thus S strongly depends on the energy derivative of the electronic density of states (DOS) at around the Fermi energy (E_F)

$$\left[\frac{\partial \mathrm{DOS}(E)}{\partial E}\right]_{E=E_F}$$

In case of bulk system, the slope of log n versus S relation should be $-\ln 10 \cdot k_B \cdot e^{-1}$ ($\equiv -198\ \mu\mathrm{V\ K^{-1}\ decade^{-1}}$) since parabolic shaped $E-k$ relation at around the conduction band minimum is generally observed. Simply say, there is a trade-off relationship between $|S|$ and n. Therefore the maximum power factor ($\mathrm{PF} = S^2 \cdot \sigma$) is determined with the DOS of the material.

In order to further enhance the PF, the DOS must be enlarged without reducing n and μ. In optoelectronic devices, two-dimensional (2D) structure usually shows exotic electron transport properties compared with bulk counterparts, due to the modified DOS near the bottom of the conduction band or top of valence band with decreasing quantum well thickness. In 1993 Hicks and Dresselhaus [7] theoretically predicted a similar 2D enhancement effect in thermoelectric materials that two-dimensional thermoelectric figure of merit, $Z_{2D}T$ of quantum well can dramatically be enhanced by using superlattices as the quantum well thickness is narrower than the de Broglie wavelength (λ_D):

$$\lambda_D = \sqrt{\frac{2\pi \cdot \hbar^2}{k_B \cdot T \cdot m^*}}$$

where $\hbar$, m^*, and k_B are reduced Planck's constant, effective mass of conductive carriers, and Boltzmann constant, respectively. The increased DOS only enhances S without sacrificing electrical conductivity, resulting in the rise in ZT.

This theory has been proved experimentally by Dresselhaus et al. [8] in 1996 as shown in Fig. 8.1. They fabricated superlattice structure of PbTe (well)/$Pb_{0.927}Eu_{0.073}Te$ (barrier) on (111) BaF_2 substrate using 200-nm-thick $Pb_{0.958}Eu_{0.042}Te$ buffer layer (Fig. 8.1A). The carrier effective mass (m^*) of PbTe is 0.2 m_0 and the de Broglie wavelength is 14 nm. The experimentally observed $S^2 \cdot n$ values are well reproduced with the theoretical results (Fig. 8.1B and C).

In order to further prove the 2D thermoelectric theory, in 2007, Ohta et al. started to verify the theory by using the artificial superlattices [9–14] and field-effect transistors [15–21]. This chapter reviews 2D thermopower based on the authors' efforts focusing on superlattices (multiple quantum wells) and filed effect transistors.

8.2 Thermopower of 2D superlattices

In order to verify the 2D thermoelectric theory, Ohta et al. selected electron-doped $SrTiO_3$ as the model system. Electron-doped $SrTiO_3$ has been considered to be a good candidate of a thermoelectric material because it shows rather large PF [22–28]. As compared with the heavy metal based thermoelectric materials such as Bi_2Te_3 and PbTe, $SrTiO_3$ has several good points including rich elemental resources, nontoxicity, and high thermal- and chemical stability. In order to verify the 2D thermoelectric theory, the most important thing is that the atomic layer superlattice film growth is possible. $SrTiO_3$ and its related oxide are well known as the material which atomic layer superlattice can be grown by pulsed laser deposition technique on perovskite oxide single crystal substrate [29,30].

In 2007 Ohta et al. fabricated $SrTi_{0.8}Nb_{0.2}O_3/SrTiO_3$ superlattices and observed an enhancement of S, which is factor of 5 larger than that of bulk [9]. As schematically shown in Fig. 8.2A,

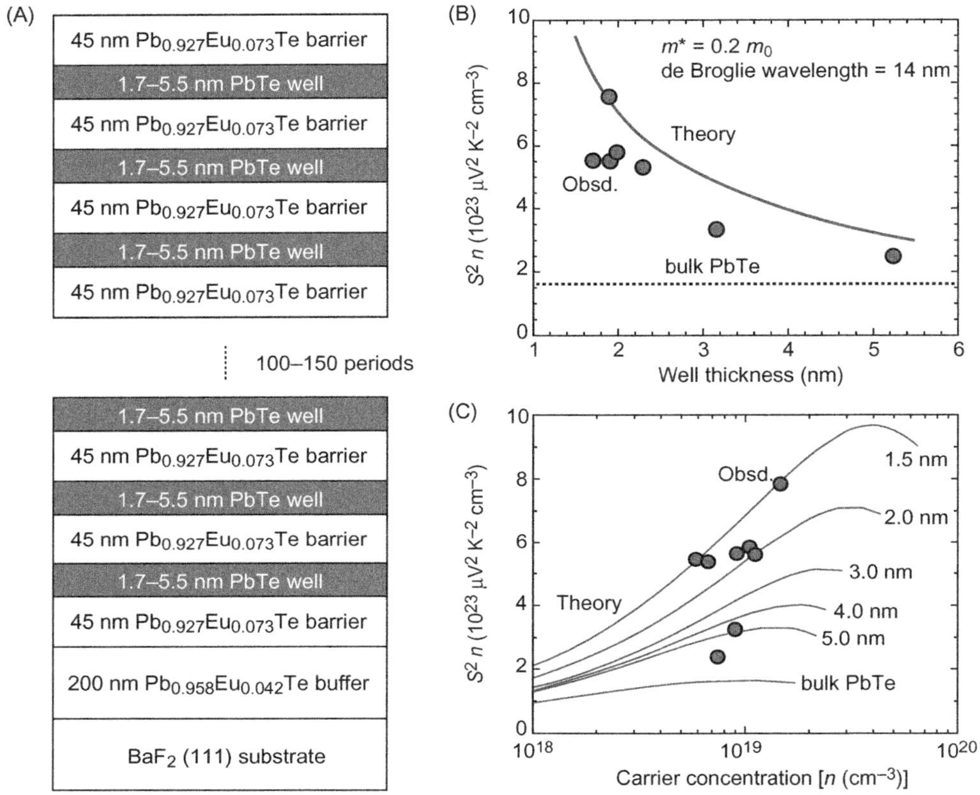

Figure 8.1 Thermopower enhancement of PbTe-based quantum well. In 1993 Hicks and Dresselhaus theoretically predicted that the $S^2 \cdot n$ dramatically increases with decreasing the well thickness when the well thickness is thinner than the de Broglie wavelength. They experimentally proved the theory in 1996. (A) Schematic superlattice structure of PbTe (well)/$Pb_{0.927}Eu_{0.073}Te$ (barrier) grown on (111) BaF_2 substrate using 200-nm-thick $Pb_{0.958}Eu_{0.042}Te$ buffer layer. (B) $S^2 \cdot n$ plots of the PbTe-based superlattice as a function of well thickness at 300 K. (C) $S^2 \cdot n$ plots of the superlattice as a function of carrier concentration at 300 K. Solid lines indicate the calculation results of each well thickness. The carrier effective mass (m^*) of PbTe is 0.2 m_0 and the de Broglie wavelength is 14 nm. *Source*: Reproduced with permission from L.D. Hicks, T.C. Harman, X. Sun, M.S. Dresselhaus, Experimental study of the effect of quantum-well structures on the thermoelectric figure of merit, *Phys. Rev. B* 53 (1996) 10493–10496. Copyright 2016, American Physical Society.

two-dimensional electron gas (2DEG) was confined in quantum well (QW) structures with well thickness ranging from 1 to 16 unit cells. With decreasing the QW thickness under 4 unit cells, enhancement in $|S|$ was observed. For 1 unit cell superlattice, enhancement factor (S_{2DEG}/S_{Bulk}) reached to ~5 (Fig. 8.2B). This large enhancement of S was attributed to a dimensional crossover of the polaron from 3D to quasi 2D [12]. Similar enhancement of S was observed in La-doped $SrTiO_3$-based

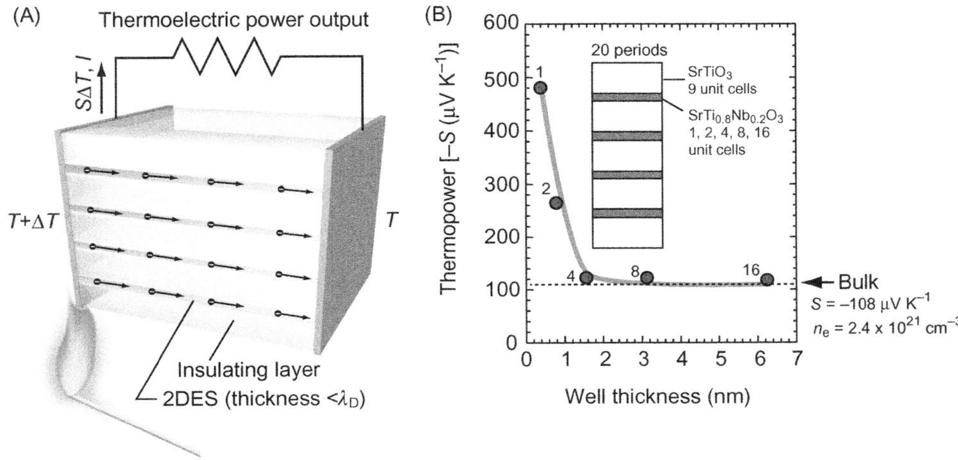

Figure 8.2 Thermopower enhancement of $SrTiO_3$-based 2DEG confined in the superlattices. In 2007 Ohta et al. experimentally clarified that the thermopower enhancement occurs in $SrTiO_3$-based 2DEG when the conducting layer thickness is thinner than 2 nm. (A) Schematic illustration of the superlattice structure. When the thickness of 2DEG is thinner than the de Broglie wavelength, the thermopower can be enhanced. (B) Experimental result. The thermopower dramatically increases when the well layer (Nb-substituted $SrTiO_3$) thickness is thinner than 4-unit cell thickness. *Source*: Reproduced with permission from H. Ohta, et al., Giant thermoelectric Seebeck coefficient of two-dimensional electron gas in SrTiO3, *Nat. Mater.* 6 (2007) 129–134. Copyright 2007, Springer Nature.

superlattices [13]. Thus enhancement of S should occur in the case of electron-doped $SrTiO_3$ superlattice, however, in order to confine the 2DEG very thick insulating $SrTiO_3$ barrier layers are indispensable [10]. When insulating layers are taken into calculations, ZT_{300K} is only $\sim$0.24 [9].

In 2016 Hung et al. [31] reported that the enhancement factor of S_{2DEG}/S_{Bulk} can be improved when the 2DEG showing longer de Broglie wavelength is used. Since the quantum size effect starts when the quantum well thickness is thinner than the de Broglie wavelength, this assumption would be reasonable. Zhang et al. examined the change in the de Broglie wavelength of $SrTiO_3$–$SrNbO_3$ solid solution system and found that the de Broglie wavelength increased from $\sim$4.1 nm to $\sim$5.3 nm with increasing Nb concentration [32]. As shown in Fig. 8.3A, the carrier effective mass (m^*) for compositions with $x < 0.3$ was calculated to be $\sim$1.1 m_0, which further decreased to $\sim$0.7 m_0 as $x > 0.3$. A step-like decease was detected at $x \sim 0.3$. The author owed the step-like decrease behavior in m^* at $x \sim 0.3$ to the conduction band transition from Ti 3d to Nb 4d. Thus the λ_D in the region A is $\sim$4.1 nm, whereas that in the region B is $\sim$5.3 nm.

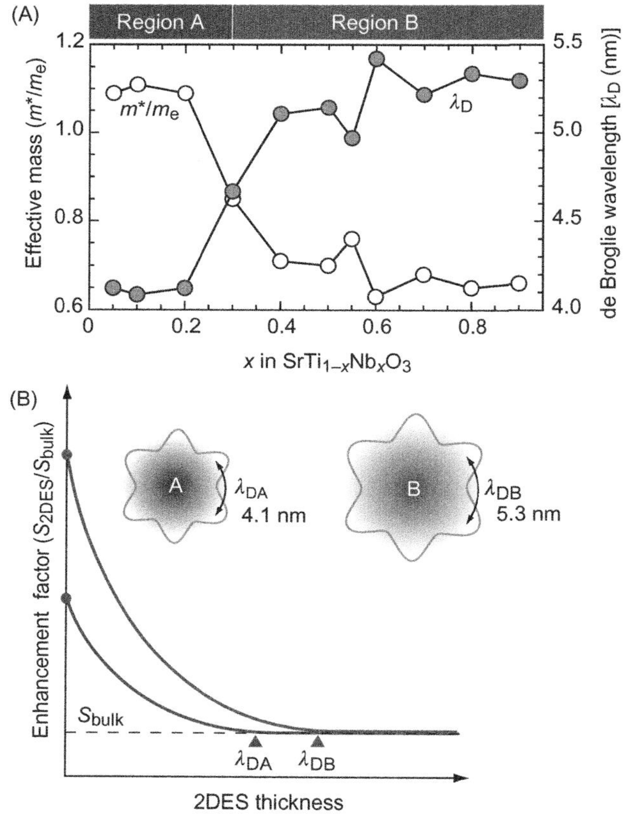

Figure 8.3 Evolution of the de Broglie wavelength of $SrTiO_3$–$SrNbO_3$ solid solution system. (A) Change in the effective mass (m^*) and the de Broglie wavelength (λ_D) as a function of x in $SrTi_{1-x}Nb_xO_3$. In the region A, the m^* is ~1.1 m_0 whereas it is ~0.6 m_0 in the region B. The λ_D in the region A is ~4.1 nm, whereas that in the region B is ~5.3 nm. (B) Hypothesis: The enhancement factor of S_{2DEG}/S_{Bulk} would be dramatically improved when well layer has longer de Broglie wavelength. This theory has been published by Hung et al. [31]. *Source*: Reproduced with permission from Y.Q. Zhang, B. Feng, H. Hayashi, C. P. Chang, Y.M. Sheu, I. Tanaka, et al., Double thermoelectric power factor of a 2D electron system, Nat. Commun. 9 (2018) 2224. Copyright 2018, Springer Nature.

In 2018 Zhang et al. hypothesized that the enhancement factor of S_{2DEG}/S_{Bulk} would be dramatically improved when the region B is confined in a unit cell layer [14]. Zhang et al. fabricated [N uc $SrTi_{1-x}Nb_xO_3$|11 uc $SrTiO_3$]$_{10}$ superlattices ($1 \leq N \leq 12$, $x = 0.2$–0.9) on (001) $LaAlO_3$ single crystal substrates and found that the region B based superlattices showed the enhancement factor of S_{2DEG}/S_{Bulk} ~10 (Fig. 8.4). For 2DEG with $x = 0.2$ and 0.3, the highest S_{2DEG}/S_{Bulk} values are around 4 and 5, which are similar with the previous report [9], respectively, whereas that for the $x = 0.8$ counterpart is ~10 (Fig. 8.4B). As hypothesized, the enhanced S_{2DEG}/S_{Bulk} should stem from the increasing λ_D with x. In this experiment, S_{2DEG}/S_{Bulk} for the $x = 0.2$ and 0.3 2DEGs are saturated around 11 uc, which is consistent with λ_D in Region A (~4.2 nm indicated by dashed line λ_{DA}). As λ_D increases in Region B, the saturation position for the $x = 0.8$ 2DEG has a thickness larger than the λ_D (~5.2 nm indicated by dashed line λ_{DB}).

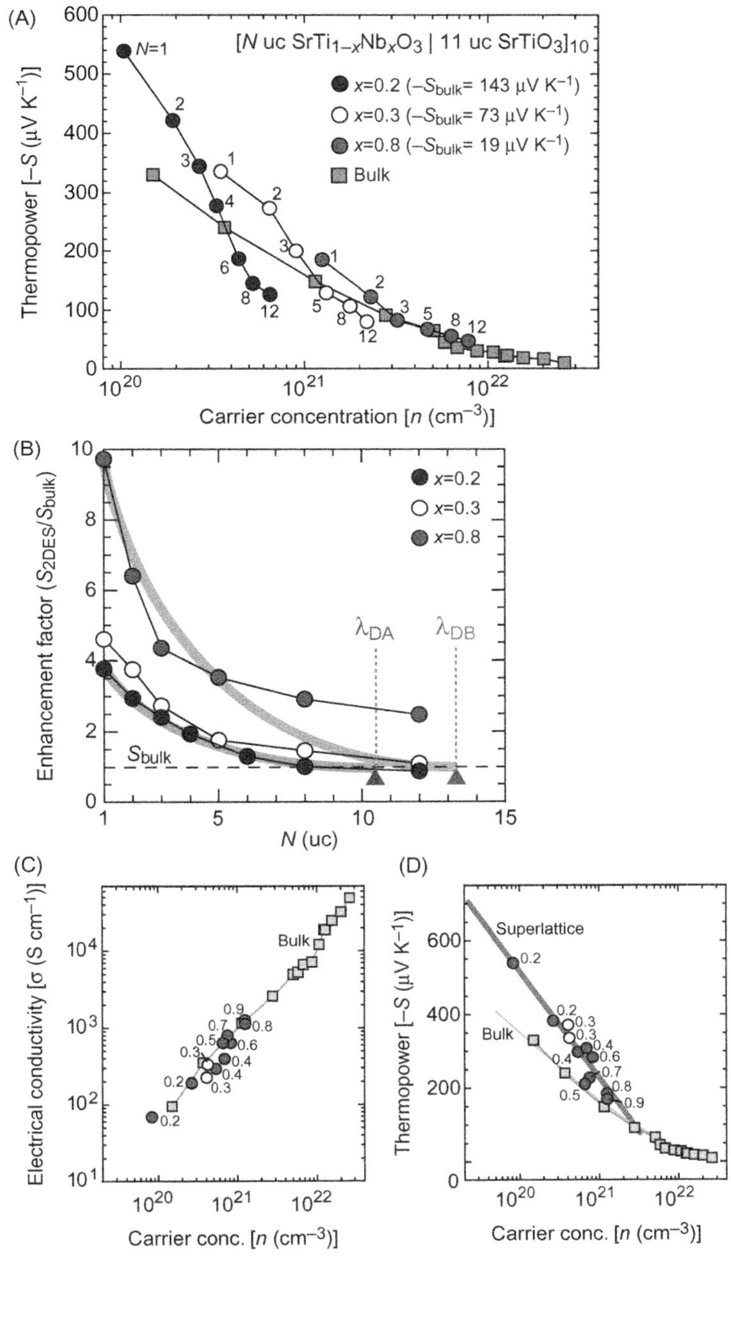

Figure 8.4 Two-dimensionality of 2DEG: a key to enhance thermopower. (A) Plots of thermopower of the 2DEGs, [*N* uc $SrTi_{1-x}Nb_xO_3$|11 uc $SrTiO_3]_{10}$ superlattices ($x=0.2$, 0.3, and 0.8), versus the carrier concentration (*n*). Compared to bulk values (gray squares), all the 2DEGs show an enhanced $-S$ as *N* is reduced under 3 uc. (B) Enhancement factors in $-S$ (S_{2DEG}/S_{Bulk}) for three sets of 2DESs. For $x=0.2$ and 0.3 2DEGs, the highest S_{2DEG}/S_{Bulk} values are obtained at $N=1$, which are 4 and 5, respectively, while that of $x=0.8$ can reach 10. (C–D) Thermoelectric properties of the [1 uc $SrTi_{1-x}Nb_xO_3$|11 uc $SrTiO_3]_{10}$ superlattices ($x=0.2$–0.9). Note that the *x*-axis is effective carrier concentration, which includes both conducting $SrTi_{1-x}Nb_xO_3$ and insulating $SrTiO_3$. The slope of the superlattice in (D) is $-300\ \mu V\ K^{-1}$, whereas that of bulk is $-200\ \mu V\ K^{-1}$, indicating the DOS of the superlattice is larger than that of bulk. *Source*: Reproduced with permission from Y.Q. Zhang, B. Feng, H. Hayashi, C.P. Chang, Y.M. Sheu, I. Tanaka, et al., Double thermoelectric power factor of a 2D electron system, Nat. Commun. 9 (2018) 2224. Copyright 2018, Springer Nature.

Following the bulk values, σ increases almost linearly with n (Fig. 8.4C), indicating that n dominates σ. In the $SrTi_{1-x}Nb_xO_3$ system, carriers are mostly due to Nb substitution. The high n of heavily Nb substituted region induces superiority in σ. The slope of the S—log n plot for the superlattice in Fig. 8.4D is $\sim -300\ \mu V\ K^{-1}$, whereas that of bulk is $\sim -200\ \mu V\ K^{-1}$, indicating the DOS of the superlattice is larger than that of bulk. Based on the conclusions above, Zhang et al. successfully improved the thermoelectric PF in $[1\ uc\ SrTi_{1-x}Nb_xO_3|11\ uc\ SrTiO_3]_{10}$ superlattices by adjusting x between $x = 0.2$ and 0.9.

8.3 Electric field thermopower modulation of two-dimensional electron gas

In addition to the superlattice structure, field-effect transistor (FET) is another method to realize the 2D confinement. In 2008 Pernstich et al. first demonstrated the field-effect modulation of thermopower of organic semiconductors [33]. They used two Peltier devices to introduce the temperature difference between the transistor channels and successfully measured the thermopower of rubrene single crystals at temperatures from 208 K to 294 K. Ohta et al. also developed the measurement system of the electric field thermopower modulation and measured that of several materials [15–20,34–37].

In this section, $SrTiO_3$ [15,17,19], $BaSnO_3$ [20], and GaN [21] based FET structures were reviewed to discuss the effectiveness of FET on modulating the 2DEG and exploring thermoelectric materials.

8.3.1 Method

First, the three-terminal transistor was prepared on the materials as schematically shown in Fig. 8.5A. The transistor characteristics were measured using a semiconductor device analyzer. For the S measurements (Fig. 8.5B), the transistor was placed in between two Peltier devices to generate a temperature difference between the source and the drain electrodes. Two thermocouples, which were mechanically attached at both edges of the 2DEG channel, monitored the temperature difference (ΔT). The thermo-electromotive force (ΔV) and ΔT values were simultaneously measured at room temperature, and the slope of the ΔV–ΔT plots yielded the S values.

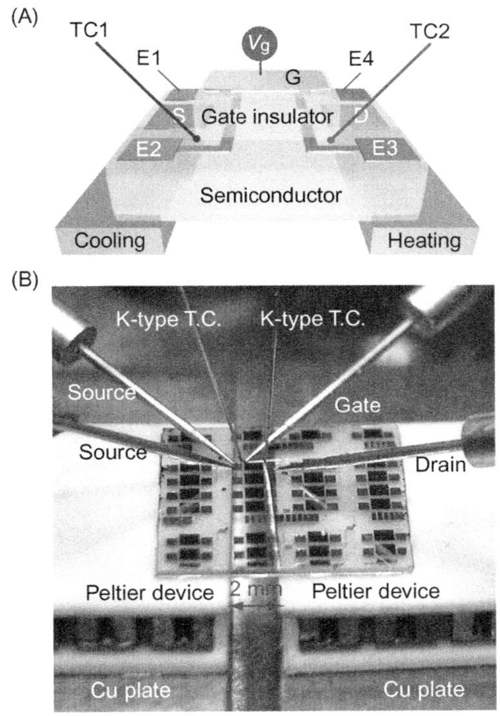

Figure 8.5 Measurement setup of the electric field modulated thermopower. (A) Schematic illustration of the three-terminal field effect transistor structure. The thermopower of the electric field modulated channel can be measured during the gate voltage application. (B) Photograph of the measurement setup.

8.3.2 $SrTiO_3$ [15,17,19]

Since electron-doped $SrTiO_3$-based superlattices showed an enhanced S when the electron-doped layer is thinner than the de Broglie wavelength, Ohta et al. expected that the electrostatically accumulated 2DEG also shows an enhanced S due to the modification of DOS (Fig. 8.6A and B). First, the authors examined the FET structure on $SrTiO_3$ single crystal [15]. The resultant FET showed good transistor characteristics at room temperature. The on/off current ratio was $\sim 10^6$. The field-effect mobility was $\sim 2\ cm^2\ V^{-1}\ s^{-1}$. The sheet carrier concentration was modulated up to $\sim 10^{13}\ cm^{-2}$. However, the S of the 2DEG channel of the FET monotonically decreased from -900 to $-600\ \mu V\ K^{-1}$ with increasing the sheet carrier concentration up to $\sim 10^{13}\ cm^{-2}$. As compared with the $S-n$ relationship of electron-doped $SrTiO_3$ bulk, the effective thickness ($\approx n_{2D}/n_{3D}$) was ~ 10 nm, thicker than the de Broglie wavelength.

In order to reduce the 2DEG thickness thinner than the de Broglie wavelength, Ohta et al. fabricated the FET structures on a $SrTiO_3$ single crystal plate using water infiltrated nanoporous glass (CAN), which contains spontaneously absorbed water, as a

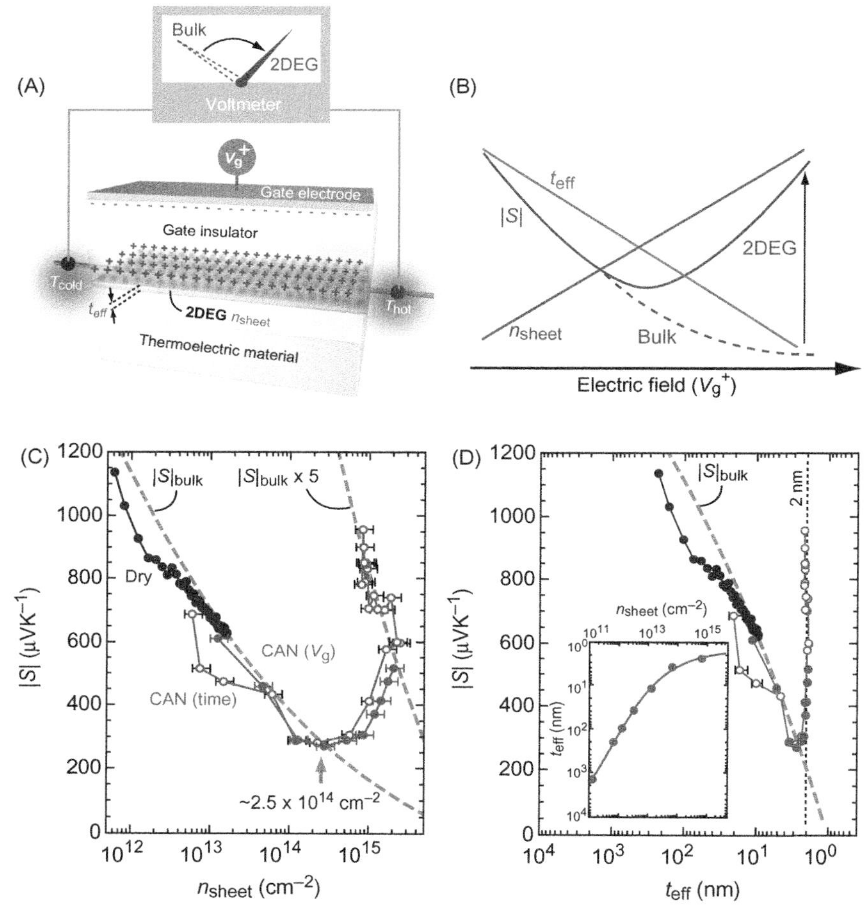

Figure 8.6 Unusually large enhancement of thermopower in $SrTiO_3$-based field-effect transistor. (A) Schematic illustration of a conventional metal-insulator-semiconductor (n-type) field-effect transistor structure on a thermoelectric material. For a positive V_g, a 2DEG is formed at the gate insulator/semiconductor interface. (B) Electric field modulation of $|S|$, n_{sheet} and t_{eff}. As the applied electric field increases, n_{sheet} of the 2DEG layer monotonically increases and the t_{eff} becomes thinner. When t_{eff} is thinner than the critical thickness, unusually large enhancement of $|S|$ is observed. $|S|$ values of the 2DEGs are plotted as a function of (C) n_{sheet} and (D) t_{eff} on a logarithmic scale. $|S|$ initially decreases with n_{sheet} from ~1150 to ~250 $\mu V\ K^{-1}$. Simultaneously, t_{eff} decreases from ~100 to ~2 nm. In this region, the n_{sheet} dependence of $|S|$ is similar to that of simulated bulk values (gray dotted line). When the n_{sheet} value exceeds $\sim 2.5 \times 10^{14}\ cm^{-2}$, $|S|$ increases drastically and is modulated from ~600 to ~950 $\mu V\ K^{-1}$, while t_{eff} remains nearly constant (~2 nm). $|S|$ versus log n_{sheet} relation is approximately five times larger than that of the bulk, clearly indicating that the electric field induced 2DEG in $SrTiO_3$ exhibits an unusually large $|S|$. *Source*: Reproduced with permission from H. Ohta, T. Mizuno, S.J. Zheng, T. Kato, Y. Ikuhara, K. Abe, et al., Unusually large enhancement of thermopower in an electric field induced two-dimensional electron gas, Adv. Mater. 24 (2012) 740–744. Copyright 2012, Wiley.

gate insulator [19]. The CAN gate insulator (H_3O^+) can accumulate carrier electrons up to $\sim 10^{15}$ cm^{-2}, which is two orders of magnitude higher than that can be accumulated by conventional solid gate insulator such as SiO_2, Al_2O_3, and dry C12A7. $|S|$ initially decreases with n_{sheet} from ~ 1150 to ~ 250 $\mu V\ K^{-1}$ (Fig. 8.6C). Simultaneously, effective thickness of the 2DEG (t_{eff}) decreases from ~ 100 to ~ 2 nm (Fig. 8.6D). In this region, the n_{sheet} dependence of $|S|$ is similar to that of simulated bulk values. When the n_{sheet} value exceeds $\sim 2.5 \times 10^{14}$ cm^{-2}, $|S|$ increases drastically and is modulated from ~ 600 (n_{sheet} $\sim 2 \times 10^{15}$ cm^{-2}) to ~ 950 $\mu V\ K^{-1}$ (n_{sheet} $\sim 8 \times 10^{14}$ cm^{-2}), while t_{eff} remains nearly constant (~ 2 nm). $|S|$ versus log n_{sheet} relation is approximately five times larger than that of the bulk, clearly indicating that the electric field induced 2DEG in $SrTiO_3$ exhibits an unusually large $|S|$. Moreover, because the present electric field thermopower modulation method is simple and effectively verifies the performance of thermoelectric materials, it may accelerate the development of nanostructures for high performance thermoelectric materials.

8.3.3 $BaSnO_3$ [20]

The electric field thermopower modulation method is useful to analyze the thickness of the FET channel experimentally. Here we explain the effectiveness using $BaSnO_3$-based FET, which is a transparent oxide semiconductor with large bandgap of 3.1 eV. After the finding of high electron mobility (~ 320 $cm^2\ V^{-1}\ s^{-1}$) of La-doped $BaSnO_3$ single crystal at room temperature by Kim et al., many researchers tried to improve the FET performance of $BaSnO_3$ [38,39]. However, the carrier mobility of La-doped $BaSnO_3$ films and FETs are still much lower than that of bulk single crystal. In order to improve the mobility, clarification of fundamental physical properties such as carrier effective mass (m^*) is very important. Although there are so many reports on the m^* of $BaSnO_3$, it is almost impossible to know the intrinsic m^* value because the reported m^* values are scattered from 0.06 to 3.7 m_0.

In 2017 Sanchela et al. fabricated the thin film transistor structure on the $BaSnO_3$ film grown by PLD. As shown in Fig. 8.7A, Sanchela et al. have clarified the intrinsic m^* value of $BaSnO_3$, $m^* = 0.40 \pm 0.01$ m_0, by the thermopower modulation clarification method [20]. The authors also found the threshold of degenerate/nondegenerate semiconductor of $BaSnO_3$. At the threshold, the thermopower value of both La-doped $BaSnO_3$ (Fig. 8.7A) and $BaSnO_3$ TFT structure (Fig. 8.7B) was 240 $\mu V\ K^{-1}$,

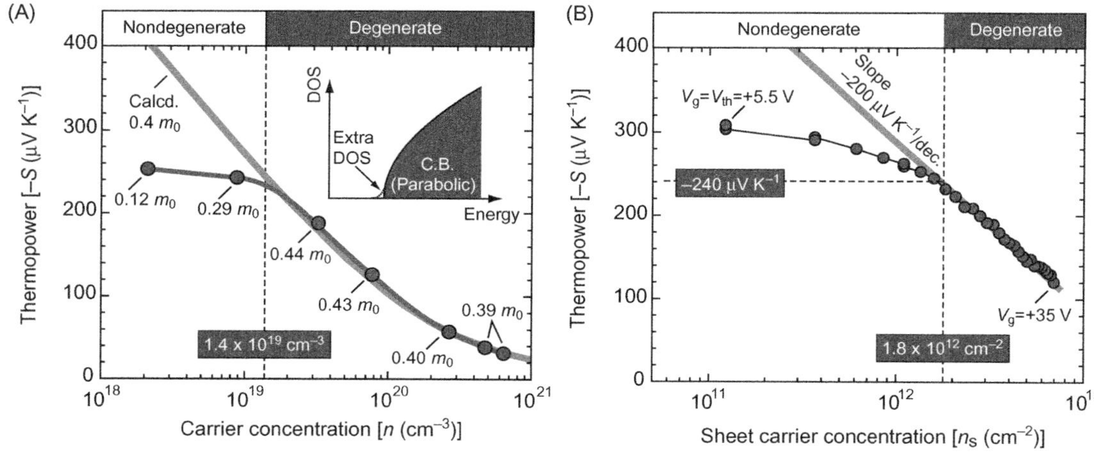

Figure 8.7 Comparison of thermopower of $BaSnO_3$: conventional chemical doping and the electric field modulation method. (A) Change in the thermopower S of La-doped $BaSnO_3$ films grown on (001) $SrTiO_3$ substrates as a function of the volume carrier concentration n at room temperature. Red line is to guide the eye. Effective mass m^* values are shown. Gray line is the $S-n$ curve calculated using $m^* = 0.4 \pm 0.1\ m_0$. Dotted line is the threshold of the degenerate/nondegenerate semiconductor around $n = 1.4 \times 10^{19}$ cm^{-3}. Inset shows a schematic explanation of the energy dependence of the DOS. (B) Electric field modulated S of the $BaSnO_3$ TFT as a function of the sheet carrier concentration at room temperature. Inflection point occurs around $(|S|, n_s) = (240\ \mu V\ K^{-1}, 1.8 \times 10^{12}\ cm^{-2})$. Slope of $S-\log n_s$ plot is $\sim$200 μV K^{-1}/decade (gray line) when n_s exceeds 1.8×10^{12} cm^{-2}. These data indicate the threshold of a degenerate/nondegenerate semiconductor is around $n = 1.8 \times 10^{12}$ cm^{-2}. The effective thickness ($t_{eff} \approx n_s/n$) is $\sim$1 nm. *Source*: Reproduced with permission from A.V. Sanchela, T. Onozato, B. Feng, Y. Ikuhara, H. Ohta, Thermopower modulation clarification of the intrinsic effective mass in transparent oxide semiconductor BaSnO3, Phys. Rev. Mater. 1 (2017) 034603. Copyright 2017, American Physical Society.

bulk carrier concentration was 1.4×10^{19} cm^{-3}, and two-dimensional sheet carrier concentration was 1.8×10^{12} cm^{-2}. Thus the effective thickness ($t_{eff} \approx n_s/n$) is $\sim$1 nm. When the E_F locates above the parabolic shaped conduction band bottom, rather high mobility was observed. On the contrary, very low carrier mobility was observed when the E_F lays below the threshold, most likely due to that the tail states suppress the carrier mobility. This information would be useful to further improve the FET performance of $BaSnO_3$-based TFTs.

8.3.4 AlGaN/GaN interface [21]

Since electrostatically accumulated 2DEG usually shows very high carrier mobility because of minimized impurity scattering, the 2DEG in conventional semiconductor would show very high PF compared with the impurity-doped one. As shown in Fig. 8.8A, the electrical conductivity would be suppressed by the

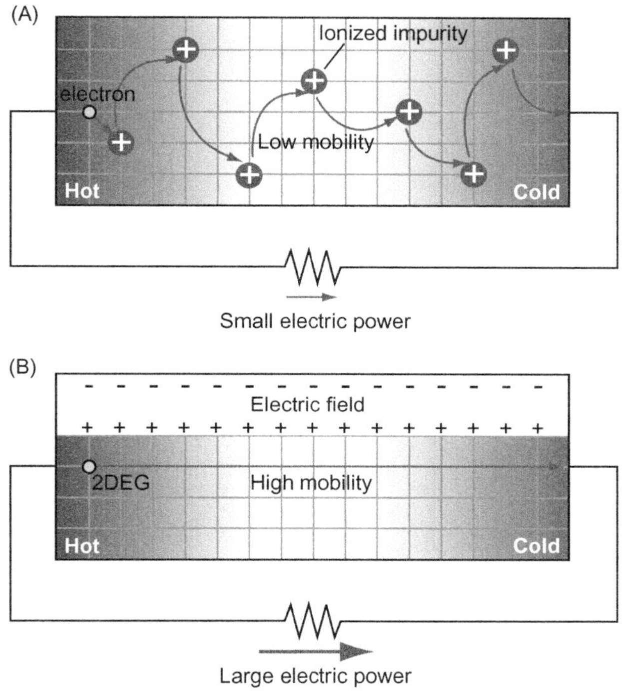

Figure 8.8 Schematic illustration of thermoelectric power generation in an n-type semiconductor. (A) Conventional impurity-doped n-type semiconductor. Carrier electron flow from the hot to the cold side due to the Seebeck effect. Low mobility of the carrier electron originates from ionized impurity scattering. (B) Electric field-induced high-mobility 2DEG. Larger electric power can be obtained compared with (A). *Source*: Reproduced with permission from H. Ohta, S.W. Kim, S. Kaneki, A. Yamamoto, T. Hashizume, High thermoelectric power factor of high-mobility 2D electron gas, Adv. Sci. 5 (2018) 1700696. Copyright 2017, Wiley.

ionized impurity in the case of conventional impurity-doped semiconductor. However, the carrier mobility would not be suppressed in the case of the 2DEG (Fig. 8.8B). Thus PF of the conventional semiconductor would be maximized by utilizing the 2DEG structure. In order to examine this hypothesis, AlGaN/GaN interface is one of the best materials systems because it has been known that high-density 2DEG is spontaneously formed between AlGaN and GaN heterointerfaces.

In 2018 Ohta et al. fabricated an Al_2O_3/AlGaN/GaN metal-oxide-semiconductor high electron mobility transistor (MOS-HEMT) and measured the thermoelectric properties of the 2DEG as schematically shown in Fig. 8.9A. The authors used a commercially available $Al_{0.24}Ga_{0.76}N$ (20 nm)/GaN (900 nm) heterostructure film, which was grown on a semi-insulating (0001) SiC substrate by metal-organic chemical vapor deposition [40]. The sheet resistance (R_s), Hall mobility (μ_{Hall}), and sheet carrier concentration (n_s) were $423\ \Omega\ sq^{-1}$, $1730\ cm^2\ V^{-1}\ s^{-1}$, and $8.53 \times 10^{12}\ cm^{-2}$, respectively, at room temperature.

The observed S values are always negative, indicating that the channel is an n-type semiconductor. As V_g increases, $|S|$ monotonically decreases from $490\ \mu V\ K^{-1}$ to $90\ \mu V\ K^{-1}$.

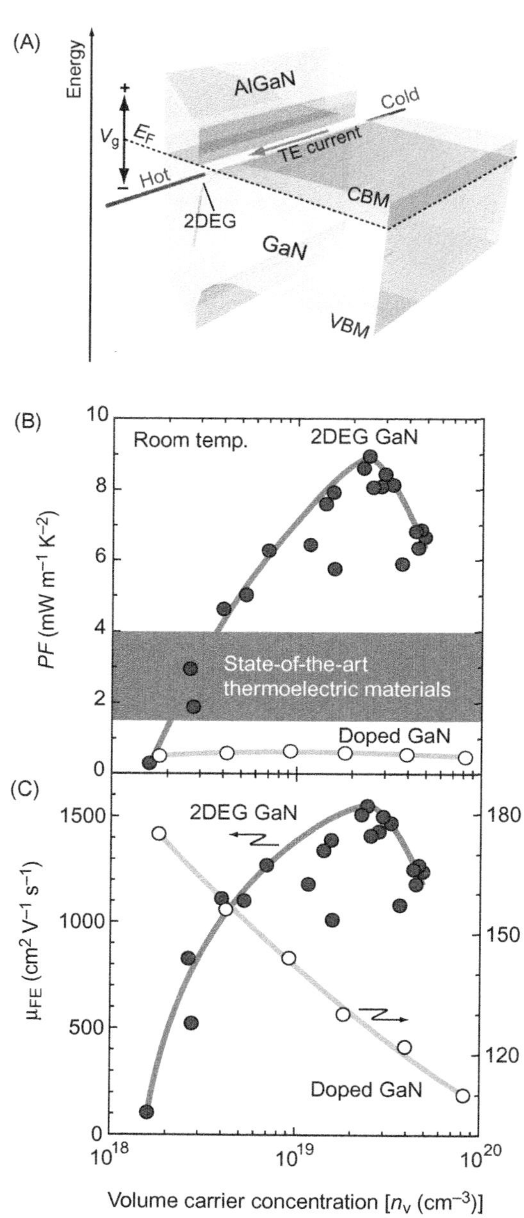

Figure 8.9 High thermoelectric power factor of high-mobility 2DEG at AlGaN/GaN interface. (A) Schematic energy band diagram of 2DEG confined at an AlGaN/GaN heterointerface. CBM and VBM denote the conduction band minimum and the valence band maximum, respectively. Thermoelectric properties can be modulated by applying a gate voltage (V_g). Carrier concentration dependence of (A) PF and (B) carrier mobility (μ) for the 2DEG and conventional impurity-doped n-type GaN. Maximized PF of the 2DEG with $n_v \sim 2.5 \times 10^{19}$ cm^{-3} is $\sim$9 mW m^{-1} K^{-2} at room temperature, which is an order magnitude greater than that of doped GaN bulk and a factor of 2–6 greater than that state-of-the-art practical thermoelectric materials (1.5–4 mW m^{-1} K^{-2}). *Source:* Reproduced with permission from H. Ohta, S.W. Kim, S. Kaneki, A. Yamamoto, T. Hashizume, High thermoelectric power factor of high-mobility 2D electron gas, Adv. Sci. 5 (2018) 1700696. Copyright 2017, Wiley.

The observed S values reflect a bulk-like energy derivation of the parabolic shaped DOS of the conduction band near the E_F Fig. 8.9A and B shows the carrier concentration dependence of (1) PF and (2) carrier mobility (μ) for the 2DEG and conventional impurity-doped n-type GaN. Maximized PF of the 2DEG with $n_V \sim 2.5 \times 10^{19}\,cm^{-3}$ is $\sim 9\,mW\,m^{-1}\,K^{-2}$ at room temperature, which is an order magnitude greater than that of doped GaN bulk and a factor of 2–6 greater than that state-of-the-art practical thermoelectric materials (1.5–4 $mW\,m^{-1}\,K^{-2}$).

8.4 Summary

In this chapter, we have discussed the effectiveness of 2DEG in enhancing thermoelectric performance with an emphasis on both multiple quantum wells (superlattices) of $SrTiO_3$ and filed effect transistors of $SrTiO_3$, $BaSnO_3$, and AlGaN/GaN. By combining volume carrier concentration dependence of thermopower, the electric field thermopower modulation results give the effective thickness of two-dimensional electron gas. This method is very useful to characterize the electronic structure of semiconductors especially the dimensionality. I hope this manuscript is useful to the materials scientists.

References

[1] T. J. Seebeck, agnetische Polarisation der Matalle und Erze durch TemperaturDifferenz. Abh. K. Akad. Wiss. (1823) 265–373.

[2] F.J. DiSalvo, Thermoelectric cooling and power generation, Science 285 (1999) 703–706.

[3] G.J. Snyder, E.S. Toberer, Complex thermoelectric materials, Nat. Mater. 7 (2008) 105–114.

[4] C.B. Vining, An inconvenient truth about thermoelectrics, Nat. Mater. 8 (2009) 83–85.

[5] Q.H. Zhang, X.Y. Huang, S.Q. Bai, X. Shi, C. Uher, L.D. Chen, Thermoelectric devices for power generation: recent progress and future challenges, Adv. Eng. Mater. 18 (2016) 194–213.

[6] M. Cutler, N.F. Mott, Observation of Anderson localization in an electron gas, Phys. Rev. 181 (1969) 1336.

[7] L.D. Hicks, M.S. Dresselhaus, Effect of quantum-well structures on the thermoelectric figure of merit, Phys. Rev. B 47 (1993) 12727–12731.

[8] L.D. Hicks, T.C. Harman, X. Sun, M.S. Dresselhaus, Experimental study of the effect of quantum-well structures on the thermoelectric figure of merit, Phys. Rev. B 53 (1996) 10493–10496.

[9] H. Ohta, et al., Giant thermoelectric Seebeck coefficient of two-dimensional electron gas in $SrTiO_3$, Nat. Mater. 6 (2007) 129–134.

[10] Y. Mune, H. Ohta, K. Koumoto, T. Mizoguchi, Y. Ikuhara, Enhanced Seebeck coefficient of quantum-confined electrons in $SrTiO_3/SrTi_{0.8}Nb_{0.2}O_3$ superlattices, Appl. Phys. Lett. 91 (2007) 192105.

[11] H. Ohta, R. Huang, Y. Ikuhara, Large enhancement of the thermoelectric Seebeck coefficient for amorphous oxide semiconductor superlattices with extremely thin conductive layers, Phys. Stat. Solid.: Rapid Res. Lett. 2 (2008) 105–107.

[12] W.S. Choi, H. Ohta, S.J. Moon, Y.S. Lee, T.W. Noh, Dimensional crossover of polaron dynamics in Nb:$SrTiO_3$/$SrTiO_3$ superlattices: possible mechanism of thermopower enhancement, Phys. Rev. B 82 (2010) 024301.
[13] W.S. Choi, H. Ohta, H.N. Lee, Thermopower enhancement by fractional layer control in 2D oxide superlattices, Adv. Mater. 26 (2014) 6701–6705.
[14] Y.Q. Zhang, B. Feng, H. Hayashi, C.P. Chang, Y.M. Sheu, I. Tanaka, et al., Double thermoelectric power factor of a 2D electron system, Nat. Commun. 9 (2018) 2224.
[15] H. Ohta, Y. Masuoka, R. Asahi, T. Kato, Y. Ikuhara, K. Nomura, et al., Field-modulated thermopower in $SrTiO_3$-based field-effect transistors with amorphous 12CaO-7Al_2O_3 glass gate insulator, Appl. Phys. Lett. 95 (2009) 113505.
[16] A. Yoshikawa, K. Uchida, K. Koumoto, T. Kato, Y. Ikuhara, H. Ohta, Electric-field modulation of thermopower for the $KTaO_3$ field-effect transistors, Appl. Phys. Express 2 (2009) 121103.
[17] H. Ohta, Y. Sato, T. Kato, S. Kim, K. Nomura, Y. Ikuhara, et al., Field-induced water electrolysis switches an oxide semiconductor from an insulator to a metal, Nat. Commun. 1 (2010) 118.
[18] T. Mizuno, Y. Nagao, A. Yoshikawa, K. Koumoto, T. Kato, Y. Ikuhara, et al., Electric field thermopower modulation analysis of an interfacial conducting layer formed between Y_2O_3 and rutile TiO_2, J. Appl. Phys. 110 (2011) 063720.
[19] H. Ohta, T. Mizuno, S.J. Zheng, T. Kato, Y. Ikuhara, K. Abe, et al., Unusually large enhancement of thermopower in an electric field induced two-dimensional electron gas, Adv. Mater. 24 (2012) 740–744.
[20] A.V. Sanchela, T. Onozato, B. Feng, Y. Ikuhara, H. Ohta, Thermopower modulation clarification of the intrinsic effective mass in transparent oxide semiconductor $BaSnO_3$, Phys. Rev. Mater. 1 (2017) 034603.
[21] H. Ohta, S.W. Kim, S. Kaneki, A. Yamamoto, T. Hashizume, High thermoelectric power factor of high-mobility 2D electron gas, Adv. Sci. 5 (2018) 1700696.
[22] T. Okuda, K. Nakanishi, S. Miyasaka, Y. Tokura, Large thermoelectric response of metallic perovskites: $Sr_{1-x}La_xTiO_3$ ($0 <= x <= 0.1$), Phys. Rev. B 63 (2001) 113104.
[23] S. Ohta, T. Nomura, H. Ohta, M. Hirano, H. Hosono, K. Koumoto, Large thermoelectric performance of heavily Nb-doped $SrTiO_3$ epitaxial film at high temperature, Appl. Phys. Lett. 87 (2005) 092108.
[24] S. Ohta, T. Nomura, H. Ohta, K. Koumoto, High-temperature carrier transport and thermoelectric properties of heavily La- or Nb-doped $SrTiO_3$ single crystals, J. Appl. Phys. 97 (2005) 034106.
[25] H. Ohta, Thermoelectrics based on strontium titanate, Mater. Today 10 (2007) 44–49.
[26] K. Kato, M. Yamamoto, S. Ohta, H. Muta, K. Kurosaki, S. Yamanaka, et al., The effect of Eu substitution on thermoelectric properties of $SrTi_{0.8}Nb_{0.2}O_3$, J. Appl. Phys. 102 (2007) 116107.
[27] M. Yamamoto, H. Ohta, K. Koumoto, Thermoelectric phase diagram in a $CaTiO_3$-$SrTiO_3$-$BaTiO_3$ system, Appl. Phys. Lett. 90 (2007) 072101.
[28] H. Ohta, K. Sugiura, K. Koumoto, Recent progress in oxide thermoelectric materials: p-type $Ca_3Co_4O_9$ and n-type $SrTiO_3$, Inorg. Chem. 47 (2008) 8429–8436.
[29] A. Ohtomo, D.A. Muller, J.L. Grazul, H.Y. Hwang, Artificial charge-modulation in atomic-scale perovskite titanate superlattices, Nature 419 (2002) 378–380.

[30] H.N. Lee, H.M. Christen, M.F. Chisholm, C.M. Rouleau, D.H. Lowndes, Strong polarization enhancement in asymmetric three-component ferroelectric superlattices, Nature 433 (2005) 395–399.
[31] N.T. Hung, E.H. Hasdeo, A.R.T. Nugraha, M.S. Dresselhaus, R. Saito, Quantum effects in the thermoelectric power factor of low-dimensional semiconductors, Phys. Rev. Lett. 117 (2016) 036602.
[32] Y.Q. Zhang, B. Feng, H. Hayashi, T. Tohei, I. Tanaka, Y. Ikuhara, et al., Thermoelectric phase diagram of the $SrTiO_3$-$SrNbO_3$ solid solution system, J. Appl. Phys. 121 (2017) 185102.
[33] K.P. Pernstich, B. Rossner, B. Batlogg, Field-effect-modulated Seebeck coefficient in organic semiconductors, Nat. Mater. 7 (2008) 321–325.
[34] Y. Nagao, A. Yoshikawa, K. Koumoto, T. Kato, Y. Ikuhara, H. Ohta, Experimental characterization of the electronic structure of anatase TiO_2: thermopower modulation, Appl. Phys. Lett. 97 (2010) 172112.
[35] T. Katase, Y. Suzuki, H. Ohta, Reversibly switchable electromagnetic device with leakage-free electrolyte, Adv. Electron. Mater. 2 (2016) 1600044.
[36] T. Katase, T. Onozato, M. Hirono, T. Mizuno, H. Ohta, A transparent electrochromic metal-insulator switching device with three-terminal transistor geometry, Sci. Rep. 6 (2016) 25819.
[37] J. Pu, K. Kanahashi, N.T. Cuong, C.H. Chen, L.J. Li, S. Okada, et al., Enhanced thermoelectric power in two-dimensional transition metal dichalcogenide monolayers, Phys. Rev. B 94 (2016) 014312.
[38] H.J. Kim, et al., High mobility in a stable transparent perovskite oxide, Appl. Phys. Express 5 (2012) 061102.
[39] H.J. Kim, et al., Physical properties of transparent perovskite oxides (Ba, La) SnO_3 with high electrical mobility at room temperature, Phys. Rev. B 86 (2012) 165205.
[40] Y. Hori, Z. Yatabe, T. Hashizume, Characterization of interface states in Al_2O_3/AlGaN/GaN structures for improved performance of high-electron-mobility transistors, J. Appl. Phys. 114 (2013) 244503.

9

Hydrogen storage in two-dimensional and three-dimensional materials

Johnny Deschamps

Department of Chemistry and Chemical Engineering (UCP), Ecole Nationale Supérieure de Techniques Avancées (ENSTA IP Paris), Palaiseau, France

9.1 Context

Nowadays renewable energies are one of the most promising alternatives to face the future rarefaction of fossil fuels and counter threats of climate changes due to emissions of greenhouse gases. However, in the current state of knowledge, each resource (solar, wind, hydraulic, geothermal, and biomass) is still subject to technological constraints and the most important of them is the intermittence. To circumvent this problem, the development and the use of energy vectors seems to be an appropriate solution and among the energy vectors known, hydrogen seems to be the ideal candidate as it is a good way to store energy massively over long periods of time that can be employed in a wide variety of systems such as mobility, heat, or industrial processes. Moreover, it has no impact on the carbon footprint and its combustion with oxygen only produces water.

Hydrogen can be used for mobile or stationary applications using fuel cells or by direct combustion but its utilization depends on its storage, which represents currently a crucial problem, especially for the mobility and the on-board applications. Currently, only two on-board storage techniques are available: the liquid state storage at very low temperature and the storage under high pressure. However, these conditions of temperature and pressure are extreme (20K and 700 bar) and an alternative medium for hydrogen storage at moderate temperature and pressure has to be developed. Moreover, the thermal properties of the storage material have to match the operation

2D Nanomaterials for Energy Applications. DOI: https://doi.org/10.1016/B978-0-12-816723-6.00009-5

conditions of a fuel cell, which means that the temperature required to release the hydrogen from the storage medium should not exceed the working temperature of the fuel cell device. Storing hydrogen in nanoporous two-dimensional (2D) and three-dimensional (3D) materials such as graphene, aerogels, and metal organic frameworks (MOFs) structures could be a promising option for achieving satisfactory gravimetric and volumetric densities, and through this chapter the following discussion will help the reader to understand the potentialities of these materials. It has to be understood that the most relevant results in terms of hydrogen storage capacity are presented and therefore the list of referenced works presented is not meant to be exhaustive.

9.2 Graphene and graphene-based structures

Graphene is a two-dimensional one atom thick crystal which results from the arrangement of carbon atoms in a honeycomb geometry and arranged hexagonally [1]. This material possesses very interesting and useful properties such as a quantum Hall effect at room temperature which can be restored by annealing [2,3]. It can be considered as a medium for hydrogen storage and can be then included in the class of hydrocarbons and N-and B-hydrides [1,4–6]. It is robust, stable, and can be transported over long distances. Moreover, its carbon chemical versatility allows it the advantage of changing from sp2 to sp3 hybridization which improves the binding of hydrogen atoms [3]. This hydrogen-carbon binding can be associated to charging/discharging strategies enabled by its mechanical flexibility [7].

The most common methods used to prepare graphene are the chemical vapor deposition (CVD) of the camphor [8], the exfoliation of graphitic oxide [9], the thermal conversion of nanodiamond [10], and the arc evaporation of silicon carbide (SiC) [11]. Subrahmanyam et al. have performed some characterizations of graphene samples synthesized by using these different methods [12]. As a conclusion of this comparative study, the exfoliation of graphitic oxide method provides the best graphene samples in terms of surface area, crystallite sizes, and number of layers (smaller number of layers). Graphene obtained by thermal conversion of nanodiamond also is of good quality and possesses the best thermal stability. Finally, the samples obtained by these two methods

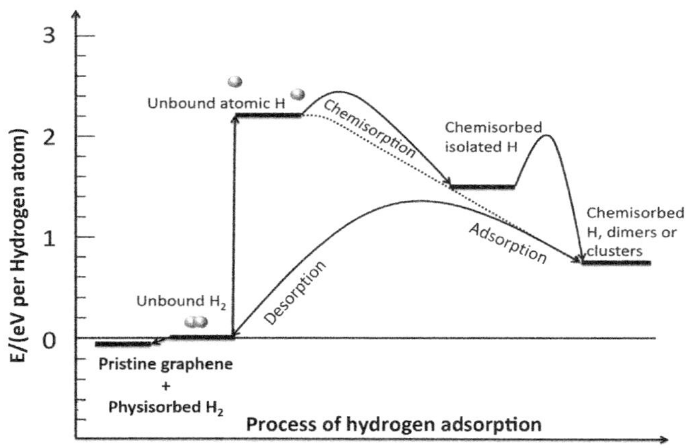

Figure 9.1 Energy profiles of hydrogen adsorption on graphene as a function of the hydrogen adsorption process. The reference level is the pristine graphene and the unbonded hydrogen.

exhibit significant hydrogen uptake capacity [2,12]. This point is detailed in the Section 9.2.2.

9.2.1 Hydrogen sorption in graphene

Hydrogen can be adsorbed on graphene by forming chemical bonds with the carbon atoms (chemisorption) or by interacting with the surface of the solid (physisorption). The energy profiles of the hydrogen adsorption on graphene are given in Fig. 9.1 [1].

Chemisorption of hydrogen on graphene is characterized by the formation of chemical bonds between hydrogen and carbon atoms. Chemisorption of atomic hydrogen is a more favorable process than that of molecular hydrogen. During the “atomic process” the H binding and the barriers of energy are approximately 0.7 and 0.3 eV [13–16], respectively while the “molecular process” is associated to more important barriers of energy estimated around 1.5 eV [17]. This increase of energy is associated to the required energy for the dissociation of the hydrogen molecule. It has been shown by theoretical studies that adsorption of the first hydrogen atom locally modifies the graphene structure providing a stabilization effect and favoring further hydrogen binding [18,19]. The maximum gravimetric hydrogen uptake reachable by chemisorption is 8.3 wt.% and corresponds

Figure 9.2 Molecular structure of graphane. The hydrogen atoms are represented in white and the carbon atoms in gray.

to a completely saturated graphene sheet with 1:1 C:H ratio [1,3,20]. This structure, represented in Fig. 9.2, is called "graphane" which is known to be a broad band-gap insulator with a controllable band-gap opening [20,21]. The maximum volumetric hydrogen uptake capacity depends on the possibility of compacting graphene sheets in specific 3D structures [1].

Physisorption is characterized by Van der Waals interactions. These forces are very low and the phenomenon happens when hydrogen is in molecular form. The binding energy of H_2 was theoretically evaluated between 0.01 and 0.06 eV [7,22] and depends on the difficulty to consider the London dispersion forces. However, it is clear that the binding energy is very weak and the physisorption phenomenon takes place at low temperature and under pressure. During the process, a compact monolayer of H_2 is formed on the surface of the graphene layer and corresponds to a maximum gravimetric hydrogen uptake capacity of 3.3 wt.% (per side). Similarly, to the chemisorption process, the maximum volumetric hydrogen uptake capacity is difficult to evaluate and depends on the possibility of compacting graphene sheets in 3D structures [1].

9.2.2 Nanostructures of graphene

The volumetric density of hydrogen stored in single graphene layer quasi-2D system is not well defined and to evaluate correctly the potentialities of graphene for hydrogen storage it is necessary to consider graphene multilayers which are nanostructures or three-dimensional assemblies of graphene [1,23].

The physisorption capabilities of a two-layers graphene material was studied from quantum calculations and it was shown that the hydrogen storage capacity (gravimetric and volumetric) depends on the interlayer distance and the best results are obtained for an interlayer separation of approximately 6.0 Å [22,24]. In fact, the attractive Van der Waals forces of the two layers combine together and the physisorption energy is almost doubled. This behavior is similar to a "pump effect" capable of increasing the internal pressure of hydrogen and maintaining the external one.

Graphene nanosheets were synthesized by Srinivas et al. and the hydrogen uptake capacities of the materials were measured at pressures up to 10 bar. The maximum hydrogen storage capacities were 1.2 and 0.1 wt.% at 77K and 298K, respectively. The room hydrogen storage capacity of the sample was estimated at 0.72 wt.% at 100 bar and the values of isosteric heat of adsorption are between 4.0 and 5.9 kJ mol^{-1}, indicating good interactions between hydrogen and graphene sheets [25].

Subrahmanyam et al. have performed some syntheses and characterizations of graphene sheets by exfoliation of graphitic oxide and thermal conversion of nanodiamond. The materials obtained by exfoliation methods exhibit hydrogen uptake capacities of 1.38 wt.% at 77K under 1 bar and 3.1 wt.% at 300K under 100 bar [12]. By using the thermal conversion of nanodiamond method, the graphene samples prepared yield storage capacities of 0.68 wt.% at 77K under 1 bar and 2.5 wt.% at 300K under 100 bar [12]. The adsorption isotherms are completely reversible and comparable to those of carbon nanotubes [25]. Gosh et al. performed the same experiments and the hydrogen storage capacities of the samples prepared by exfoliation method are of 1.4 wt.% at 77K under 1 bar and 3.1 wt.% at 298K under 100 bar. The samples obtained by thermal conversion of nanodiamond exhibit hydrogen uptake capacity of 2.5 wt.% at 298K under 100 bar [2,12].

Klechikov et al. also have prepared graphene structures by thermal exfoliation of graphite oxide. The materials prepared were then activated by using a specific procedure based on potassium hydroxide treatment followed by high temperature hydrogen annealing. The graphene structures obtained are three-dimensional and contain strongly defective and interconnected graphene layers. They are very porous (surface area of 3400 $m^2 g^{-1}$ and pore volume of 2.2 $cm^3 g^{-1}$), contain essentially micropores, and their hydrogen uptake capacities measured under 120 bar of pressure are 7.5 wt.% at 77K, 4.23 wt.% at 193K, 1.61 wt.% at 273K, and 1.25 wt.% at 293K [26].

9.2.3 Functionalization of graphene

Although layered or nanostructured graphene can reach reasonable hydrogen uptake capacities though physisorption, the DOE targets [27] in terms of volumetric (4.0 vol% for 2025 and 5.0 vol% ultimate) and gravimetric (5.5 wt.% for 2025 and 6.5 wt.% ultimate) capacities can only be reached at cryogenic temperature (77K) and under pressure (100 bar or more). Among the various possibilities to enhance the hydrogen adsorption at ambient conditions, the chemical functionalization can be considered as a promising alternative and the chemical decoration by using hydroxyl and other groups [28–31] or metal atoms [32] such as palladium (Pd) [33,34], platinum (Pt) [33], nickel (Ni) [35], iron (Fe) [36], and aluminum (Al) [37].

Kim et al. have studied the hydrogen storage in multilayered graphene oxide (GO) which is a single layer of carbon atoms with hydroxyl, epoxy and carboxyl surface groups [28]. The layered structure of (GO) possesses a mean interlayer distance between 6.0 and 12.0 Å which depends on the interlamellar water content and the extent of the oxidation process [38–40]. This interlayer distance is controlled by thermal annealing to prevent any introduction of moisture and an optimum distance of 6.5 Å which corresponds to the maximum hydrogen uptake, was found. In terms of hydrogen storage, the graphene oxide has a hydrogen uptake capacity of 4.8 and 0.5 wt.% at 77K and 298K, respectively, up to a pressure of 90 bar [28]. Similarly, Hudson et al. also studied the hydrogen uptake capacity of graphene oxide (GO) and thermally reduced graphene oxide (GO-TR). The structures of these two materials are shown in the Transmission Electron Microscopy (TEM) images given in Fig. 9.3 [36]. The hydrogen storage capacities of the (GO) and

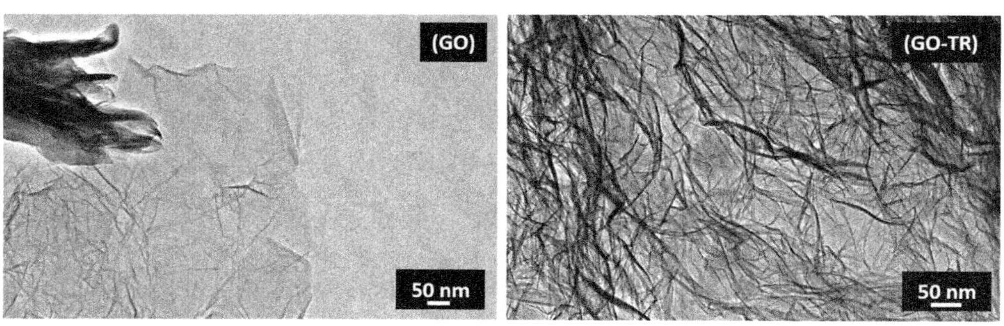

Figure 9.3 Transmission electron microscopy (TEM) images of graphene oxide (GO) and thermally reduced graphene oxide (GO-TR).

the (GO-TR) materials at 300K and 50 bar are 0.17 and 0.32 wt.%, respectively. Moreover, the (GO-TR) exhibits a hydrogen uptake capacity of 2.07 wt.% at 77K under 50 bar and kinetically speaking the saturation of the adsorption takes place in few minutes (measurements performed at 77K and 27, 40, and 47 bar) [36].

Huang et al. studied the hydrogen storage in Pd and Pt nanoparticles decorated graphene prepared by using electroless deposition technique [33]. It was shown that the hydrogen uptake capacity of graphene is significantly enhanced by the decoration with Pd and Pt. The hydrogen storage capacities of the Pd decorated (G-Pd) and the Pt decorated (G-Pt) materials at 303K and 57 bar are 0.16 and 0.15 wt.%, respectively. Compared to the pristine graphene, the hydrogen storage capacities of G-Pd and G-Pt materials are enhanced by factors 2.32 and 2.23, respectively [33]. If the increase of hydrogen capacity of the G-Pd material is totally due to the hydrogen adsorbed in Pd cluster, the hydrogen content can be estimated at 4.3H/Pd and is larger than the H/Pd ratio of Pd bulk (0.6). This difference is related to the spillover effect and the small size of Pd particles. Similarly, the calculated H/Pt ratio is 0.33 whereas the enhancing factor of hydrogen uptake is 2.23. The spillover effect associated to the G-Pt material can be then attributed to the mass of the Pt particles (almost 50% of the sample) and the reduced surface area of the sample [33]. This result is confirmed by the system Pt-Activated Carbon which has the same behavior [41]. The hydrogen adsorption in doped Pd and Pt graphene is then governed by the amount and the size and the metal particles.

The nickel decoration of graphene was reported by Zhou et al. and it was shown that the nickel particles (10 nm size) are well dispersed in the graphene structure and they presumably are linked with Ni–O–C and Ni–C bonds. Basically, hydrogen atoms migrate into graphene by a Ni-graphene interface which can be assumed as linker and foster the spillover process [35,42]. The composite material obtained yield a hydrogen uptake capacity of 0.14 wt.% under atmospheric temperature and pressure conditions. This uptake reaches 1.18 wt.% at ambient temperature and a pressure up to 60 bar. The hydrogen release occurs at 150°C and is completed at 250°C [35].

In the same vein, Hudson et al. studied the hydrogen uptake capacity of iron (Fe) decorated graphene nanosheets and the capacities measured are of 2.16 wt.% at 77K and 50 bar, and 0.27 wt.% at 300K under 50 bar [36].

Aluminum (Al) also is an interesting metal for graphene doping. The hydrogen storage in Al-adsorbed graphene was studied by using functional density theory by Ao and Peeters [37]. It was found that excess hydrogen storage uptake capacity is 6.0 wt.% under atmospheric temperature and pressure conditions. This behavior is due to the adsorbed atoms that act as bridges between the electron cloud of hydrogen and the graphene layer, and a two layers arrangement of hydrogen molecules is formed on each side of the Al-adsorbed graphene layer [37].

9.3 3D structures from nanostructured 1D and 2D materials

9.3.1 Aerogels

Aerogels are very porous materials with large pore volumes and low bulk densities which can be used in a wide range of applications [43–45]. The typical types of aerogel structures are metal oxide (Al_20_3 and SIO_2) [43,46,47], carbon materials (carbon, graphene, carbon nanotubes, and glucose/graphene) [43,48–52], semiconducting chalcogenide (CdS, PbTe, and CdSe) [43,53,54], polysaccharide [43,55], and protein [43,56]. They are usually obtained by liquid extraction of a gel by using critical point drying (CPD) technique to keep the gel network intact. The gel networks are then assembled by using sol–gel process in presence of nanoparticle suspensions and gelling agent precursor [43]. Basically, nanoparticle suspensions are formed and the nanoparticles are crosslinked into 3D network by the gelling agent precursor. However, the crosslinking reactions are very specific and the material types of aerogels are limited [43]. It is then possible to create a wide range of aerogels by assembling gels via the crosslinking of the material itself without the precursor [43]. Jung et al. have developed a specific methodology to make gels without precursor in order to enable the synthesis of aerogel over a wider range. In this method, 1D colloidal particles (nanotubes or nanowires) or 2D nanosheets are dispersed in a liquid (solvent) and when the concentration of liquid reaches a specific level, the 1D or 2D objects interlink with each other, forming a gel in a 3D network. The resulting aerogel material is then formed by liquid extraction via critical point drying (CPD) [43].

Singh et al. have studied experimentally hydrogen storage in carbon aerogel (CA) and platinum doped carbon aerogel (CA-Pt) [57]. These materials, obtained from resorcinol, formaldehyde, and sodium carbonate contain a large density of micropores

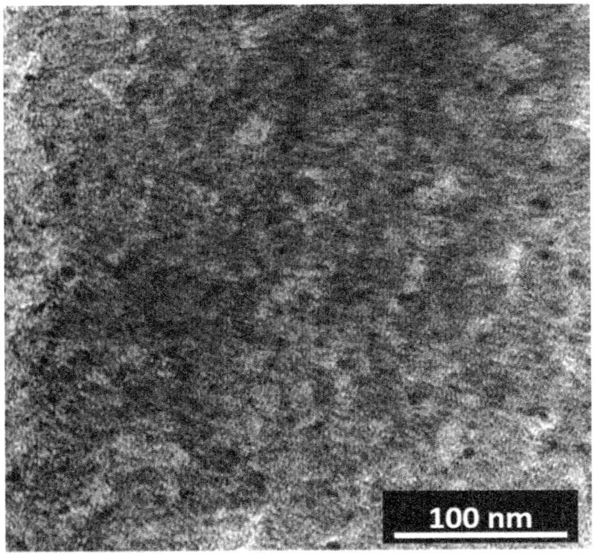

Figure 9.4 Transmission electron microscopy (TEM) image of cobalt doped carbon aerogel (CA-Co). The metal particles (*black dots*) are totally dispersed inside the material.

with a pore size majority between 0.30 and 1.46 nm. At room temperature and at pressure up to 22 bar, the hydrogen storage capacities of CA and CA-Pt are 0.40 and 0.33 wt.%, respectively. Under the same conditions of pressure but at cryogenic temperature (77K), CA and CA-Pt have hydrogen storage capacities of 5.65 and 5.15 wt.%, respectively. Similarly, Tian et al. performed some hydrogen storage measurements in carbon aerogel (CA) and cobalt doped carbon aerogel (CA-Co) prepared from resorcinol, furfural, and hexamethylene tetramine. The dispersion of the metal inside the carbon aerogel material is illustrated by the microscopy image given in Fig. 9.4 [58].

At 77K under 46 bar, CA and CA-Co have hydrogen uptake capacities of 4.88 and 4.38 wt.%, respectively. The hydrogen uptake capacity of CA-Co is slightly reduced compared to pristine carbon aerogel (CA) due to a small reduction in surface area associated to the cobalt doping [58].

9.3.2 Doped metal organic frameworks

Metal organic frameworks (MOFs) are crystalline porous materials composed of inorganic subunits and rigid organic linkers. They consist of metal ions or metal clusters and organic ligands as linkers and possess very low densities, extremely high surface area values and adjustable pore size which make them suitable for a wide range of application including, separation, catalysis, drug delivery, sensing, and gas storage [59]. Considering their high specific surface areas, tunable pore sizes and functionalized pores walls, these materials have attracted the attention of the researchers in the field of hydrogen storage [60,61]. However, the best hydrogen uptake capacities reported in MOFs are mainly at cryogenic temperatures only, and the room temperature hydrogen uptake capacities are too low and need improvements [62,63]. Different modification strategies such as framework catenation, organic linkers functionalization, or impregnation have been tested to enhance their hydrogen uptake capacities [64]. The MOFs doping with carbonaceous materials like activated carbon, carbon nanotubes, graphite oxide, or graphite nanofibers leads to an an enhanced composite performance which possesses usual hydrophobic and mechanical properties [65–67]. These hybrid composites also are more thermally stable and have better moisture resistance compared to bare MOFs [66] which contain large void spaces not beneficial for retention of hydrogen molecules at ambient conditions owing to weak dispersive forces [59]. The formation of composite increases the dispersive forces in MOFs via the presence of carbon materials inside the pores [65,67] and the new pore space available in the composite results in increased dispersive forces, and contributes to enhance adsorption uptake capacity [65,67].

The hydrogen uptake capacities in MOFs can also be increased by the improvement of the hydrogen binding energy which can be induced by the possibility of doping light metal cations such as Mg^{2+}, Na^{+}, Li^{+}, and Al^{3+} into the porous framework of MOFs [68–71]. Among these metal cations, lithium is very promising due to its low atomic weight and high affinity by charge induced dipole interactions [68,70]. Simulation studies have shown that doped lithium ions in MOF provide approximately 0.9 electron to the MOF linker which corresponds to a hydrogen binding of about 12.0kJ mol^{-1} and a strong binding to lithium near MOF corners [70,72,73]. Goufhi, Deschamps, and Maurin reported a significant increase of hydrogen storage capacity up to 10 wt.% in Lithium doped MIL-101 at 77K by grand canonical Monte Carlo (GCMC) simulation [74]. Theoretical studies have shown that lithium doped MOFs can store more than 6.0 wt.% of hydrogen at ambient conditions [70,75]. Experimental studies also were performed and they report that lithium doped MOFs have shown increased hydrogen uptake capacity at 77K [70,76] and 298K [70]. Lithium doped MIL-101 materials exhibit higher hydrogen sorption capacity compared to bare MIL-101 [70,76,77]. Karikkethu and Deschamps have reported that the hydrogen uptake capacity obtained for lithium doped MIL-101 at 77K and 298K at 100 bar were 79.20 and 6.55 mg g^{-1}, respectively compared to 72.50 and 5.50 mg g^{-1}, respectively for the undoped MIL-101 in the same conditions [70]. Similarly, aluminum (Al^{3+}) also is a good candidate for MOFs doping to increase the hydrogen storage capacity of the material. Karikkethu, Catoire, and Deschamps have reported experimentally that the hydrogen uptake capacity obtained for aluminum-doped MIL-101 at 298K under 100 bar was 6.90 mg g^{-1} compared to 5.50 mg g^{-1} for the undoped MIL-101 in the same conditions [59].

Another approach for increasing hydrogen uptake capacities is the metal doping of hybrid composite MOF to get the synergic effect of both carbonaceous material incorporation and metal doping. The modification of MIL-101 both by activated carbon incorporation and lithium doping significantly improves the hydrogen uptake capacity of the material [70]. Karikkethu and Deschamps have shown experimentally that the doping of activated carbon incorporated MIL-101 by using a little amount of lithium (727 ppm) almost double the hydrogen uptake capacity of the material both at 77K and 298K up to 100 bar. In terms of performances, this lithium doped composite material possesses a maximum hydrogen uptake capacity of 143.9 and 11.4 mg g^{-1} at 77K and 298K, respectively, up to a pressure of 100 bar [70]. The enhanced hydrogen uptake can be attributed to the combined effect of three factors which are the activation of unsaturated metal sites by carbon incorporation, the formation of ultramicropores by activated carbon incorporation, and the strong interaction of lithium inside the pores of the MIL-101 material towards hydrogen [70].

Similarly, the same authors demonstrate experimentally that the doping of activated carbon incorporated MIL-101 with aluminum greatly enhance the hydrogen uptake capacity. The activated carbon incorporated MIL-101 material doped with a small amount of aluminum (925 ppm) exhibits a hydrogen uptake capacity of 17.40 mg g^{-1} at 298K up to 100 bar, which is more than three times higher than the hydrogen uptake capacity of pristine MIL-101 under the same conditions of temperature and pressure. This result is shown in Fig. 9.5 [59].

In addition to the activation of unsaturated metal sites by carbon incorporation and the formation of ultramicroporosity by activated carbon incorporation, the confined aluminum nanoparticles finely dispersed inside MOF framework contribute positively to the hydrogen adsorption at room temperature and under pressure. The dispersion of the aluminum nanoparticles is illustrated by the transmission electron microscopy (TEM) image given in Fig. 9.6.

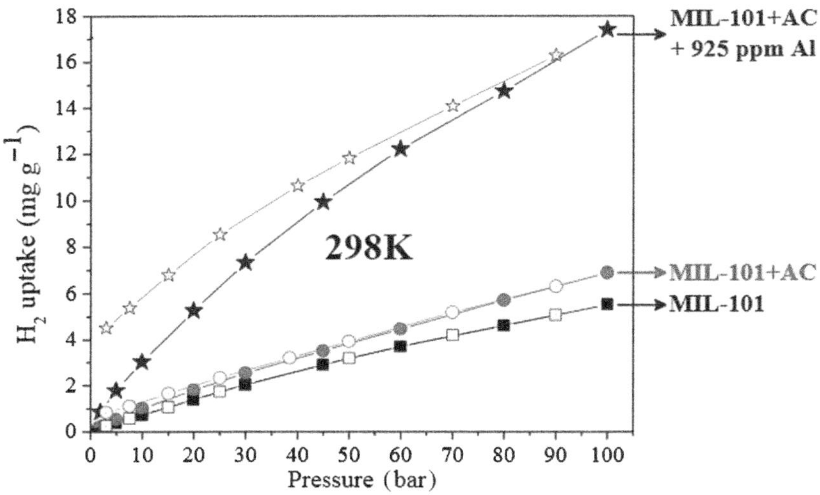

Figure 9.5 Hydrogen adsorption-desorption isotherms (gravimetric measurements) of MIL-101 material, Activated carbon doped MIL-101 material (MIL-101 + AC), and aluminum-doped activated carbon incorporated MIL-101 material (MIL-101 + AC + 925 ppm Al) at 298K up to 100 bar (*close symbols* = adsorption, *open symbols* = desorption).

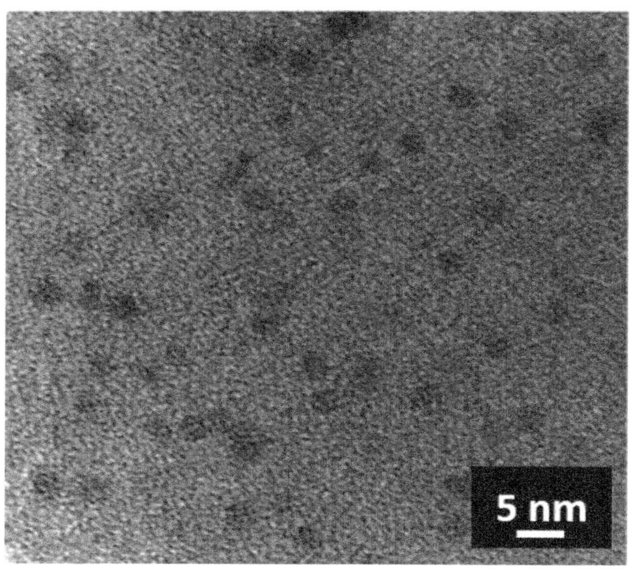

Figure 9.6 Transmission electron microscopy (TEM) image of aluminum-doped activated carbon incorporated MIL-101 material (MIL-101 + AC + 925 ppm Al).

9.4 Conclusion

Hydrogen seems to be an ideal solution to circumvent the problems related to the energy such as the intermittence of the renewable energies, the rarefaction of fossil fuels, and the climate problems. It can be used for mobile or stationary applications using fuel cells or by direct combustion but its utilization depends on its storage which represents currently a crucial problem. A medium for hydrogen storage at moderate temperature and pressure has to be then developed and through this chapter the potentialities of several 2D–3D materials like graphene, aerogels, and metal organic frameworks (MOFs) structures were mentioned and discussed. These materials can reach reasonable hydrogen uptake capacities though physisorption at ambient temperature and moderate pressure (100 bar), but the DOE targets in terms of volumetric and gravimetric capacities can only be reached at cryogenic temperature (77K) and up to a pressure of 100 bar. However, even if the DOE targets are not reached at ambient temperature and moderate pressure, the synthesis procedures of the 2D–3D materials have been considerably improved over the past decade and the hydrogen storage performances of the materials have been significantly enhanced. Finally, solid state storage in porous nanostructures such as 2D and 3D materials could be a promising solution but before that improvements in terms of hydrogen uptake capacity of storage at ambient conditions are necessary, and for that fundamental research is still needed to better understand the potential of this technology and to contemplate its development.

References

[1] V. Tozzini, V. Pellegrini, Prospects for hydrogen storage in graphene, Phys. Chem. Chem. Phys. 15 (2013) 80–89.
[2] A. Ghosh, K.S. Subrahmanyam, K.S. Krishna, S. Datta, A. Govindarj, S.K. Pati, et al., Uptake of H_2 and CO_2 by graphene, J. Phys. Chem. C 112 (2008) 15704–15707.
[3] D.C. Elias, R.R. Nair, T.M. Mohiuddin, S.V. Morozov, et al., Science 323 (2009) 610–613.
[4] M.E. Bluhm, M.G. Bradley, R. Butterick, U. Kusari, L.G. Snedon, Amineborane-based chemical hydrogen storage: enhanced ammonia borane dehydrogenation in Ionic Liquids, J. Am. Chem. Soc. 128 (2006) 7748–7749.
[5] K. Müller, K. Stark, B. Müller, W. Arlt, Amineborane based hydrogen carriers: an evaluation, Energy Fuels 26 (2012) 3691–3696.
[6] H.W. Li, Y. Yan, S.I. Orimo, A. Züttel, C.M. Jensen, Recent progress in metal borohydrides for hydrogen storage, Energies 4 (2011) 185–214.
[7] V. Tozzini, V. Pellegrini, Reversible hydrogen storage by controlled buckling of graphene layers, J. Phys. Chem. C 115 (2011) 25523–25528.

[8] P.R. Somani, S.P. Somani, M. Umeno, Planer nano-graphene from camphor by CVD, Chem. Phys. Lett. 430 (2006) 56–59.
[9] H.C. Schniepp, J.-L. Li, M.J. MacAllister, et al., Functionalized single graphene sheets derived from splitting graphite oxide, J. Phys. Chem. B 110 (2006) 8535–8539.
[10] O.E. Andersson, B.L.V. Prasad, H. Sato, T. Enoki, Structure and electronic properties of graphite nanoparticles, Phys. Rev. B 58 (1998) 16387–16395.
[11] Y. Li, S. Xie, W. Zhou, D. Tang, X.P. Zou, Z. Liu, et al., Nanographite ribbons grown from a SiC arc-discharge in a hydrogen atmosphere, Carbon 39 (2001) 626–628.
[12] K.S. Subrahmanyam, S.R.C. Vivekchand, A. Govindaraj, C.N.R. Rao, A study of graphenes prepared by different methods: characterization, properties and solubilization, J. Mater. Chem. 18 (2008) 1517–1523.
[13] A. Zuttel, P. Sudon, Ph Mauron, et al., Hydrogen storage in carbon nanostructures, Int. J. Hydrogen Energy 27 (2002) 203–212.
[14] Z. Sljivancanin, E. Rauls, L. Hornekaer, et al., Extended atomic hydrogen dimer configurations on the graphite (0001) surface, J. Chem. Phys. 131 (2009) 084706.
[15] N. Rougeau, D. Teillet-Billy, V. Sidis, Double H atom adsorption on a cluster model of a graphite surface, Chem. Phys. Lett. 431 (2006) 135–138.
[16] X. Sha, B. Jackson, First principles study of the structural and energetic properties of H atoms on a graphite (0001) surface, Surf. Sci. 496 (2002) 318–330.
[17] Y. Miura, W. Dino, H. Nakanishi, First principles studies for the dissociative adsorption of H_2 on graphene, J. Appl. Phys. 93 (2003) 3395–3400.
[18] D. Stojkovic, P. Zhang, P.E. Lammert, V.H. Crespi, Collective stabilization of hydrogen chemisorption on graphemic surfaces, Phys. Rev. B: Condens. Matter Mater Phys. 68 (2003). 195406-1–195406-5.
[19] S. Casolo, O.M. Lovik, R. Martinazzo, G.F. Tantardini, Understanding adsorption of hydrogen atoms on graphene, J. Chem. Phys. 130 (2009). 054704-1–054704-10.
[20] J.O. Sofo, A.S. Chaudhari, G.D. Barber, Graphane: a two-dimensional hydrocarbon, Phys. Rev. B: Condens. Matter Mater Phys. 75 (2007). 153401-1–153401-4.
[21] R. Balog, B. Jorgensen, L. Nilson, M. Andersen, et al., Bandgap opening in graphene induced by patterned hydrogen sorption, Nat. Mater. 9 (2010) 315–319.
[22] S. Patchkovskiim, J.S. Tse, S.N. Yurchenko, L. Zhechkov, T. Heine, G. Seifer, Graphene nanostructure as tunable storage media for molecular hydrogen, Proc. Natl Acad. Sci. U.S.A. 102 (2005) 10439–10444.
[23] S.R. Peurifoy, E. Castro, F. Liu, X.Y. Zhu, et al., Three-dimensional graphene nanostructures, J. Am. Chem. Soc. 140 (2018) 9341–9345.
[24] I. Cambria, M.J. Lopez, J.A. Alonso, The optimum average nanopore size for hydrogen storage in carbon nanoporous materials, Carbon 45 (2007) 2649–2658.
[25] G. Srinivas, Y. Zhu, R. Piner, N. Skipper, M. Ellerby, R. Ruoff, Synthesis of graphene-like nanosheets and their hydrogen adsorption capacity, Carbon 48 (2009) 630–635.
[26] A. Klechikov, G. Mercier, T. Sharifi, I.A. Baburin, G. Seifert, A.V. Talyzin, Hydrogen storage in high surface area graphene scaffolds, Chem. Commun. 51 (2015) 15280–15283.
[27] Department of Energy. Target explanation document: onboard hydrogen storage for light-duty fuel cell vehicles. <https://www.energy.gov/sites/prod/files/2017/05/f34/fcto_targets_onboard_hydro_storage_explanation.pdf>, 2017.

[28] B.H. Kim, W.G. Hong, H.Y. Yu, Y.-K. Han, et al., Thermally modulated multilayered graphene oxide for hydrogen storage, Phys. Chem. Chem. Phys. 14 (2012) 1480–1484.
[29] H.G. Shiraz, O. Tavakoli, Investigation of graphene-based systems for hydrogen storage, Renew. Sustain. Energy Rev. 74 (2017) 104–109.
[30] T. Xu, J. Chen, W. Yuan, B. Li, et al., Preparation and hydrogen storage characteristics of surfactant-modified graphene, Polymers 10 (2018) 1220. Available from: https://doi.org/10.3390/polym10111220.
[31] A. Ariharan, B. Viswanathan, V. Nandhakumar, Nitrogen doped graphene as potential material for hydrogen storage, Graphene 6 (2017) 41–60.
[32] D.P. Broom, in: J. Amouroux (Ed.), Hydrogen Storage Materials – The Characterization of Their Storage Properties, Bentham Science Publishers, 2011.
[33] C.C. Huang, N.-W. Pu, C.-A. Wang, J.-C. Huang, Y. Sung, M.-D. Ger, Hydrogen storage in graphene decorated with Pd and Pt nano-particles using an electroless deposition technique, Sep. Purif. Technol. 82 (2011) 210–215.
[34] C. Zhou, J.A. Szpunar, Hydrogen storage performance in Pd/graphene nanocomposites, ACS Appl. Mater. Interfaces 8 (2016) 25933–25940.
[35] C. Zhou, J.A. Szpunar, X. Cui, Synthesis of Ni/graphene nanocomposite for hydrogen storage, ACS Appl. Mater. Interfaces 8 (2016) 15232–15241.
[36] M.S.L. Hudson, H. Raghubanshi, S. Awasthi, T. Sadhasivam, Hydrogen uptake of reduced graphene oxide and graphene sheets decorated with Fe nanoclusters, Int. J. Hydrogen Energy 39 (2014) 8311–8320.
[37] Z.M. Ao, F.M. Peeters, High-capacity hydrogen storage in Al-adsorbed graphene, Phys. Rev. B 81 (2010). 205406-1–205406-7.
[38] D.R. Dreyer, S. Park, C.W. Bielawski, R.S. Ruoff, The chemistry of graphene oxide, Chem. Soc. Rev. 39 (2010) 228–240.
[39] A. Buchsteiner, A. Lerf, J. Pieper, Water dynamics in graphite oxide investigated with neutron scattering, J. Phys. Chem. B 110 (2006) 22328–22338.
[40] N.V. Medhekar, A. Ramasubramaniam, R.S. Ruoff, V.B. Shenoy, Hydrogen bond networks in graphene oxide composite paper: structure and mechanical properties, ACS Nano 4 (2010) 2300–2306.
[41] N.P. Stadie, J.J. Purewal, C.C. Ahn, B. Fultz, Measurements of hydrogen spillover in platinum doped superactivated carbon, Langmuir 26 (2010) 15481–15485.
[42] L. Wang, R.T. Yang, New sorbents for hydrogen storage by hydrogen spillover-a review, Energy Environ. Sci. 1 (2008) 268–279.
[43] S.M. Jung, H.Y. Jung, M.S. Dresselhaus, Y.J. Jung, J. Kong, A facile route for 3D aerogels from nanostructured 1D and 2D materials, Sci. Rep. 2 (2012) 849.
[44] R.T. Olsson, Making flexible magnetic aerogels and stiff magnetic nanopaper using cellulose nanofibrils as templates, Nat. Nanotechnol. 5 (2010) 584–588.
[45] A.E. Aliev, et al., Giant-stroke, superelastic carbon nanotube aerogel muscles, Science 323 (2009) 1575–1578.
[46] M. Moner-Girona, A. Roig, E. Molins, J. Llibre, Sol-gel route to direct formation of silica aerogel microparticles using supercritical solvents, J. Sol-Gel Sci. Technol. 26 (2003) 645–649.
[47] T.F. Baumann, et al., Synthesis of high-surface-area alumina aerogels without the use of alkoxide precursors, Chem. Mater. 17 (2005) 395–401.

[48] T. Horikawa, J. Hayashi, K. Muroyama, Size control and characterization of spherical carbon aerogel particles from resorcinol–formaldehyde resin, Carbon 42 (2004) 169–175.
[49] A. Worsley, et al., Synthesis and characterization of monolithic carbon aerogel nanocomposites containing double-walled carbon nanotubes, Langmuir 24 (2008) 9763–9766.
[50] A.W. Marcus, et al., Synthesis of graphene aerogel with high electrical conductivity, J. Am. Chem. Soc. 132 (2010) 14067–14069.
[51] J. Mao, J. Locozzia, J. Huang, K. Meng, Y. Lai, Z. Lin, Graphene aerogels for efficient energy storage and conversion, Energy Environ. Sci. 11 (2018) 772–799.
[52] K.K. Liu, B. Jin, L.Y. Meng, Glucose/graphene-based aerogels for gas adsorption and electric double layer capacitors, Polymers 11 (2019) 40. Available from: https://doi.org/10.3390/polym11010040.
[53] J.L. Mohanan, I.U. Arachchige, S.L. Brock, Porous semiconductor chalcogenide aerogels, Science 307 (2005) 397–400.
[54] B. Santanu, N.T. Pantelis, J.C. Peter, S.A. Gerasimos, G.K. Mercouri, Porous semiconducting gels and aerogels from chalcogenide clusters, Science 317 (2007) 490–493.
[55] C.A. García-González, M. Alnaief, I. Smirnova, Polysaccharide-based aerogels-Promising biodegradable carriers for drug delivery systems, Carbohydr. Polym. 86 (2011) 1425–1438.
[56] M. Betz, C.A. García-González, R.P. Subrahmanyamb, I. Smirnova, U. Kulozik, Preparation of novel whey protein-based aerogels as drug carriers for life science applications, J. Supercrit. Fluids 72 (2012) 111–119.
[57] S. Singh, A. Bhatnagar, V. Dixit, V. Shukla, et al., Synthesis, characterization and hydrogen storage characteristics of ambient temperature dried carbon aerogel, Int. J. Hydrogen Energy 41 (2016) 3561–3570.
[58] H.Y. Tian, C.E. Buckley, D.A. Sheppard, M. Paskevicius, N. Hanna, A synthesis method for cobalt doped carbon aerogels with high surface area and their hydrogen storage properties, Int. J. Hydrogen Energy 35 (2010) 13242–13246.
[59] P. Karikkethu Prabhakaran, L. Catoire, J. Deschamps, Aluminium doping composite metal organic framework by alane nanoconfinement: impact on the room temperature hydrogen uptake, Micropor. Mesopor. Mater. 243 (2017) 214–220.
[60] M. Hirscher, B. Panella, B. Schmitz, Metal organic framework for hydrogen storage, Micropor. Mesopor. Mater. 129 (2010) 335–339.
[61] L.J. Murray, M. Dinca, J.R. Long, Hydrogen storage in metal organic frameworks, Chem. Soc. Rev. 38 (2009) 1294–1314.
[62] D. Zhao, D. Yuan, H.C. Zhou, The current status of hydrogen storage in metal organic frameworks, Energy Environ. Sci. 1 (2008) 222–235.
[63] M.P. Suh, H.J. Park, T.K. Prasad, D.W. Lim, Hydrogen storage in metal organic frameworks, Chem. Rev. 112 (2012) 782–835.
[64] J.L.C. Rowsell, O.M. Yaghi, Strategies for hydrogen storage in metal organic frameworks, Angew. Chem. Int. Ed. 44 (2005) 4670–4679.
[65] K.P. Prasanth, P. Rallapalli, M.C. Raj, H.C. Bajaj, R.V. Jasra, Enhanced hydrogen sorption in single walled carbon nanotube incorporated MIL-101 composite metal organic framework, Int. J. Hydrogen Energy 36 (2011) 7594–7601.
[66] S.J. Yang, J.Y. Choi, H.K. Chae, J.H. Cho, K.S. Nahm, C.R. Park, Preparation and enhanced hydrostability and hydrogen storage capacity of CNT@MOF-5 composite, Chem. Mater. 21 (2009) 1893–1897.

[67] C. Petit, T.J. Bandosz, Enhanced adsorption of ammonia on metal organic framework/graphite oxide composites: analysis of surface interactions, Adv. Funct. Mater. 20 (2010) 111–118.
[68] P.K. Prabhakaran, J. Deschamps, Hydrogen adsorption in lithium doped MIL-101 and MIL-53 Al at 77 K and 298 K up to 100 bar: effect of lithium concentration, J. Porous Mater. 4 (2015) 1073–1081.
[69] P. Karikkethu Prabhakaran, J. Deschamps, Room temperature hydrogen uptake in single walled carbon nanotubes incorporated MIL-101 doped with lithium: effect of lithium doping, J. Porous Mater. 6 (2015) 1635–1642.
[70] P. Karikkethu Prabhakaran, J. Deschamps, Doping activate carbon incorporated composite MIL-101 using lithium: impact on hydrogen uptake, J. Mater. Chem. A 13 (2015) 7014–7021.
[71] M. Barbatti, G. Jalbert, M.A.C. Nascimento, Clustering of hydrogen molecules around a molecular cation: the $Li_3^+(H_2)_n$ clusters (n = 1–6), J. Phys. Chem. A 106 (2002) 551–555.
[72] A. Blomqvist, C.M. Araujo, P. Srepusharawoot, R. Ahuja, Li-decorated metal-organic framework 5: a route to achieving a suitable hydrogen storage medium, Proc. Natl Acad. Sci. U.S.A. 104 (2007) 20173–20176.
[73] P. Dalach, H. Frost, R.Q. Snurr, D.E. Ellis, Enhanced hydrogen uptake and the electronic structure of lithium-doped metal organic frameworks, J. Phys. Chem. C 112 (2008) 9278–9284.
[74] A. Ghoufi, J. Deschamps, G. Maurin, Theoretical hydrogen cryostorage in doped MIL-101 (Cr) metal organic frameworks, J. Phys. Chem. C 116 (2012) 10504–10509.
[75] S.S. Han, W.A. Goddard, Lithium doped metal organic frameworks for reversible H_2 storage at ambient temperature, J. Am. Chem. Soc. 129 (2007) 8422–8423.
[76] M. Kubo, A. Shimojima, T. Okubo, Effect of lithium doping into MIL-53 (Al) through thermal decomposition of anion species on hydrogen adsorption, J. Phys. Chem. C 116 (2012) 10260–10265.
[77] K.L. Mulfort, O.K. Farha, C.L. Stern, A.A. Sarjeant, J.T. Hupp, Post-synthetic alkoxide formation within metal organic framework materials: a strategy for incorporating highly coordinatively unsaturated metal ions, J. Am. Chem. Soc. 131 (2009) 3866–3868.

Further reading

B.E. Conway, Electrochemical Supercapacitors, Kluwer Academic, Plenum, New York, 1999.

10

2D nanomaterials for electrokinetic power generation

Si Qin, Dan Liu and Weiwei Lei
Institute for Frontier Materials, Deakin University, Waurn Ponds Campus, Geelong, Victoria, Australia

10.1 Introduction

10.1.1 Electrokinetic effect and streaming current

Ions are ubiquitously presented in aqueous environment. The transport of ions and the charges they carry are essential to many electrochemical reactions and biological process, for example, in batteries, supercapacitors, fuel cells, and cellular membranes [1]. In the recent years, research on ion transportation in nanoscale has attracted increasing interest since ions behave very differently in nanoscale than those in the bulk solution [2,3]. The knowledge on ion transportation not only deepens our understanding on many fundamental electrochemical and biological processes, and guides the development of modern batteries and supercapacitors, it also enables numerous new applications such as lab-on-chip devices, DNA manipulation [4,5], ion transistors [6–8], water desalination, and energy harvesting [9–11].

In classic fluid dynamics, solution is generally considered to be a homogenous media where equal densities of positive and negative ions are uniformly dispersed and can move freely. This assumption usually works well in microfluidic and macrofluidic systems. However, for a system consisted of nanomaterials, a large amount of the materials surface is exposed to the environment. This high surface-to-volume ratio, on one hand, improves heat and mass transfer between the nanomaterial and the environment. In a nanofluidic system, on the other hand, it provides a large amount of interface for the ions to interact with the materials. In fact, nanofluidic ion transport is largely governed by the surface properties of channel walls which leads to various unique transport phenomena compared to that in the bulk

2D Nanomaterials for Energy Applications. DOI: https://doi.org/10.1016/B978-0-12-816723-6.00010-1

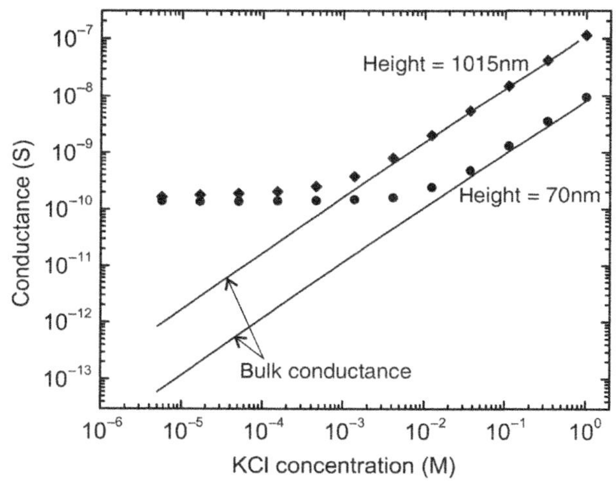

Figure 10.1 Conductance of two channels in KCl aqueous solution made from silica with channel heights of 1015 and 70 nm. The lines reflect the conductance expected in the channel from the bulk solution. *Source*: Adapted with permission from D. Stein, M. Kruithof, C. Dekker, Surface-charge-governed ion transport in nanofluidic channels, Phys. Rev. Lett. 93 (3) (2004) 035901. Copyright 2004 the American Physical Society.

form (Fig. 10.1). More importantly, the interaction between ions and the charged channel wall can no longer be neglected when the channel size approaches the Debye screening length or the thickness of the electric double layer (EDL).

Surface charge on a water/solid interface is common on most materials. The surface charge can be positive or negative, depending on the materials' surface functionality and the pH value of the solution. The amount of positive or negative charge per unit area is called the surface charge density. At a water/solid interface, counterions concentrate in the vicinity of the charged surface and forms what is called the Stern layers. Further away from the interface the distribution of the counter-ions (diffusion) could be described by the Gouy–Chapman model. Within the EDL, co-ions are repelled and excluded from the vicinity of the surface. The thickness of the EDL (Debye screening length) is inversely proportional to the square of the bulk ionic concentration ($\lambda_D \propto \sqrt{c}$) where c is the ionic concentration. As a result, under a lower concentration where the Debye length is equal or larger than the channel size, only counterions can enter and exist in the channel, while co-ions can only present when the concentration is high.

The electrokinetic effect describes the phenomenon where transport of fluid and ion is under the effect of surface charge. For microfluidic and macrofluidic systems, equal amount of positive and negative ions are transported simultaneously, resulting in no net charge transport, while in nanofluidic system, a net charge transport (streaming current) will occur from the transportation of a selective ions across the channel. This streaming current can become prominent in nanofluidic channels, when surface-to-volume ratio is greatly increased. The typical characteristic in the nanofluidic system is that the conductivity through the channels will be governed by surface charge and become independent to the concentration change at low concentration. A comprehensive review on the research and theory on nanofluidic transportation is discussed in a few review papers. This chapter will focus more on the current research progress on energy harvesting using structures or devices fabricated from two-dimensional (2D) nanomaterials.

10.1.2 Harvesting energy from streaming current

In nature, some animals such as electric eels can generate bioelectricity with high voltage through their cell membranes [12,13]. The idea to create artificial devices to mimic this process and harvest energy from seawater has attracted much interest. Theoretically, a remarkable amount of energy is available from water with salinity difference between seawater (~0.6 M NaCl) and fresh water. In theory, up to 2.8 MJ (0.8 kW h^{-1}) of energy is available per cubic meter of seawater could be extracted (counting both positive and negative ions)—equivalent to the energy generated from the same amount of water falling over a dam more than 280 m high [14,15]. Power generation by using the electrokinetic devices had been proposed long ago [16]. In principle, the streaming current can be used to power electric loads when paired with an external circuit. When pumping an electrolyte into a fine channel, the ions in the electrolyte travel through the charged channel. The applied pressure is converted in electric energy (streaming current). This feasibility of harvesting energy from the streaming current (or the so-call blue energy) was explored by Osterle [16] in the 1960s. One way of harvesting this energy is to streaming current which occurs when the electrolyte is driven through nanofluidic channels either by a mechanical pressure gradient or by a salt concentration gradient [17,18].

10.2 Energy harvesting and 2D materials

10.2.1 Fabrication of nanofluidic channels

A number of technologies have been proposed and explored to prepare nanofluidic channels with suitable size [10,17–20]. Conventional methods like lithography are common for microfabrication which can fabricate 1D nanofluidic channel with outstanding precision in terms of shape and length of the channel [19,20]. For example, charged channels can be fabricated by drilling holes in silicon nitride thin films by focused ion beam [21], electron beam lithography on silica substrate [3], and track-etch technique on polymers such as polyimide (PI) and poly(ethylene terephthalate) (PET) [22,23]. However, the procedure requires costly and specialized equipment, materials, and multistep procedure. The number of the channels produced is usually limited to one or a few reported in the research but could be potentially scaled up [17]. Membranes developed for pressure retarded osmosis (PRO) and reversed electrodialysis have seen great development in the recent years, but they are currently still expensive to make and suffer from low selectivity and efficiency [15]. A family of 2D layered materials has recently been explored as a platform to provide large amount of nanofluidic channels with relatively lost fabrication cost. This approach may provide an innovative solution to the problem.

Tremendous efforts have been devoted to the development of synthetic methods of functional 2D nanomaterials and assembly techniques of membrane for nanofluidic devices in terms of low cost, large scale, simplicity in processing, and high performance in practical nanofluidic application. It is also found that the interplay of the various nanoscales, controllable distance of nanochannels, geometric, and surface charges factors, dominates the performance of such nanofluidic devices based on 2D nanomaterials membranes.

10.2.1.1 Graphene oxide

Typically, graphene oxide (GO) is a water-soluble nanosheet synthesized by a modified Hummers' method and post washing. In view of the structure of GO, it remains as a one-atom-thick planar sheet with aromatic sp^2-bonded carbon structure while derivatized with various oxygen-containing functional groups (e.g., hydroxyl, epoxy, carbonyl, carboxyl, etc.) decorated on the basal plane and the sheet edges. This feature allows for assembling of GO membrane unprecedentedly facile for using in energy storage and conversion, sensors, and water purification

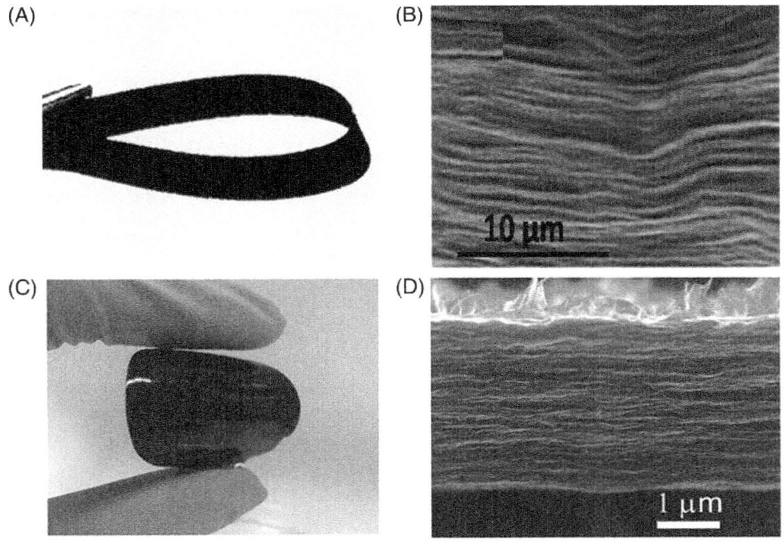

Figure 10.2 (A) A flexible GO strip with lamellar structure and (B) the cross-sectional SEM image of GO strip. (C) A photograph of the freestanding graphene hydrogel membrane and (D) the SEM image of the cross section of the graphene hydrogel membrane after freeze-dried. *Source*: (A and B) Adapted and reprinted with permission from K. Raidongia, J. Huang, Nanofluidic ion transport through reconstructed layered materials, J. Am. Chem. Soc. 134 (40) (2012) 16528–16531. Copyright 2012, American Chemical Society. (C and D) Adapted and reprinted with permission from W. Guo, C. Cheng, Y. Wu, Y. Jiang, J. Gao, D. Li, et al., Bio-inspired two-dimensional nanofluidic generators based on a layered graphene hydrogel membrane, Adv. Mater. 25 (42) (2013) 6064–6068. Copyright 2013, Wiley-VCH.

applications. In 2012, Huang et al. first demonstrated that a GO membrane with 2D laminated structure is an ideal platform for nanofluidic ion transport with novel physical and chemical phenomena, as shown in Fig. 10.2A and B [24]. They showed that the anomalous transport phenomena occurring in the massive arrays of nanofluidic channels with the distance of sub-nanometer range, giving rise to hydrodynamic flow, surface-charge-governed regime, and precise ionic selectivity between neighboring GO layers. The highly negative surface charged nature of GO nanosheets leads to a high affinity to the cations while anions are repelled, supporting cation unipolar selective ion transport. In addition, the nanofluidic channels within GO membrane are very robust due to the high flexibility of the membrane, which can be high mechanical folding for 30 cycles. In order to improve the ion transport performance of GO membranes, subsequently, several strategies including efficient chemical modification and tuning channel spacing have been proposed. For example, Li et al. reported that by using chemically converted graphene as building blocks, the interlayer spacing of the layered graphene hydrogel membrane can be precisely tuned in a wide range from sub-nanometer to more than 10 nm. Because of the surface charged nanocapillaries and the enlarged interlayer distance, the membrane exhibited an improved nanofluidic performance with facilitating selective transport of counterions transport of cations under the pressure-driven ionic flow for energy conversion [10,25]. Compared to the pristine GO sheets with negatively charged (n-GO), the surface of GO sheets can be functionalized and modified to positively charges (p-GO) by the conversion of surface groups from carboxylic acid groups to alkyl-substituted imidazolium groups [26]. Subsequently, both negatively and positively charged 2D nanofluidic ion channels have been assembled in one pairs of oppositely charged GO membrane. The surface charge density of the 2D nanochannels can be efficiently tuned from -123 to $+147$ mC m^{-2}. By coupling the GO membrane pair into a single-unit reverse electrodialysis cell, cations are preferentially transported by the n-GO membrane toward the anode, while anions are migrated to the cathode through the p-GO membrane, resulting in efficient charge separation and creating superposed ionic flux and electrochemical potential for osmotic energy conversion.

10.2.1.2 2D transition metal dichalcogenides

Following the discovery of unique properties of graphene, the family of 2D transition metal dichalcogenides (TMDs) has also garnered a renewed interest. TMDs are the group of hexagonal structured and layered materials with strong in-plane bonding along and weak out-of-plane van der Waals bonding, which comprise over 40 compounds with the molecular formula MX_2 (M: transition metals such as Mo, W, Ta, Ti, Nb, and X: chalcogens such as S, Se, and Te) [27]. As a typical example, molybdenum disulfide (MoS_2) comprises planes wherein two S − Mo − S layers composed of edge sharing MoS_6 trigonal prisms and the neighboring planes bonded each other by van der Waals forces. Therefore MoS_2 nanosheets could be easily exfoliated along these planes of the bulk materials. Interestingly, the MoS_2 nanosheets display unique features such as wide band gaps around 1–2 eV, high mobility of charge carriers and 2D planar advantage, which present a promising application in photovoltaic, chemical sensors, thermal management, and energy storage and conversion that are not necessarily observed in bulk MoS_2 materials [28–31].

Tremendous efforts have been devoted to develop methods for obtaining MoS_2 nanosheets dispersed in organic solvents or aqueous without layer aggregation for easily fabrication as membrane or film for range of applications. Liquid-phase exfoliation is the most efficient way to produce the large quantities of monolayer or few-layer MoS_2 nanosheets by ultrasonic waves in a suitable solvent, which is a mechanical force with appropriate surface energy. Coleman et al. first reported that liquid exfoliation technique through sonication of bulk MoS_2 materials in organic solvents such as *N*-methyl-pyrrolidone (NMP) yielded few-layered MoS_2 nanosheets dispersion [32]. Similarly, Smith et al. proposed that exfoliation of few-layered MoS_2 nanosheets can be obtained in an aqueous containing surfactants (such as sodium cholate, bovine serum albumin) by sonication [33]. Recently, Kang et al. reported that functionalized 2D TMDs including MoS_2, WS_2, WSe_2, and $MoSe_2$ with glucan multivalency were exfoliated using suitable polymers in an aqueous solution by pulsed sonication process (Fig. 10.3A) [34].

Intercalation method is another optional way to exfoliate bulk MoS_2 materials through some atoms or molecules intercalation between the MoS_2 layers and enlarge their interlayer distance. Specifically, the bulk MoS_2 materials in Li containing organic solution of the solution of *n*-butyllithium (C_4H_9Li, BuLi) in hexane yield an intercalation compound Li_xMoO_2. Then, by reacting the intercalated Li_xMoO_2 with water, a colloidal stable dispersion of almost exclusively monolayer MoS_2 was formed (Fig. 10.3B) [35,36].

10.2.1.3 Boron nitride nanosheets

Boron nitride nanosheets (BNNSs) also called "white graphene," consist of one to a few layers of alternating boron and nitrogen atoms in a hexagonal arrangement. BNNSs are ideal candidate in many applications, such as composite reinforcement, electrocatalysts, water purification, thermal management, and energy storage and conversion due to their intrinsic properties including wide band gap, extremely high resistance to oxidation, as well as good chemical and thermal stability [37–41]. However, it is difficult to form a colloidal aqueous suspension of BNNSs due to the strong "$\pi-\pi$" interaction between the BN layers [42].

Up to date, considerable effort has been developed for fabricating functionalize BNNSs to achieve good dispersibility in

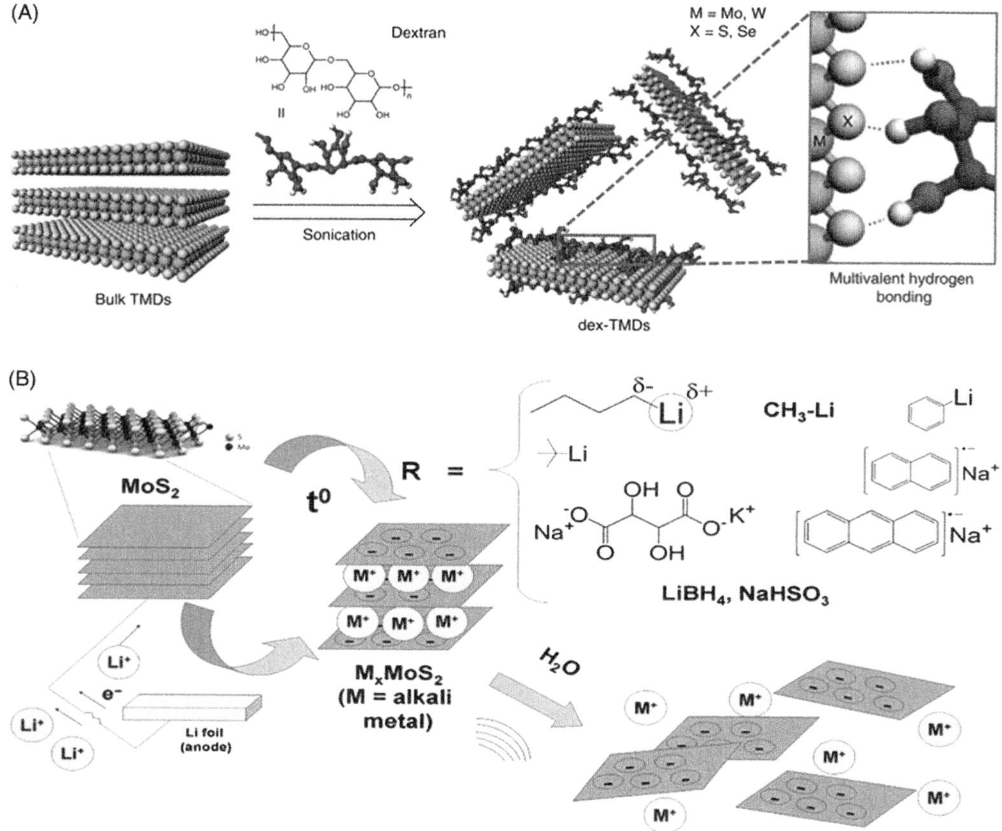

Figure 10.3 (A) Schematic illustration of the liquid-phase exfoliation of TMDs via multivalent hydrogen bonding. (B) Schematic illustration of intercalation exfoliation method toward colloidal MoS_2 nanosheets. *Source*: (A) Adapted and reprinted under Creative Commons Attribution 4.0 International License: http://creativecommons.org/licenses/by/4.0/ from T.W. Kang, J. Han, S. Lee, I.J. Hwang, S.J. Jeon, J.M. Ju, et al., 2D transition metal dichalcogenides with glucan multivalency for antibody-free pathogen recognition, Nat. Commun. 9 (2018) 2549. Copyright 2018 Nature. (B) Adapted and reprinted with permission from E.D. Grayfer, M.N. Kozlova, V.E. Fedorov, Colloidal 2D nanosheets of MoS2 and other transition metal dichalcogenides through liquid-phase exfoliation, Adv. Colloid Interface Sci. 245 (2017) 40–61. Copyright 2017 Elsevier.

water or different solvents via introducing hydrophilic functional groups on the basal plane and edges of BNNSs by free radical covalent function [43], liquid exfoliation [44], Lewis acid function [45], sonication-centrifugation [46], mechanochemical ball milling [47], and so on. For example, Coleman et al. obtained functionalization BNNSs by decorating OH groups on the BNNSs edge via the oxidation of exfoliated h-BN nanosheets by a two-step functionalization procedure, and the concentration of the BNNSs solution was 0.107 mg mL^{-1}, even after long

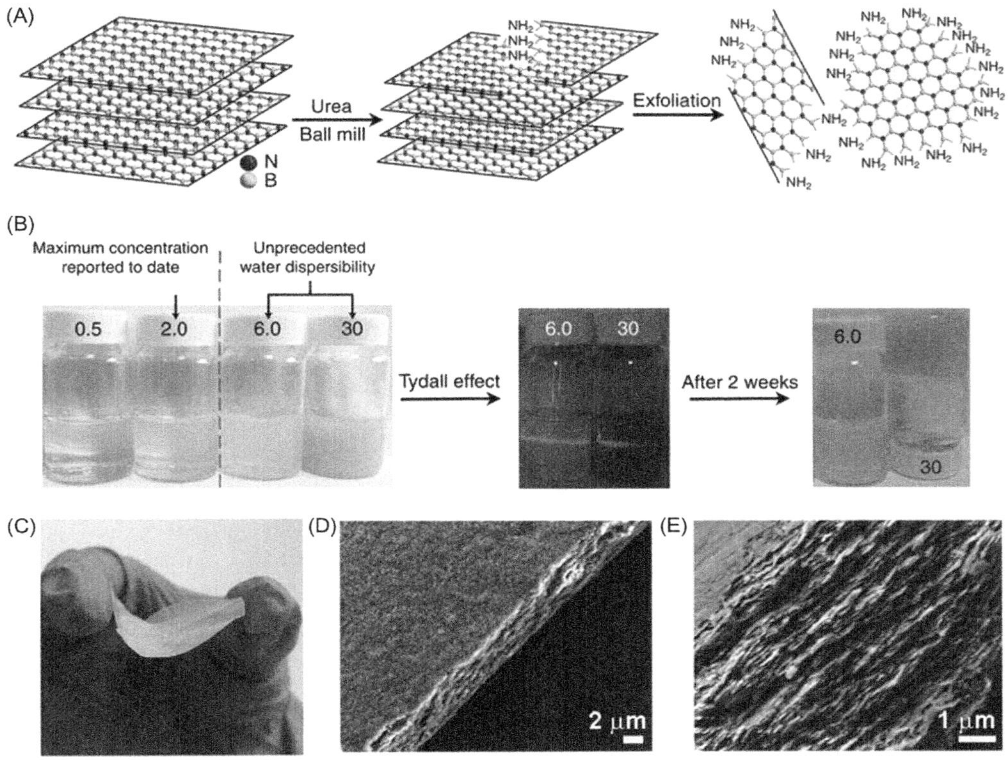

Figure 10.4 (A) Schematic illustration of the exfoliation. (B) Photos of as-prepared colloidal solutions of few-layer BN with concentrations from 0.5 to 30 mg mL^{-1}, and demonstration of the Tyndall effect and stability of the colloidal solutions. (C) Photo of the freestanding BN membrane. (D and E) SEM images of the basal fracture surface of the BN membrane, respectively. *Source*: Adapted and reprinted under Creative Commons Attribution 4.0 International License: http://creativecommons.org/licenses/by/4.0/ from W.W. Lei, V.N. Mochalin, D. Liu, S. Qin, Y. Gogotsi, Y. Chen, Boron nitride colloidal solutions, ultralight aerogels and freestanding membranes through one-step exfoliation and functionalization, Nat. Commun. 6 (2015) 8849. Copyright 2015 Nature.

times of intense bath-sonication [43]. Fu and coworkers have prepared few-layers BNNSs dispersion with a production yield of 19% by a facile liquid exfoliation under the assistance of molten sodium hydroxide [48]. Recently, Lei et al. has developed a simple and efficient one-step method for the preparation and functionalization of few-layer BN nanosheets by mechanochemical ball milling of commercially available h-BN and urea powder, as shown in Fig. 10.4 [47]. The results showed that the concentration of stable water-soluble BNNSs suspensions was up to 30 mg mL^{-1} due to a high concentration of negative $-NH_2$ groups. In addition, the yield of the exfoliated BNNSs was as high as 85%. Such high stable BNNSs suspensions can be directly fabricated to BN membranes using simple vacuum-assisted filtration for 2D nanofluidic applications.

10.2.1.4 Clay (kaolinite and vermiculite)

Clays are the group of naturally occurring phyllosilicate minerals, composed of tetrahedral silicate layers and octahedral hydroxide layers by comparatively weak van der Waals, hydrogen bonding, and Coulomb attraction forces between the two adjacent layers [49]. It can be classified into two types: layered 1:1 (such as kaolinite and nacrite) and 2:1 (such as vermiculite and illite) phyllosilicate structures based on the ratio of tetrahedron and octahedron stacked layers [50]. Such distinctive lamellar structure allows clay minerals to be exfoliated into 2D nanosheets, endowing them with high surface area and rich surface functionality properties to lead diverse applications in catalysis [51], drug delivery [52], and energy conversion [53]. Sun et al. summarized three different exfoliation, including direct delamination, intercalation-assisted exfoliation, and ion exchange–assisted exfoliation methods according to various stacking structures of clays, as shown in Fig. 10.5 [54]. Shear mixing and/or ultrasonication are the direct delamination methods to exfoliate clays. For example, Samori et al. reported that ultrathin vermiculite nanosheets with a yield of 27% were successfully produced by using ultrasound-assisted exfoliation in aqueous hydrochloric acid solutions over 3 h [55].

Intercalation of various guest species is a promising strategy to swelling the layered structure of clay minerals for their further delamination. This method is most used for exfoliating of kaolinite due to their tighter bound stacked layers by van der Waals interactions and a network of hydrogen than the 2:1 type clays. In the past few years, tremendous efforts have been devoted to prepare internal surface modification of kaolinite through the intercalation of different reactive guest species with various chain lengths including dimethyl sulfoxide, carboxylic acids, and methanol (short-chain) and hexylamine and 3-aminopropyltriethoxysilane (medium-chain), and dodecylamine (long-chain), aiming to swell the basal spacing and weak the cohesive force between of clay layers for facilitating the delamination of clay minerals [56]. In order to further exfoliate ultrathin clay nanosheets, subsequently, certain agitation methods such as stir, ultrasonication, and thermal shock treatment were involved to help fully separate the loosened clay layers, forming 2D nanosheets. Recently, ultrathin kaolinite nanosheets with

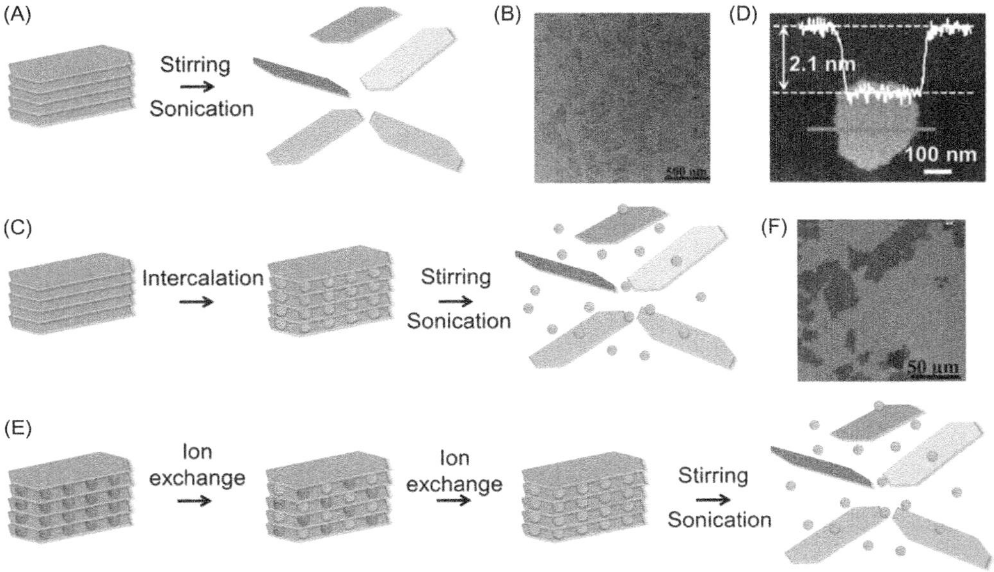

Figure 10.5 Representative exfoliation methods and exfoliated nanosheets. (A) Direct delamination. (B) SEM image of the exfoliated MMT nanosheets by intercalation-assisted exfoliation. (C) Intercalation-assisted exfoliation. (D) Atomic force microscopy (AFM) image of exfoliated kaolinite nanosheets with a combination of multiple methods. (E) Ion exchange–assisted exfoliation and (F) SEM image of exfoliated VMT nanosheets with ion exchange–assisted exfoliation [54]. *Source*: Adapted with permission from Y. Zhou, A.M. LaChance, A.T. Smith, H. Cheng, Q. Liu, L. Sun, Strategic design of clay-based multifunctional materials: from natural minerals to nanostructured membranes, Adv. Funct. Mater. 29 (16) (2019) 1807611. Copyright 2019 Advanced Functional Materials.

the thickness from 2 to 20 nm with 75.7% of which were thinner than 12 nm were successfully exfoliated by Jiang et al. using intercalation and post ultrasonication method [53]. Then the kaolinite-based Janus-like nanobuilding blocks were successfully fabricated into lamellar nanofluidic devices with high capability in osmotic and hydraulic energy conversion.

Ion exchange–assisted exfoliation method is another effective route to get 2D clay nanosheets. For example, a representative form of smectite clays, typically Si^{4+} and Al^{3+} ions located at the central of tetrahedral and octahedral lattice, respectively, can be substituted by the lower-valence metal ions (such as Al^{3+}, Fe^{2+}, and Mg^{2+}), inducing a charge imbalance as each sheet carry a negative layer charge. Usually, the hydratable cations, such as Na^+ or Ca^{2+} used to compensate the negative layer charge, consequently reduce the strong electrostatic forces between of clay layers. In order to achieve further exfoliation, additional ion-exchange treatment by larger cations to significantly weak or even break the interlayer bonding and expand interlayer spacing is required for facilitating exfoliation [57,58]. Recently, Huang et al. reported that the bilayer vermiculite nanosheets with few microns lateral dimension have been successfully exfoliated by two consecutive steps of ion exchange [59]. The results showed that the concentration of stable water-soluble vermiculite nanosheets dispersion was 1.5 mg mL^{-1}, which can be readily reassembled to form membrane with lamellar nanofluidic channels, exhibiting high surface charge-governed proton conductivity and extraordinary thermal stability for nanofluidic device [59].

10.2.2 Nanofluidic channels constructed by 2D nanomaterials

First observed on graphene oxide (GO) membrane [60], it was found that the nanochannels from the restacking of 2D nanomaterials are large enough to allow water molecules and ions to pass through with little impedance. This has opened up new avenues for material design and large-scale integration of artificial nanofluidic channels. Later, nanofluidic ion transport was observed on many other 2D nanomaterials such as boron nitride (BN), metal dichalcogenides (MoS_2 and WS_2) [61,62], Ti_3C_2 MXene and layered metal oxide (kaolinite and vermiculite) [53,63–65]. The restacking of 2D nanomaterials are usually achieved by filtration of a colloidal dispersion of the 2D nanomaterials, as the shearing force during the filtration process helps the nanosheets to stack in an orderly manner. The interlayer space between restacked individual nanosheets forms lamellar channels for fluidic and ionic transport.

Compared with conventionally methods, fabricating nanofluidic channels from 2D nanomaterials has many advantages. The first advantages is the facile fabrication. In the case of filtration, the fabrication only requires simple equipment setup, minimal processing technique, and no chemical modification process. The filtration-based method also means that the fabrication is easy to scale up. While this is beneficial for lowing the fabrication cost of the membrane, the main cost of the membrane is associated with synthesizing and preparation of the required 2D nanomaterials. Up to date, the cost for manufacturing 2D nanomaterials are still high and generally uneconomical for membrane manufacturing. For example, (although price of commercially available GO flakes varies depending on the quality), highly quality GO (in dispersion form) is currently available at a price around USD $40,000 per kilogram [66], so it is estimated that the cost for fabricating a membrane at a density 1 mg cm^{-2} is higher than USD $400 per square meter at the moment. However, the price is anticipated to come down dramatically with manufacturing capacity increasing in the future, meanwhile, other 2D nanomaterials such as layered metal oxides which are derived from natural mineral cost much less to produce. It is anticipated that the cost for the manufacture of 2D nanomaterials membranes will make financial sense in the near future.

The second advantage of membrane based on 2D nanomaterials is that the whole process is highly customizable. The order structure of the nanosheets can be modulated by using different filtration methods (Fig. 10.6). The thickness and size of the membrane can be easily adjustable by changing the size of the filter and amount of 2D nanomaterials for filtration. The sheet size of the 2D nanomaterial can be controlled by controlling the synthesis process and by using posttreatment process such as sonication and mechanical/chemical exfoliation. Simulation and experimental results have shown that nanofluidic channels within 2D material membrane are dependent on the properties (e.g. lateral size, thickness, and surface functionalities) of the nanosheet [68,69]. For example, GO membranes fabricated from small sheets usually offer higher flux, while those with larger nanosheets offer higher selectivity [70]. When adjusting the size of GO nanosheets and the thickness of the membrane, the effective channel length for ion transport can be controlled. However, there is often a trade-off between

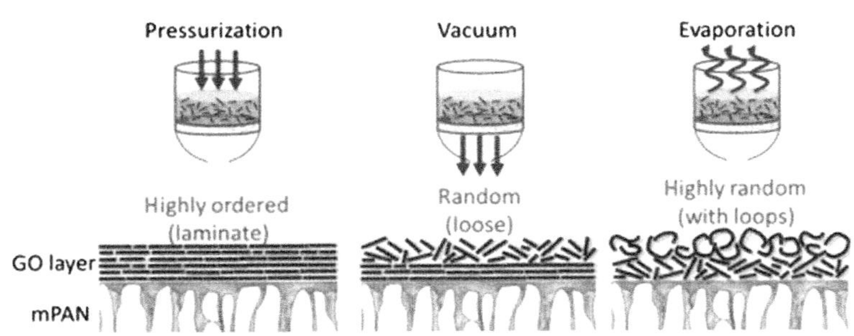

Figure 10.6 Schematic diagrams showing the difference in GO membrane structure fabricated by pressure filtration, vacuum filtration, and evaporation [67]. *Source*: Reprinted with permission from C.-H. Tsou, Q.-F. An, S.-C. Lo, M. De Guzman, W.-S. Hung, C.-C. Hu, et al., Effect of microstructure of graphene oxide fabricated through different self-assembly techniques on 1-butanol dehydration, J. Membr. Sci. 477 (2015) 93–100. Copyright 2015 Elsevier.

permeability and selectivity, increasing the selectivity is often at the cost of permeability. In general, the easily adjustable 2D nanomaterial membrane will provide some degree of control on some parameters such as effective size and length of the channels, and flux through the membrane, which enables modulating the ion transport across the membrane.

More importantly, as nanofluidic ion transport is governed by surface charge on the channel wall, the high aspect ratio of 2D materials provides a large contact area, and thus dramatically increases the interaction between ions and channel wall. In addition, chemical modification/functionalization of the 2D materials prior to the fabrication is a feasible way to modulate the surface charge, For example, it has been reported that new surface functional groups can be introduced onto Ti_3C_2 MXene by simple solution treatment to change the surface charge of the MXene sheets [71,72]; there also are many reports on surface functionalization of GO [73–75]. This strategy enables the design of channels suitable for specific ion species and concentration, and the possibility of achieving a higher efficiency.

However, many surface functional groups show different dissociation level under different pH, which means that the ion transport behavior will be affected when there a change in pH. For example, in highly acidic or basic conditions ($pH < 2$ or > 11), the GO membranes are nearly impermeable to water [76]. Moreover, highly basic conditions may lead to quick deoxygenation of GO [77–79]. Highly basic conditions can even devastate the ionic channels in GO membranes. In general, there are still many challenges to use the GO membranes in practical industry application and there is an urgent need to explore better membranes.

10.3 Energy harvesting using 2D materials

10.3.1 Pressure-driven energy harvesting

Pressure-driven energy conversion in nanofluidic system was proposed long ago [16]. When electrolyte is driven through charged nanofluidic channels buy external pressure, a streaming current can be observed and this energy can be harvested to power electric loads. A schematic illustration can be found in Fig. 10.7. This advantage of the pressure-driven devices is that the same electrolyte can be used on both sides of the membrane—it does not require electrolyte with different concentration, which is different from the salt gradient-driven system

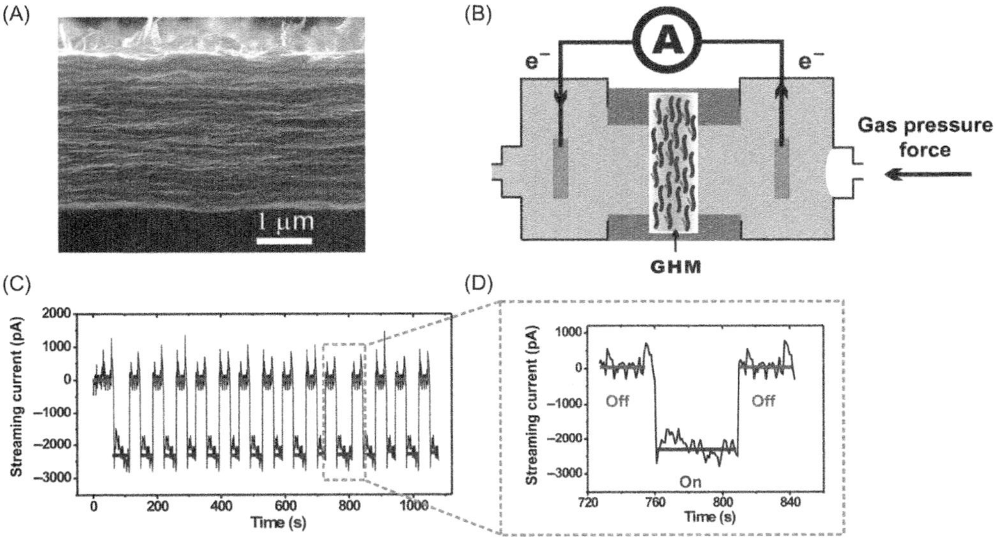

Figure 10.7 (A) Cross-section SEM image of a freeze-dried GO membrane shows a highly order layered structure. (B) Schematic showing the experimental design and the streaming current resulted from the pressure. (C and D) Synchronized current was observed when the gas pressure (5 kPa) was toggled. The current output was stable and reproducible for many cycles. *Source*: Adapted and reprinted with permission from W. Guo, C. Cheng, Y. Wu, Y. Jiang, J. Gao, D. Li, et al., Bio-inspired two-dimensional nanofluidic generators based on a layered graphene hydrogel membrane, Adv. Mater. 25 (42) (2013) 6064–6068. Copyright 2015 Wiley-VCH.

discussed in detail in Section 10.3.2. This characteristic makes the pressure-driven device potentially useful for harvesting mechanical energy from sources such as footsteps, doors, and other physical movements. Early research mainly focused on achieving a charged fluidic channel through porous material [80,81]. It was not until the recent years that fabricating fluidic channels in the nanometer scale through controllable process became feasible [18,82]. Daiguji et al. investigated the theoretical transport phenomena inside nanotube under external pressure gradient and the resulting current-potential ($I-V$) characteristics, and concluded that the maximum efficiency is reached when the Debye length of the solution is about half of the channel height [9]. Later, van der Heyden et al. experimentally studied the electrokinetic power generation in individual rectangular silica nanofluidic channel [20], and found that the generated streaming potential across the nanotube is proportional to the applied pressure.

Guo et al. was the first to report power generation using restacked 2D nanomaterial (GO) membrane [10]. The graphene

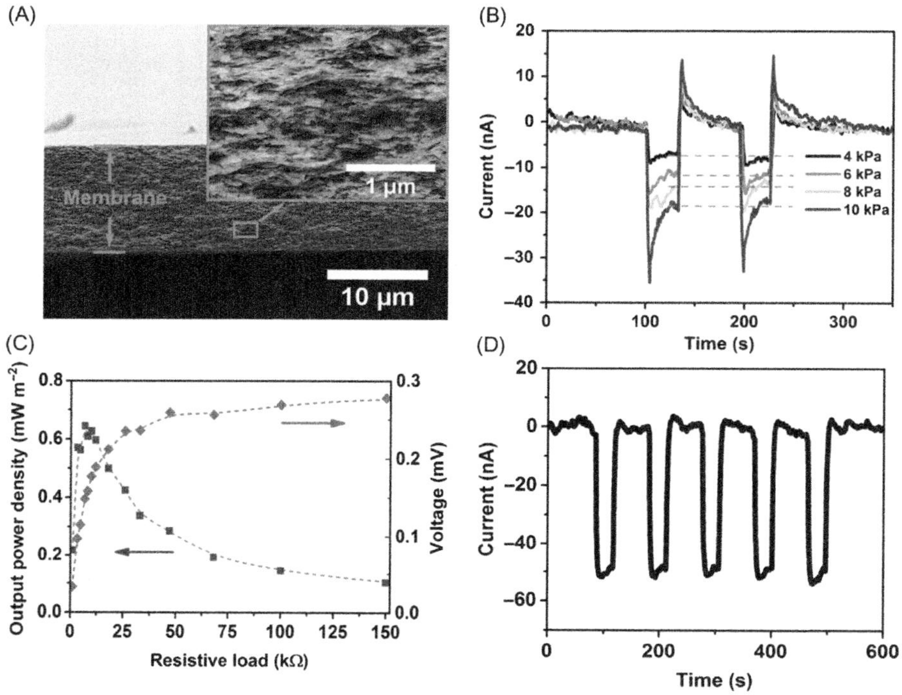

Figure 10.8 (A) Cross-section SEM images of a BN membrane showing the closely stacked BN nanosheets. (B) Current response across the BN membrane from different pressures. The current scales with applied pressure proportionally. (C) Output power and voltage when the system was powering resistive external loads. (D) Current response of a device with five BN membrane devices in a parallel connection. *Source*: Adapted and reprinted with permission from S. Qin, D. Liu, Y. Chen, C. Chen, G. Wang, J. Wang, J. M. Razal, W. Lei, Nanofluidic Electric Generators Constructed from Boron Nitride Nanosheet Membranes, Nano Energy 47 (2018) 368–373. Copyright 2018 Elsevier.

hydrogel membrane was fabricated by reconstructing chemically converted graphene nanosheets (Fig. 10.7). A streaming current of approximately 2 nA was observed with a gas pressure of 5 kPa across the GO membrane ($\sim 3\ mm^2$).

Qin et al. reported nanofluidic ion transport behavior on BN membrane (Fig. 10.8) [64,83]. Being highly chemical inert, the membrane fabricated from BN nanosheets shows stable ion transport characteristic in acidic and alkaline solutions. The streaming current across BN membrane ($\sim 5\ mm^2$) was up to 13 nA when driven by a 5 kPa gas pressure. It was found that current is proportional to the applied pressure, in consistent with the prediction made on a single channel nanofluidic system [9]. More importantly, it was demonstrated in this work that the streaming current scales proportionally with the number of nanofluidic devices connected together, suggesting that higher power can be achieved using multiple small devices.

Both theoretical prediction and experimental results suggest that streaming current increases with the increasing driving pressure. However, the challenge with pressure-driven system is to design membrane robust enough to withstand the physical pressure applied on the membrane. Currently, the amount of pressure that a 2D nanomaterials membrane can withstanding is still low (usually around 5 kPa), and it often requires a supporting membrane to achieve sufficient mechanical strength required for the application. Fabrication of a strong 2D material membrane which truly reflects the high mechanical strength of the 2D nanomaterials remains a challenge for further improving the power output of the pressure-driven nanofluidic generators.

10.3.2 Osmotic energy harvesting from salinity gradient

Unlike pressure-driven devices, devices driven by salinity gradient rely on the concentration difference across the two sides of the membrane. The membrane is the only component that provides the ionic pathway between the two electrolytes. This configuration generates an osmotic pressure which is dependent on the salinity gradient within the membrane [80]. The osmotic pressure generated within the membrane can reach a very high level. For example, a 1 M salinity difference (e.g., NaCl) at 25°C could generate an osmotic pressure in the order of ~5000 kPa, which is around 1000 times greater than typical pressure used in the pressure-driven system. This tremendous osmotic pressure usually translates to a much higher power output capability in the salinity-driven nanofluidic generators. In addition, salinity difference (e.g., between seawater and fresh water) is a naturally occurring "resources", which is widely available around the globe.

Siria et al. reported giant osmotic power through single BN nanotube with inner diameter in the range of 15–40 nm and a length of ~1 μm [18]. The maximum power density per unit tube surface reaches 4 kW m^{-2} for the single BNNT, which is several orders of magnitude higher than the power densities (usually around a few tens of watts per square meter) of classical reversed electrodialysis.

Feng et al. reported and demonstrated a single-layer molybdenum disulfide (MoS_2) nanopores as osmotic nanopower generator (Fig. 10.9) [53]. Nanopores with size in the range of

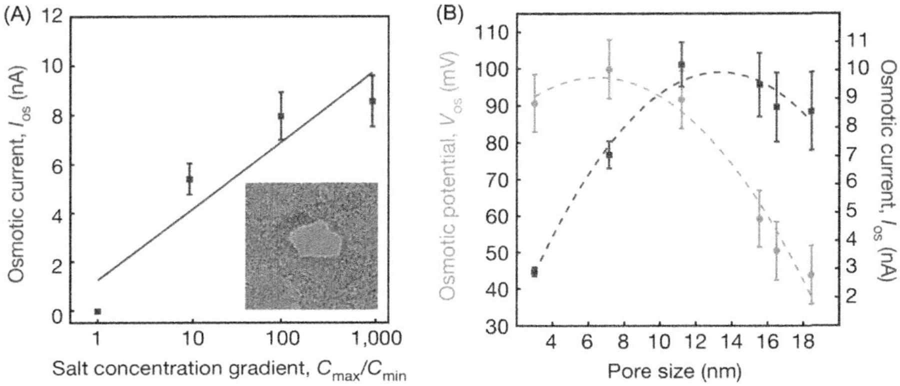

Figure 10.9 (A) Osmotic current as a function of salt gradient. The solid line is a linear fitting of the results. Inset showing the MoS_2 with a nanopore around 5 nm drilled by using TEM. (B) Osmotic potential and current as a function of pore size. The dashed lines are a guide to the eye and show the trend as the pore size changes. *Source*: Adapted and reprinted with permission from J. Feng, M. Graf, K. Liu, D. Ovchinnikov, D. Dumcenco, M. Heiranian, et al., Single-layer MoS2 nanopores as nanopower generators, Nature 2016, advance online publication. Copyright 2016 Springer Nature.

2–25 nm were "drilled" on a 0.65-nm-thick, single-layer MoS_2 using either transmission electron microscope or electrochemical reaction. A large, osmotically induced current was observed with an estimated power density up to 106 $W\,m^{-2}$—when normalized to the atomically thin MoS_2 membrane. This work has demonstrated osmotic energy harvesting at an atomic level, and, more importantly, shown the tremendous amount of energy that can potentially be extracted from osmotic pressure.

The abovementioned work demonstrates at an atomic level the potential energy available from osmotic pressure. Guo and coworkers demonstrated the energy harvesting using membrane constructed from 2D nanomaterials (Fig. 10.10) [26]. In the work, GO was treated prior to membrane assembly to achieve different surface charge polarity on the surface. This procedure enables the efficiently tuning of the surface charge density of the GO from negative ($-123\,mC\,m^{-2}$) to positive ($+147\,mC\,m^{-2}$). GO membrane (GOM) with different charges can be used as either cation- or anion-selective membrane which could improve the efficiency when they are paired together in a nanofluidic cell. The device fabricated from the GOM could deliver a maximum output power density of 0.77 $W\,m^{-2}$ with an electrolyte (NaCl) concentration gradience of 0.5 versus 0.01 M. This power density is about 54% higher than that using commercial ion exchange membranes. Furthermore, this work demonstrated tandem connected

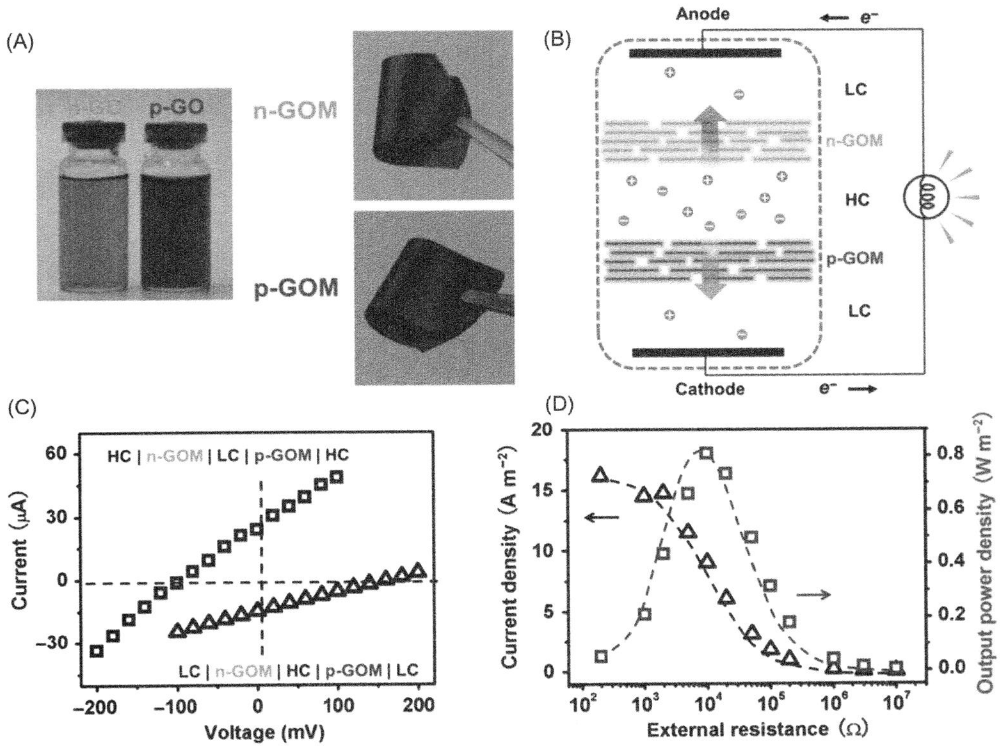

Figure 10.10 (A) Negatively and positively charge GO dispersion and assembled GOM. (B) Schematic illustration of the device and mechanism. (C) Current–voltage responses for a single nanofluidic reverse electrodialysis device measured with the two configurations. (D) Output current and power density of the GOM device under different external load. *Source*: Adapted and reprinted with permission from J. Ji, Q. Kang, Y. Zhou, Y. Feng, X. Chen, J. Yuan, et al., Osmotic power generation with positively and negatively charged 2D nanofluidic membrane pairs, Adv. Funct. Mater. 27 (2) (2017) 1603623. Copyright 2016 Wiley-VCH.

alternating GOM pairs can extend the practical operating voltage of the device, more specifically, fifteen connected devices with an output voltage of 2.7 V were capable to power electronic devices such as LED and electronic calculator.

Later, Guo and coworkers demonstrated osmotic energy harvesting using membrane fabricated from kaolinite (RKM) (Fig. 10.11) [53]. The kaolinite composes of a silicon tetrahedral sheet (STS) and an aluminum octahedral sheet (AOS) separated by a plane of oxygen atoms. The STS and AOS layers exhibit different reactivity for chemical modification due to the presence of hydrophilic μ-Al_2-OH groups on the STS surface. Through chemical functionalization, the charge on two sides of the nanosheets can be modified. This asymmetric crystal structure

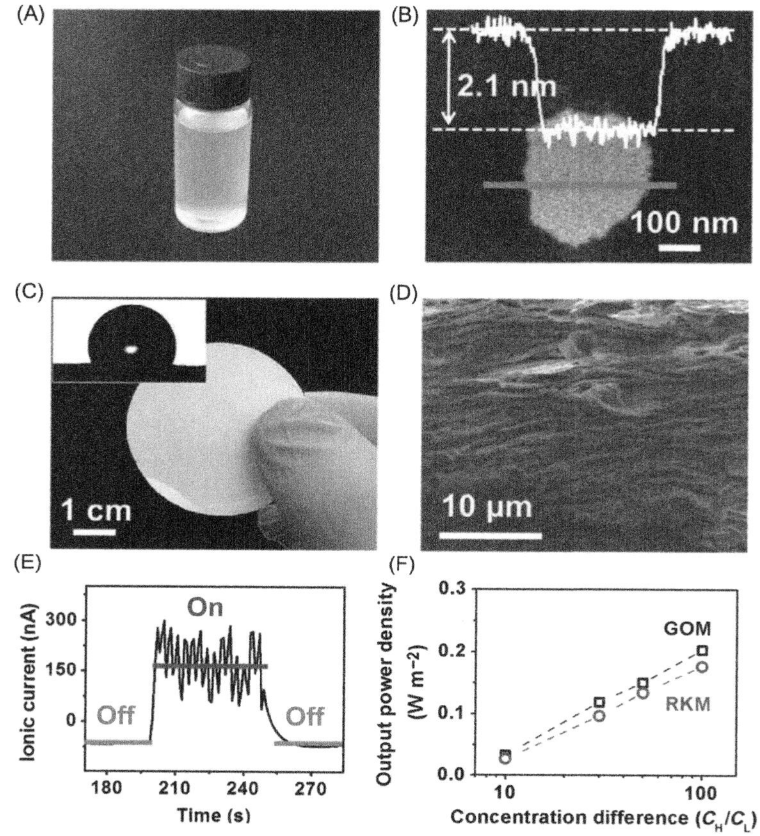

Figure 10.11 (A) Photo showing the nanokaolinite dispersion (0.24 mg mL^{-1}). (B) AFM image and height profile of the nanokaolinite flake with lateral size of ~320 nm and thickness of 2.1 nm (three layers). (C) A photograph of the white-colored free-standing RKM. (D) SEM image of the cross-section of the RKM showing order and lamellar structure. *Source*: Adapted and reprinted with permission from H. Cheng, Y. Zhou, Y. Feng, W. Geng, Q. Liu, W. Guo, et al., Electrokinetic energy conversion in self-assembled 2D nanofluidic channels with janus nanobuilding blocks, Adv. Mater. 29 (23) (2017) 1700177. Copyright 2017 Wiley-VCH.

could create large superposed dipoles and cohesive energy in the lamellar structure. The reconstructed RKM contains two types of nanofluidic channels with sizes of about 6.8 and 13.8 Å. The nanogenerator fabricated using RKM can deliver an output power density of ~0.18 W m^{-2}, which is slightly lower compared to that obtained on the GOM-based device. The kaolinite as a naturally occurring mineral provides a cost-effective alternative for membrane based on GO.

10.4 Challenges and outlook

Research on 2D nanomaterials membranes for electrokinetic power generation began few years ago, and the progress has been very rapid. In this review chapter, we have highlighted some important studies on these areas. The interstitial space between opposite nanosheets can be treated as lamellar channels for fluidic, ionic, and molecular transport. Therefore 2D nanomaterials membranes with nanofluidic channels can be used to harvest electrokinetic power. In such a system, when ions move through the nanoscale fluidic channels, an electric field is formed due to the surface charges of the channel walls and their interaction with ions can be converted into electrokinetic energy (electricity). In addition, theoretically, the 2D nanomaterials membranes with nanofluidic channels function as a fence, allowing certain molecules and ions to pass through easily while blocking others depending upon the size of nanochannels. Therefore precise and uniform control of the size and surface charge of nanochannel is beneficial for specific power generation and water treatment applications. There are still some critical research issues including the low generated voltage, low efficiency, and poor scalability, and other components of the electrokinetic power generation devices that need to be solved before they can be utilized in practical devices. With continued research and development in this field, we can expect that the 2D nanomaterials membranes for practical applications in electrokinetic power generation, energy storage (separator for batteries), and environmental applications (molecular separation and water treatment) will soon be in sight.

Acknowledgment

This work was financially supported by the Australian Research Council Discovery Program (DP190103290) and Australian Research Council Discovery Early Career Researcher Award scheme (DE150101617 and DE140100716).

References

[1] C. Miller, Ionic channels of excitable membranes., Cell 69 (4) (1992) 579.
[2] H. Daiguji, P. Yang, A. Majumdar, Ion transport in nanofluidic channels, Nano Lett. 4 (1) (2004) 137–142.
[3] D. Stein, M. Kruithof, C. Dekker, Surface-charge-governed ion transport in nanofluidic channels, Phys. Rev. Lett. 93 (3) (2004) 035901.
[4] H. Liu, J. He, J. Tang, H. Liu, P. Pang, D. Cao, et al., Translocation of single-stranded DNA through single-walled carbon nanotubes, Science 327 (5961) (2010) 64–67.
[5] K.-G. Wang, S. Yue, L. Wang, A. Jin, C. Gu, P.-Y. Wang, et al., Manipulating DNA molecules in nanofluidic channels, Microfluid. Nanofluid. 2 (1) (2006) 85–88.
[6] K. Tybrandt, R. Forchheimer, M. Berggren, Logic gates based on ion transistors, Nat. Commun. 3 (2012) 871.

[7] K. Tybrandt, K.C. Larsson, A. Richter-Dahlfors, M. Berggren, Ion bipolar junction transistors, Proc. Natl. Acad. Sci. U.S.A. 107 (22) (2010) 9929–9932.
[8] Z. Zhang, X.-Y. Kong, K. Xiao, G. Xie, Q. Liu, Y. Tian, et al., A bioinspired multifunctional heterogeneous membrane with ultrahigh ionic rectification and highly efficient selective ionic gating, Adv. Mater. 28 (1) (2016) 144–150.
[9] H. Daiguji, P. Yang, A.J. Szeri, A. Majumdar, Electrochemomechanical energy conversion in nanofluidic channels, Nano Lett. 4 (12) (2004) 2315–2321.
[10] W. Guo, C. Cheng, Y. Wu, Y. Jiang, J. Gao, D. Li, et al., Bio-inspired two-dimensional nanofluidic generators based on a layered graphene hydrogel membrane, Adv. Mater. 25 (42) (2013) 6064–6068.
[11] Y. Xu, P. Chen, H. Peng, Generating electricity from water through carbon nanomaterials, Chem. Eur. J. 24 (24) (2018) 6287–6294.
[12] M. Altamirano, Electrical properties of the innervated membrane of the electroplax of electric eel, J. Cell. Comp. Physiol. 46 (2) (1955) 249–277.
[13] J. Xu, D.A. Lavan, Designing artificial cells to harness the biological ion concentration gradient, Nat. Nanotechnol. 3 (11) (2008) 666–670.
[14] R.E. Pattle, Production of electric power by mixing fresh and salt water in the hydroelectric pile, Nature 174 (4431) (1954). 660.
[15] O. Schaetzle, C.J.N. Buisman, Salinity gradient energy: current state and new trends, Engineering 1 (2) (2015) 164–166.
[16] J.F. Osterle, Electrokinetic energy conversion, J. Appl. Mech. 31 (2) (1964) 161–164.
[17] F.H.J. van der Heyden, D. Stein, C. Dekker, Streaming currents in a single nanofluidic channel, Phys. Rev. Lett. 95 (11) (2005) 116104.
[18] A. Siria, P. Poncharal, A.-L. Biance, R. Fulcrand, X. Blase, S.T. Purcell, et al., Giant osmotic energy conversion measured in a single transmembrane boron nitride nanotube, Nature 494 (7438) (2013) 455–458.
[19] F.H.J. van der Heyden, D.J. Bonthuis, D. Stein, C. Meyer, C. Dekker, Power generation by pressure-driven transport of ions in nanofluidic channels, Nano Lett. 7 (4) (2007) 1022–1025.
[20] F.H.J. van der Heyden, D.J. Bonthuis, D. Stein, C. Meyer, C. Dekker, Electrokinetic energy conversion efficiency in nanofluidic channels, Nano Lett. 6 (10) (2006) 2232–2237.
[21] A.J. Storm, J.H. Chen, X.S. Ling, H.W. Zandbergen, C. Dekker, Fabrication of solid-state nanopores with single-nanometre precision, Nat. Mater. 2 (8) (2003) 537–540.
[22] Y. Tian, Z. Zhang, L. Wen, J. Ma, Y. Zhang, W. Liu, et al., A biomimetic mercury(II)-gated single nanochannel, Chem. Commun. 49 (91) (2013) 10679–10681.
[23] K. Xiao, L. Chen, G. Xie, P. Li, X.-Y. Kong, L. Wen, et al., A bio-inspired dumbbell-shaped nanochannel with a controllable structure and ionic rectification, Nanoscale 10 (15) (2018) 6850–6854.
[24] K. Raidongia, J. Huang, Nanofluidic ion transport through reconstructed layered materials, J. Am. Chem. Soc. 134 (40) (2012) 16528–16531.
[25] X. Yang, C. Cheng, Y. Wang, L. Qiu, D. Li, Liquid-mediated dense integration of graphene materials for compact capacitive energy storage, Science 341 (2013) 534–537.
[26] J. Ji, Q. Kang, Y. Zhou, Y. Feng, X. Chen, J. Yuan, et al., Osmotic power generation with positively and negatively charged 2D nanofluidic membrane pairs, Adv. Funct. Mater. 27 (2) (2017) 1603623.

[27] M.A. Lukowski, A.S. Daniel, F. Meng, A. Forticaux, L. Li, S. Jin, Enhanced hydrogen evolution catalysis from chemically exfoliated metallic MoS_2 nanosheets, J. Am. Chem. Soc. 135 (2013) 10274–10277.
[28] J. Zhang, W.W. Lei, J. Schutz, D. Liu, B. Tang, C.H. Wang, et al., Improving the gas barrier, mechanical and thermal properties of poly(vinyl alcohol) with molybdenum disulfide nanosheets, J. Polym. Sci. Part B: Polym. Phys. 57 (7) (2019) 406–414.
[29] S. Qin, W.W. Lei, D. Liu, Y. Chen, In-situ and tunable nitrogen-doping of MoS_2 nanosheets, Sci. Rep. 4 (2014) 7582.
[30] D. Jariwala, V.K. Sangwan, L.J. Lauhon, T.J. Marks, M.C. Hersam, Emerging device applications for semiconducting two-dimensional transition metal dichalcogenides, ACS Nano 8 (2014) 1102–1120.
[31] X. Yu, M.S. Prévot, K. Sivula, Multiflake thin film electronic devices of solution processed 2D MoS_2 enabled by sonopolymer assisted exfoliation and surface modification, Chem. Mater. 26 (20) (2014) 5892–5899.
[32] J.N. Coleman, M. Lotya, A. O'Neill, S.D. Bergin, P.J. King, U. Khan, et al., Two dimensional nanosheets produced by liquid exfoliation of layered materials, Science 331 (6017) (2011) 568–571.
[33] R.J. Smith, P.J. King, M. Lotya, C. Wirtz, U. Khan, S. De, et al., Largescale exfoliation of inorganic layered compounds in aqueous surfactant solutions, Adv. Mater. 23 (34) (2011) 3944–3948.
[34] T.W. Kang, J. Han, S. Lee, I.J. Hwang, S.J. Jeon, J.M. Ju, et al., 2D transition metal dichalcogenides with glucan multivalency for antibody-free pathogen recognition, Nat. Commun. 9 (2018) 2549.
[35] E.D. Grayfer, M.N. Kozlova, V.E. Fedorov, Colloidal 2D nanosheets of MoS_2 and other transition metal dichalcogenides through liquid-phase exfoliation, Adv. Colloid Interface Sci. 245 (2017) 40–61.
[36] Z. Zeng, T. Sun, J. Zhu, X. Huang, Z. Yin, G. Lu, et al., An effective method for the fabrication of few-layer-thick inorganic nanosheets, Angew. Chem. Int. Ed. 51 (2012) 9052–9056.
[37] C. Chen, J. Wang, D. Liu, C. Yang, Y. Liu, R.S. Ruoff, et al., Functionalized boron nitride membranes with ultrafast solvent transport performance for molecular separation, Nat. Commun. 9 (2018) 1902.
[38] C. Chen, D. Liu, J. Wang, L. Wang, J. Sun, W.W. Lei, Functionalized boron nitride membranes with multipurpose and super-stable semi-permeability in solvents, J. Mater. Chem. A 6 (2018) 21104–21109.
[39] Y.P. Wu, Y. Xue, S. Qin, D. Liu, X. Wang, X. Hu, et al., BN nanosheet/polymer films with highly anisotropic thermal conductivity for thermal management application, ACS Appl. Mater. Interfaces 9 (49) (2017) 43163–43170.
[40] J. Wang, Y. Wu, Y. Xue, D. Liu, X. Wang, X. Hu, et al., Super compatible functional BN nanosheets/polymer films with excellent mechanical property and ultra-high thermal conductivity for thermal management, J. Mater. Chem. C 6 (2018) 1363–1369.
[41] J. Hao, J. Wang, S. Qin, D. Liu, Y. Li, W.W. Lei, B/N co-doped carbon nanosphere frameworks as high-performance electrodes for supercapacitors, J. Mater. Chem. A 6 (2018) 8053–8058.
[42] J.H. Ding, H.R. Zhao, H.B. Yu, High-yield synthesis of extremely high concentrated and few-layered boron nitride nanosheet dispersions, 2D Mater. 5 (2018) 045015.
[43] T. Sainsbury, A. Satti, P. May, Z.M. Wang, I. McGovern, Y.K. Gun'ko, et al., Covalently functionalized hexagonal boron nitride nanosheets by nitrene addition, Chem. Eur. J 18 (35) (2012) 10808–10812.

[44] L. Fu, T. Wang, J. Yu, W. Dai, H. Sun, Z. Liu, et al., An ultrathin high-performance heat spreader fabricated with hydroxylated boron nitride nanosheets, 2D Mater. 4 (2017) 025047.
[45] A. Nag, K. Raidongia, K.P.S.S. Hembram, R. Datta, U.V. Waghmare, C.N.R. Rao, Graphene analogues of BN: novel synthesis and properties, ACS Nano 4 (3) (2010) 1539–1544.
[46] C. Zhi, Y. Bando, C. Tang, H. Kuwahara, D. Golberg, Large-scale fabrication of boron nitride nanosheets and their utilization in polymeric composites with improved thermal and mechanical properties, Adv. Mater. 21 (28) (2009) 2889–2893.
[47] W.W. Lei, V.N. Mochalin, D. Liu, S. Qin, Y. Gogotsi, Y. Chen, Boron nitride colloidal solutions, ultralight aerogels and freestanding membranes through one-step exfoliation and functionalization, Nat. Commun. 6 (2015) 8849.
[48] T. Sainsbury, A. Satti, P. May, Z. Wang, I. McGovern, Y.K. Gun'ko, J. Coleman, Oxygen Radical Functionalization of Boron Nitride Nanosheets, J. Am. Chem. Soc. 134 (45) (2012) 18758–18771.
[49] L. Fu, H. Yang, A. Tang, Y. Hu, Engineering a tubular mesoporous silica nanocontainer with well-preserved clay shell from natural halloysite, Nano Res. 10 (8) (2017) 2782–2799.
[50] J.I. Dawson, R.O.C. Oreffo, Clay: new opportunities for tissue regeneration and biomaterial design, Adv. Mater. 25 (30) (2013) 4069–4086.
[51] F. Song, X. Hu, Exfoliation of layered double hydroxides for enhanced oxygen evolution catalysis, Nat. Commun. 5 (2014) 4477–4486.
[52] A.C.S. Alcântara, P. Aranda, M. Darder, E. Ruiz-Hitzky, Bionanocomposites based on alginate–zein/layered double hydroxide materials as drug delivery systems, J. Mater. Chem. 20 (2010) 9495–9504.
[53] H. Cheng, Y. Zhou, Y. Feng, W. Geng, Q. Liu, W. Guo, et al., Electrokinetic energy conversion in self-assembled 2D nanofluidic channels with janus nanobuilding blocks, Adv. Mater. 29 (23) (2017) 1700177.
[54] Y. Zhou, A.M. LaChance, A.T. Smith, H. Cheng, Q. Liu, L. Sun, Strategic design of clay-based multifunctional materials: from natural minerals to nanostructured membranes, Adv. Funct. Mater. 29 (16) (2019) 1807611.
[55] I. Janica, S.D. Buffa, A. Mikołajczak, M. Eredia, D. Pakulski, A. Ciesielski, et al., Thermal insulation with 2D materials: liquid phase exfoliated vermiculite functional nanosheets, Nanoscale 10 (2018) 23182–23190.
[56] Y. Zhang, M. Long, P. Huang, H. Yang, S. Chang, Y. Hu, et al., Intercalated 2D nanoclay for emerging drug delivery in cancer therapy, Nano Res. 10 (8) (2017) 2633–2643.
[57] C.H. Zhou, L.Z. Zhao, A.Q. Wang, T.H. Chen, H.P. He, Current fundamental and applied research into clay minerals in China, Appl. Clay Sci. 119 (2016) 3–7.
[58] T. Xiao, Q. Liu, Q. Zhang, Z. Liu, J. Zhai, Temperature and voltage dual-responsive ion transport in bilayer-intercalated layered membranes with 2D nanofluidic channels, J. Phys. Chem. C. 121 (34) (2017) 18954–18961.
[59] J.-J. Shao, K. Raidongia, A.R. Koltonow, J. Huang, Self-assembled two-dimensional nanofluidic proton channels with high thermal stability, Nat. Commun. 6 (2015) 7602.
[60] R.R. Nair, H.A. Wu, P.N. Jayaram, I.V. Grigorieva, A.K. Geim, Unimpeded permeation of water through Helium-leak–tight graphene-based membranes, Science 335 (6067) (2012) 442–444.
[61] Y. Feng, L. Ding, D. Ji, L. Wang, W. Guo, Highly rectified ion transport through 2D WSe_2/MoS_2 bi-layered membranes, Chin. Chem. Lett 29 (6) (2018) 892–894.

[62] J. Feng, M. Graf, K. Liu, D. Ovchinnikov, D. Dumcenco, M. Heiranian, et al., Single-layer MoS_2 nanopores as nanopower generators, Nature 536 (7615) (2016) 197–200.
[63] J. Lao, R. Lv, J. Gao, A. Wang, J. Wu, J. Luo, Aqueous stable Ti_3C_2 MXene membrane with fast and photoswitchable nanofluidic transport, ACS Nano 12 (12) (2018) 12464–12471.
[64] S. Qin, D. Liu, G. Wang, D. Portehault, C.J. Garvey, Y. Gogotsi, et al., High and stable ionic conductivity in 2D nanofluidic ion channels between boron nitride layers, J. Am. Chem. Soc. 139 (18) (2017) 6314–6320.
[65] C. Cheng, G. Jiang, G.P. Simon, J.Z. Liu, D. Li, Low-voltage electrostatic modulation of ion diffusion through layered graphene-based nanoporous membranes, Nat. Nanotechnol. 13 (8) (2018) 685–690.
[66] Graphenea, The price of graphene. <https://www.graphenea.com/pages/graphene-price> (accessed April).
[67] C.-H. Tsou, Q.-F. An, S.-C. Lo, M. De Guzman, W.-S. Hung, C.-C. Hu, et al., Effect of microstructure of graphene oxide fabricated through different self-assembly techniques on 1-butanol dehydration, J. Membr. Sci. 477 (2015) 93–100.
[68] L. Cao, F. Xiao, Y. Feng, W. Zhu, W. Geng, J. Yang, et al., Anomalous channel-length dependence in nanofluidic osmotic energy conversion, Adv. Funct. Mater. 27 (9) (2017). 1604302.
[69] C. Cheng, G. Jiang, C.J. Garvey, Y. Wang, G.P. Simon, J.Z. Liu, et al., Ion transport in complex layered graphene-based membranes with tuneable interlayer spacing, Sci. Adv. 2 (2) (2016) e1501272.
[70] J. Chen, Y. Li, L. Huang, N. Jia, C. Li, G. Shi, Size fractionation of graphene oxide sheets via filtration through track-etched membranes, Adv. Mater. 27 (24) (2015) 3654–3660.
[71] F. Kong, X. He, Q. Liu, X. Qi, Y. Zheng, R. Wang, et al., Improving the electrochemical properties of MXene Ti_3C_2 multilayer for Li-ion batteries by vacuum calcination, Electrochim. Acta 265 (2018) 140–150.
[72] C. Chen, M. Boota, X. Xie, M. Zhao, B. Anasori, C.E. Ren, et al., Charge transfer induced polymerization of EDOT confined between 2D titanium carbide layers, J. Mater. Chem. A 5 (11) (2017) 5260–5265.
[73] A. Wang, W. Yu, Z. Huang, F. Zhou, J. Song, Y. Song, et al., Covalent functionalization of reduced graphene oxide with porphyrin by means of diazonium chemistry for nonlinear optical performance, Sci. Rep. 6 (2016) 23325.
[74] D. Chen, H. Feng, J. Li, Graphene oxide: preparation, functionalization, and electrochemical applications, Chem. Rev. 112 (11) (2012) 6027–6053.
[75] A. Navaee, A. Salimi, Efficient amine functionalization of graphene oxide through the Bucherer reaction: an extraordinary metal-free electrocatalyst for the oxygen reduction reaction, RSC Adv. 5 (74) (2015) 59874–59880.
[76] H. Huang, Y. Mao, Y. Ying, Y. Liu, L. Sun, X. Peng, Salt concentration, pH and pressure controlled separation of small molecules through lamellar graphene oxide membranes, Chem. Commun. 49 (53) (2013) 5963–5965.
[77] T. Kar, R. Devivaraprasad, R.K. Singh, B. Bera, M. Neergat, Reduction of graphene oxide - a comprehensive electrochemical investigation in alkaline and acidic electrolytes, RSC Adv. 4 (101) (2014) 57781–57790.
[78] X. Fan, W. Peng, Y. Li, X. Li, S. Wang, G. Zhang, et al., Deoxygenation of exfoliated graphite oxide under alkaline conditions: a green route to graphene preparation, Adv. Mater. 20 (23) (2008) 4490–4493.

[79] J.P. Rourke, P.A. Pandey, J.J. Moore, M. Bates, I.A. Kinloch, R.J. Young, et al., The real graphene oxide revealed: stripping the oxidative debris from the graphene-like sheets, Angew. Chem. Int. Ed. 50 (14) (2011) 3173–3177.
[80] M.-C. Lu, S. Satyanarayana, R. Karnik, A. Majumdar, C.-C. Wang, A mechanical-electrokinetic battery using a nano-porous membrane, J. Micromech. Microeng. 16 (4) (2006) 667–675.
[81] J. Gao, W. Guo, D. Feng, H. Wang, D. Zhao, L. Jiang, High-performance ionic diode membrane for salinity gradient power generation, J. Am. Chem. Soc. 136 (35) (2014) 12265–12272.
[82] A. Esfandiar, B. Radha, F.C. Wang, Q. Yang, S. Hu, S. Garaj, et al., Size effect in ion transport through angstrom-scale slits, Science 358 (6362) (2017) 511–513.
[83] S. Qin, D. Liu, Y. Chen, C. Chen, G. Wang, J. Wang, et al., Nanofluidic electric generators constructed from boron nitride nanosheet membranes, Nano Energy 47 (2018) 368–373.

11

2D materials for solar fuels production

Konstantinos C. Christoforidis

Department of Environmental Engineering, Democritus University of Thrace, Xanthi, Greece

11.1 Introduction

Sustainable supply of clean energy is one of the top technological challenges mankind is facing nowadays. This mainly originates from the rapid global rising demand for energy and the progressive depletion of fossil fuels. An additional significant contribution has been admitted in the last decades. This is the severe environmental impact and in particular climate change, which has emerged as a consequence of traditional fossil fuels combustion and the production of greenhouse gas (GHG). Carbon dioxide (CO_2) is one of the main GHG and anthropogenic activities are the major contributors of CO_2 emissions in the atmosphere [1]. The current levels of CO_2 in the atmosphere are stabilized at levels higher than 400 ppm, undoubtedly indicating that the natural carbon cycle has been destabilized. It is therefore unavoidable to search for alternative clean and renewable energy sources in our effort to enter into a "carbon-neutral" society.

Among the different proposed processing technologies including biological, chemical, thermochemical, and electrolytic approaches, the heterogeneous photocatalytic process using solar light has been identified as a suitable approach to deal with both sustainable energy production and environmental protection. There are several factors that contribute to this end. The surface of the Earth receives annually solar energy power of 1.3×10^5 TW that exceeds by orders of magnitude the estimated future global energy demands. This is valid even if nonconservative estimates are considered, for example, 28–35 TW global energy demand per year by 2050 [2]. The photocatalytic approach is a stand-alone process that does not require external

2D Nanomaterials for Energy Applications. DOI: https://doi.org/10.1016/B978-0-12-816723-6.00011-3

energy input if solar energy (a free and endless energy source) and renewable substrates are used [3]. Photocatalysis can harvest solar energy and store it in the form of chemical energy by the production of fuels, the so-called *solar fuels*, a process that does not require the immediate consumption of the produced energy.

Due to the above-mentioned reasons the photocatalytic process has emerged as one of the most promising technologies for the sustainable production of clean energy, mainly in the form of hydrogen and hydrocarbon fuels. However, there are several drawbacks that must be solved before large-scale applications can take place. These are not related with the fundamentals of the process itself regarding reactors design and the actual catalytic mechanism, but rather with the efficiency in fuels production. Although several aspects may affect the efficiency, such as the actual photocatalytic set-up or the substrate used, the central component that defines efficiency is the photocatalytic material (i.e., the photocatalyst). Different types of photocatalysts have been developed and efficiently applied in the production of solar fuels. The first and still the most investigated photocatalyst in the production of solar fuels is TiO_2 and TiO_2-based materials [4]. The reason for this lies on the unique characteristics of TiO_2 polymorphs such as the chemical and biological inertness, low-toxicity, and low cost. In addition, TiO_2 meets the essential properties that a photoactive material must possess for any practical application, that is, favorable electronic structure (band edge alignment), ability to form charges, and many different pathways available for modification (phase, composition, and morphology).

Apart from the composition, the structure and the morphology of the photocatalyst plays a critical role in the photocatalytic activity. Among the different types of photocatalysts tested in solar fuel production, two-dimensional (2D) materials are particularly attractive and have therefore gained considerable attention in recent years. 2D materials are characterized by a sheet-like structure with a lateral size varying from nanometers up to micrometers, while their thickness is ranging from a single- to few-atomic thick layers. Due to their size constrains in the thickness and the electron confinement in two dimensions, they have unique physical, electronic, and chemical properties that are not present in their bulk structures [5]. First, 2D materials present larger specific surface area as compared with their bulk counterparts. Since heterogeneous photocatalytic reactions are surface-related reactions, the increase of the surface area offers more available active sites, greater efficiency in light harvesting, and mass transport and therefore, enhanced activity. Furthermore, control of the thickness of the 2D structure may alter the electronic properties of the material affecting the light absorption, and the actual electronic band structure that is essential for performing a specific reaction. Decrease of the thickness may also improve the abundance of the photogenerated charges (electron and hole pairs, e^-/h^+) on the surface of the catalyst where the photocatalytic reactions take place. This is due to the shorter pathways offered for electron transfer from the bulk to the surface of the material (i.e., shorter migration distance) that allow faster transfer and easier availability of charges. Finally, 2D structures at the nanoscale can also serve as excellent supports (i.e., platforms) for the development of multicomponent photocatalysts, offering, for example, fine catalysts and/or cocatalysts dispersion. These characteristics emerged 2D structures as promising materials and sparked intensive research in the field of photocatalysis.

In this chapter we will discuss the recent progress on the development of typical photoactive 2D materials, composites, and heterojunctions focusing in their application for sustainable energy production. Special attention will be given in photocatalytic H_2 production since it is the most mature process in the field of solar fuels production. However, a profound understanding of the fundamental principles, drawbacks, and challenges of the photocatalytic processes in solar fuels production is essential for the rational design of efficient photocatalysts. Therefore before discussing in detail recent contributions in materials development, a brief introduction on the elementary steps of photocatalytic reactions will be given.

11.2 Basic principles of photocatalysis for solar fuel production

In heterogeneous photocatalysis electron and hole pairs (e^-/h^+) are formed when semiconductors (SCs) are excited with incident light of energy greater than their bandgap energy (E_g). Electrons from the valence band (VB) are excited to the conduction band (CB) leaving behind holes (h^+). These charges, once generated, may follow several paths such as recombination in the bulk state releasing heat. However, when separated and transferred to the surface of the catalyst, they can perform reduction and oxidation reactions, respectively. Ideally, for practical applications, the excitation of the SC should take place under solar light irradiation. This means that the SC must

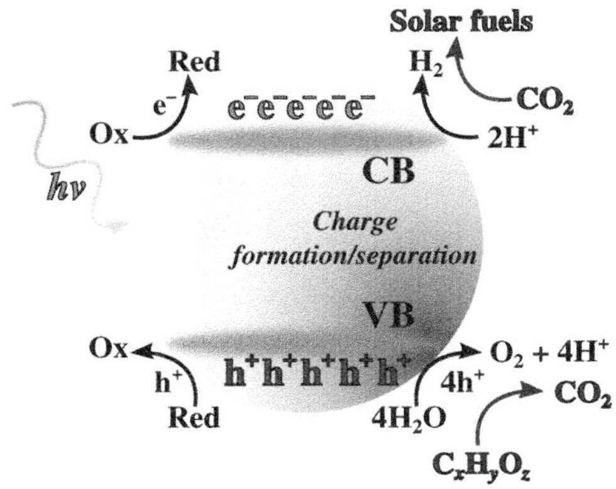

Figure 11.1 Schematic illustration of e^-/h^+ pairs formation upon irradiation of an SC with incident light of energy greater than the bandgap energy and their participation in redox reactions for solar fuels production. *Source*: Adapted from K. C. Christoforidis, P. Fornasiero, Photocatalysis for hydrogen production and CO_2 reduction: the case of copper-catalysts, ChemCatChem 11 (2018) 368–382.

absorb light in the visible region of light. Fig. 11.1 summarizes the chain of events that take place in a photocatalytic process. An important parameter that must be always taken into account is the energy level of the CB and VB edges. This is the main factor that defines the realization or not of a specific photocatalytic reaction.

The production of solar fuels using the photocatalytic route is based on two distinct processes: (1) production of H_2 and (2) production of hydrocarbons via CO_2 conversion (i.e., CO_2 photoreduction). Both reactions share the same fundamental mechanism that is based on the utilization of photogenerated e^- in the CB. To close the overall redox cycle, a reduction agent (i.e., a scavenger) must be provided to deliver e^- for the consumption of h^+ in the VB. H_2 is an ideal energy carrier for the future since it is a carbon-free fuel. On the other hand, CO_2 conversion gains continuously more attention in the scientific community, since it combines the utilization of a cheap feedstock that significantly contributes to climate change and the utilization of solar energy for the production of fuels. Although they have been both demonstrated in the same period, CO_2 photoreduction is a much less studied process as compared to H_2 production. There are several factors that contribute to this end. First, CO_2 reduction is energetically more demanding reaction. Second, CO_2 reduction proceeds through a multielectron transfer process and therefore, besides carbon reduction, it also requires hydrogenation and C–coupling reactions. Besides

Table 11.1 Required redox potential of selected reactions [1].

Reactions	E^0 (SHE, pH 7)
$H_2O + 4h^+ \rightarrow 4H^+ + 2O_2 + 4e^-$	+ 0.81
$H_2 + 2e^- \rightarrow H_2$	− 0.42
$CO_2 + e^- \rightarrow CO_2^{\bullet -}$	− 1.90
$CO_2 + 2H^+ + 2e^- \rightarrow HCOOH$	− 0.61
$CO_2 + 2H^+ + 2e^- \rightarrow CO + H_2O$	− 0.53
$CO_2 + 4H^+ + 4e^- \rightarrow HCHO + H_2O$	− 0.48
$CO_2 + 6H^+ + 6e^- \rightarrow CH_3OH + H_2O$	− 0.38
$CO_2 + 8H^+ + 8e^- \rightarrow CH_4 + 2H_2O$	− 0.24

thermodynamic and kinetic barriers, one should also mention the low availability of adsorbed CO_2 on the surface of the catalyst that hinders the overall CO_2 conversion efficiency. This originates from the limited CO_2 solubility in water (i.e., the ideal solvent for a sustainable photocatalytic process) and the low capacity of traditional SCs to adsorb CO_2 (CO_2 uptake). Furthermore, in the CO_2 conversion reaction there is always the uncertainty of the origin of the substrate converted into fuels. This requires the performance of catalytic reactions using labeled CO_2 (i.e., $^{13}CO_2$ isotopes) in order to demonstrate the catalytic conversion of CO_2. The potential production of several products from CO_2 conversion generates an additional problem related with their analytical detection. Finally, photocatalytic H_2 production and CO_2 conversion are competitive reactions. They both occur at similar potentials (Table 11.1) and they both make use of photogenerated electrons. However, comparing the two process, H_2 evolution is a more favorable reaction. A potential solution to this issue would be the development of a highly selective catalysts for CO_2 reduction but this is not a trivial task. These are the main reasons for the less studies presented in the literature for CO_2 photoconversion into fuels.

As stated above, one of the critical factors that define the realization of a specific reaction in a photocatalytic process is the level of the CB and VB potentials. In the case of H_2 photoproduction, the bottom of the CB must be more negative than the H^+/H_2 redox couple (0 V vs NHE at pH 0). On the other hand, the CB energy level plays a more prominent role in the case of CO_2 photocatalytic reduction. Since the CB edge potential defines the reduction ability of e^-, the formation of a specific product in the CO_2 reduction reaction is strictly related with the CB energy level (Table 11.1).

H_2 production through photocatalysis is achieved through the water splitting (WS) process and the photoreforming of oxygenated organic substrates. The fundamental reactions for both processes are given in Eqs. (11.1)–(11.2) and are summarized in Fig. 11.1. The WS process is consisted by two half-reactions (Table 11.1): the proton (H^+) reduction and the highly demanding 4-e^- water oxidation reaction (i.e., oxygen evolution) which is the bottleneck of the process. As an alternative to the WS process, an organic substrate is oxidized in the photoreforming process. This provides two additional benefits. First, the organic substrate acts also as proton source. Second, it may be oxidized easier than water bypassing the rate-limiting step of pure WS process. However, in the latter case, the origin of the organic substrate defines the entire sustainability of the process. For example, the method is considered sustainable only if bio-available oxygenates are used as sacrificial agents. It must be highlighted that photocatalysts are practically inactive in photocatalytic H_2 production without the presence of a cocatalyst. Expensive noble metal nanoparticles such as Pt are the most used cocatalysts in H_2 evolution reaction. However, this approach is not considered sustainable for large-scale applications and alternatives must be considered:

$$H_2O \rightarrow H_2 + 1/2O_2 \quad \text{overall waters plitting} \tag{11.1}$$

$$C_xH_yO_z + (2x - z)H_2O \rightarrow (2x - z + y/2)H_2 + xCO_2 \quad \text{photoreforming} \tag{11.2}$$

11.3 Heterojunctions

Many examples in the literature have demonstrated the importance of developing heterojunctions as catalysts in solar fuel production reactions. This is also highlighted in specific examples given in the following. The coupling of different SCs offers several advantages for improving photoactivity. The light absorption of a catalyst in the visible region can be extended via the coupling of wide E_g with narrow E_g SCs. This approach may offer better control of the overall efficiency compared with the widely used approach of doping an SC with metallic or non-metallic elements. Efficient charge separation has been also

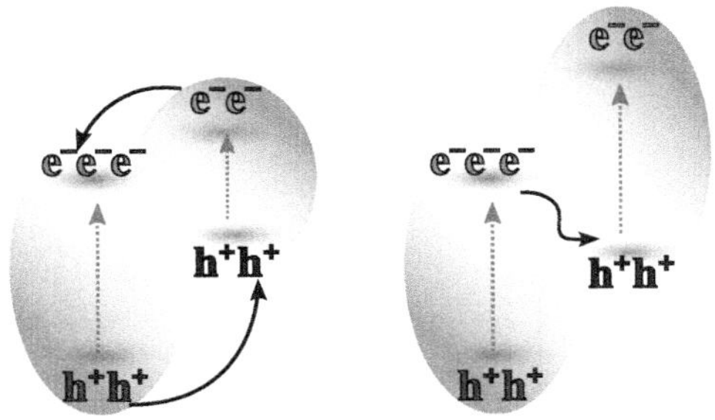

Figure 11.2 Charge transfer in a type-II (left side) and a Z-scheme (right side) heterojunction. *Source*: Adapted from K.C. Christoforidis, P. Fornasiero, Photocatalysis for hydrogen production and CO_2 reduction: the case of copper-catalysts, ChemCatChem 11 (2018) 368–382.

documented in heterojunctions. This originates from the coupling of SCs with correctly aligned band structures such as type-II heterojunctions. Charge migration is driven by the difference in the band energy levels of the two SCs (Fig. 11.2). Therefore special attention must be given on the choice of the coupled SCs in order to direct e^- and h^+ in the opposite direction. In this approach, the oxidation and reduction efficiency of the catalyst can be tuned toward the overall WS reaction, that is, both H_2 evolution and water oxidation half-reactions. However, one problem that occurs in this case is the decrease of the energy of the photogenerated charges since they are spontaneously transferred from the band located at higher energy to the one at lower energy levels. On the contrary, when e^- are transferred from the CB of the first to the second SC via the VB of the second, the power of the photogenerated electron is not reduced (Fig. 11.2). In fact, the e^- may be transferred to a CB of higher energy level and therefore the driving force for performing a specific reduction reaction is improved. This mechanism is called "Z-scheme" transfer mechanism and imitates the natural photosynthetic reaction centers.

In order for these advantages to occur in composite materials an interface must be formed between the two phases. For obvious reasons, the larger contact area on the interface is beneficial. Comparing the different possibilities offered by the variety of morphologies at the nanoscale, that is, 0D, 1D, 2D, and 3D materials, the 2D–2D coupled interfaces offer larger contact area when the coupling occurs between the lateral surfaces. In addition, the intimate interaction in a 2D–2D heterojunction favors exciton dissociation and even if there is lattice mismatch between the two phases is it easier to form the interface is such couplings [6]. We have recently investigated the effect of the

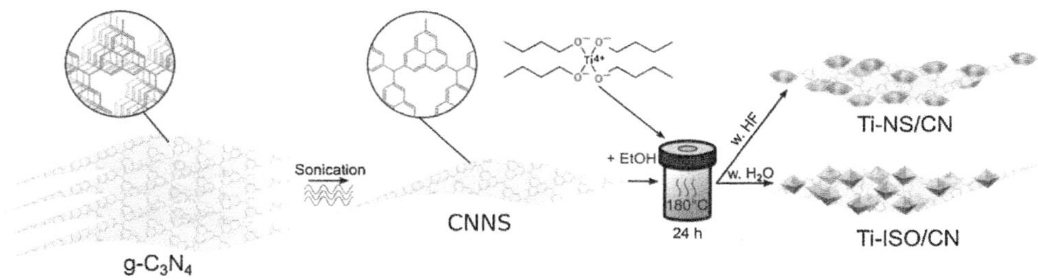

Figure 11.3 Schematic representation of the synthesis process used to produce composites by the coupling of CN nanosheets (CNNS) with TiO_2 nanoparticles of controlled 2D structure (Ti-NS) or unshaped morphology (Ti-ISO). *Source*: Reprinted with permission from A. Crake, K.C. Christoforidis, R. Godin, B. Moss, A. Kafizas, S. Zafeiratos, et al., Titanium dioxide/carbon nitride nanosheet nanocomposites for gas phase CO_2 photoreduction under UV–visible irradiation, Appl. Catal. B: Environ. 242 (2019) 369–378.

materials morphology in the coupling of TiO_2 and carbon nitride (CN) nanosheets and its impact on CO_2 photocatalytic reaction (Fig. 11.3) [7]. Bare TiO_2 nanosheets with exposed {001} facets presented lower activity in CO_2 photoreduction than non-shaped TiO_2 nanoparticles. However, when both TiO_2 nanoparticles were coupled with CN nanosheets, the heterojunction containing TiO_2 nanosheets presented superior activity than the corresponding material containing TiO_2 without crystal shape control. Based on transient absorption spectroscopy, it was demonstrated that formation of TiO_2 nanosheets improved significantly interfacial charge transfer at the TiO_2/CN junction (i.e., 2D–2D junction) by increasing the contact between the two phases. This enhanced the abundance and stability of the photogenerated charges resulting in higher photoactivity. The larger contact between TiO_2 nanosheets and MoS_2 was also the reason for the improved H_2 photocatalytic production from H_2O/methanol solution than TiO_2 bearing Pt as cocatalyst [8].

11.4 2D photocatalysts for solar fuels production

Up-to-date tremendous efforts have been devoted to the development of different types of catalysts applied in photoassisted H_2 production and CO_2 reduction. Among the variety of photocatalysts tested, TiO_2 is the most investigated due to its inherent properties. 2D TiO_2 nanosheets have also received considerable attention among the different TiO_2 morphologies. In TiO_2 the nanocrystal shape is controlled by the different orientation of the crystal facets. The surface energy determines the relative energy of each facet. This factor regulates the morphology of the crystal. TiO_2 nanocrystal morphology is mostly achieved through the use of capping agents and surfactants. The morphology of the anatase TiO_2 (i.e., the most used TiO_2 polymorph) presents a slightly truncated tetragonal bipyramid structure with {101} surfaces predominately exposed, that is, the facet with the lowest surface energy. Yang et al. demonstrated experimentally for the first time the development of anatase TiO_2 single crystals 2D nanosheets with preferentially exposed {001} facets [9]. After this pioneering work, several groups worked on the development of TiO_2 nanosheets. More importantly, it has been shown that the morphology of the nanocrystals affects significantly activity. For example, in many cases the exposed {001} facets on TiO_2 nanosheets have led to improved photoactivity. This is probably attributed the undercoordination character of the terminated species on {001 facets} (i.e., five-coordinated Ti-centers) [10]. However, regarding activity of shape-controlled anatase TiO_2 nanostructures with preferential exposed facets, controversial results have been reported in the literature. This must be related with the actual synthesis process applied and clearly suggests the need for more studies in this field. Perfect control of the shape is not trivial and usually the conditions applied affect other important parameters in parallel. For example, anatase TiO_2 nanosheets with exposed {001} facets may present higher photoactivity; however, the growth of this specific facet results in a significant reduction of the specific surface area, a critical parameter in heterogeneous catalysis. Therefore for an in-depth understanding of the effect of the crystal morphology on photocatalytic activity, the study should be always accompanied by full characterization of the catalyst.

An interesting study was performed by Liu et al. that reported the combined effect of {001} facets in anatase TiO_2 and the particle size on H_2 evolution from WS [11]. Materials with smaller particle size presented superior activity. In addition, the positive effect of removing the capping agent (i.e., fluorine) from the surface of the catalyst was indicated. The improved photoactivity was attributed to the synergistic effects of the increased specific surface area in nanosized crystals and the more negative level of the CB minimum.

One of the drawbacks of using anatase TiO_2 is the wide bang gap energy ($\sim$3.2 eV) that requires UV-light irradiation. One of the well-known approaches to improve the electronic properties related with light absorption in such materials is doping with metallic or nonmetallic elements. Anatase TiO_2 sheets with dominant {001} facets were effectively doped with nitrogen as a way to expand light absorption to the visible region [12]. This resulted in improved H_2 evolution under UV–vis irradiation. In addition, while nondoped materials did not present any activity under pure visible light irradiation (420–770 nm), N-doped TiO_2 sheets showed considerable activity under these conditions. The development of black TiO_2 photocatalysts through hydrogenation treatment at high temperatures is another interesting approach to improve light absorption in the visible region. Applying this methodology Chen et al. have shown that TiO_2 nanosheets with preferentially exposed {001} facets present higher H_2 evolution rates [13]. This was attributed to the strong morphology dependent effects on the surface and/or subsurface defects formation upon hydrogenation.

Layered titanate structures ($Ti_4O_9^{-2}$) have been also studied as photocatalysts [14,15]. Photocatalytic activity in H_2 evolution is significantly altered by the chemical exfoliation and the cocatalyst used [15]. For example, exfoliated titanate nanosheet showed an impressive photocatalytic H_2 evolution when CdS quantum dots were deposited on the surface acting as charge separators [14].

Besides TiO_2 other 2D oxides and perovskite nanosheets have been also tested as photocatalysts [16–18]. $HCa_2Nb_3O_{10}$ nanosheet structures extracted from tetrabutylammonium hydroxide assisted exfoliation presented a 16-fold improved H_2 photocatalytic production than the bulk counterpart [18]. Similar results were obtained for Rh-doped $KCa_2Nb_3O_{10}$ nanosheet when tested in aqueous methanol solution presenting approximately 165 times greater H_2 production than bulk Rh-doped layered oxide [19]. Regarding these structures, Maeda et al. recently demonstrated that their band structure is tunable by changing the composition through the introduction of Sr and Ta in the structure [20]. Introducing Sr into the nanosheets resulted in a positive shift of the CB with a negative contribution in H_2 production. This is expected due to the decrease of the driving force for H^+ reduction (less energetic CB e^-). On the contrary, the presence of Ta increased photocatalytic H_2 production. This was attributed to the shift of the CB potential to more negative values.

Layered double hydroxides (LDHs) have been also tested for solar fuels production. These are layered materials with high compositional flexibility and have been mostly used for water oxidation. Due to their high anion exchange capacity they can act as suitable substrates for anchoring various chemical moieties. This provides the benefit of tuning their properties toward

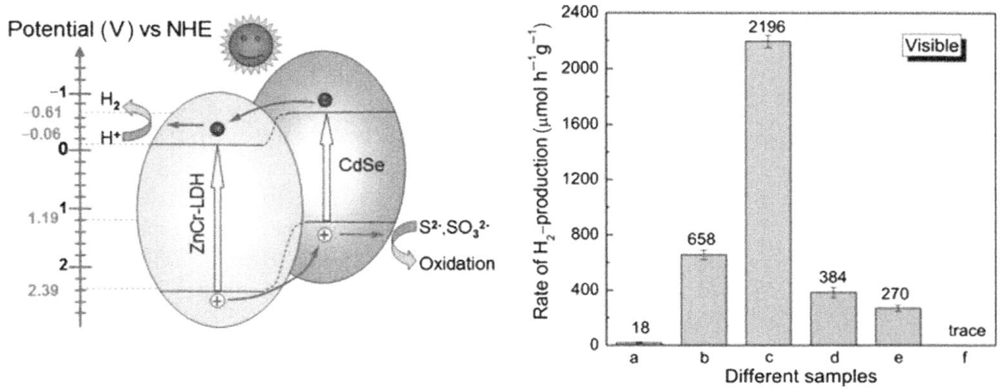

Figure 11.4 Schematic model of ZnCr-LDH and CdSe band structures and charge separation mechanism (top panel) and comparison of H_2 evolution activity of ZnCr-LDH with varying CdSe content under visible light ($\lambda > 420$ nm). *Source*: Reprinted with permission from G. Zhang, B. Lin, Y. Qiu, L. He, Y. Chen, B. Gao, Highly efficient visible-light-driven photocatalytic hydrogen generation by immobilizing CdSe nanocrystals on ZnCr-layered double hydroxide nanosheets, Int. J. Hydrogen Energy 40 (2015) 4758–4765.

a specific reaction. The immobilization of TiO_6 octahedra units on MTi-LDH (M: Ni, Zn, Mg) resulted in enhanced H_2 evolution rates [21]. This was ascribed to the significantly suppressed recombination process in the Ti-containing LDH materials and to the modification of the bandgap structure. Anchoring CdSe nanocrystals on the ZnCr-LDX nanosheets resulted in impressive H_2 photocatalytic production under pure visible light irradiation from an aqueous Na_2S/Na_2SO_3 solution (Fig. 11.4) [22]. Photoactivity was drastically affected by the CdSe content and the improvement of the heterostructure in comparison to the pure LDH was assigned to the formation of a heterojunction that allowed efficient charge separation (Fig. 11.4).

Metal chalcogenides is another class of 2D layered structures applied as photocatalysts. This group of materials is particularly interesting in photocatalytic applications mostly because they present visible light absorption due to their short bandgap energy. In addition, their composition (transition metals and chalcogens) allows modification through for example ion-exchange reactions toward the development of new and ternary or even quaternary chalcogenides. However, the difficulty to control perfectly the stoichiometry and phase structure leads to a challenging synthesis process. Considering solely photoactivity, CdS has been identified as an exceptional photocatalysts but unfortunately is susceptible in photocorrosion phenomena. Among the different chalcogenides developed, molybdenum and copper chalcogenides are particularly interested. MoS_2 is considered to be one of the most promising alternative to noble metal (i.e., Pt) cocatalysts. Since photocatalytic activity in H_2 evolution reaction is significantly low without the use of a cocatalyst, cocatalysts made from cheap and abundant materials are of great importance in the field. Chang et al. systematically investigated the effect of the MoS_2 thickness as cocatalyst deposited on CdS on the photocatalytic H_2 generation [23]. The photocatalytic activities in H_2 generation increased significantly with the decrease of MoS_2 layers number. In particular, the highest H_2 evolution activity (2.01×10^{-3} m h^{-1}) was observed for the single-layered MoS_2. However, it was still lower than the corresponding catalyst bearing Pt as cocatalyst. Recently, Yin et al. investigated the effect of MoS_2 layer thickness as well as the lateral size as cocatalyst in H_2 photocatalytic production [24]. The structure morphology in terms of thickness and lateral size was controlled by a liquid-phase exfoliation of bulk MoS_2 suitable for large-scale production of MoS_2 monolayers followed by consecutive centrifugation steps. When MoS_2 monolayers were deposited on CdS as cocatalyst, a gradual increase in H_2 evolution rate was detected with decreasing average lateral sizes. The enhanced photoactivity was attributed to the improved charge separation efficiency (electrons are transferred to MoS_2 and holes to CdS) and to the lateral size engineering that (1) increased the density of active edge sites (i.e., exposed active edge S atoms) and (2) shortened the migration distance of the photogenerated electrons from the inert basal plane to the active sites at the edges.

MoS_2 nanosheets have been also deposited on Cu_2S of varying morphology including 2D sheets and coupled with TiO_2 [25]. The prepared materials showed H_2 photocatalytic evolution under UV–vis light irradiation from a water/methanol solution. 2D–2D nanojunction through the deposition of MoS_2 nanosheets on TiO_2 2D nanosheets with preferentially exposed {001} facets have been also developed and applied in H_2 production [26]. The selective deposition of MoS_2 on {101} facets of TiO_2 nanosheets maximized the H_2 evolution rates (Fig. 11.5). This was attributed to the presence of exposed {001} facets with higher surface energy on the TiO_2 NSs and the reduced charge carrier recombination due to (1) the formation of a surface heterojunction within single TiO_2 NS (presence of both {001} and {101} facets) and (2) the formation of a heterojunction through the coupling of MoS_2 with TiO_2. TiO_2 nanobelts coated with MoS_2 nanosheets presented high photocatalytic H_2 production from aqueous solution containing Na_2S/Na_2SO_3 as sacrificial

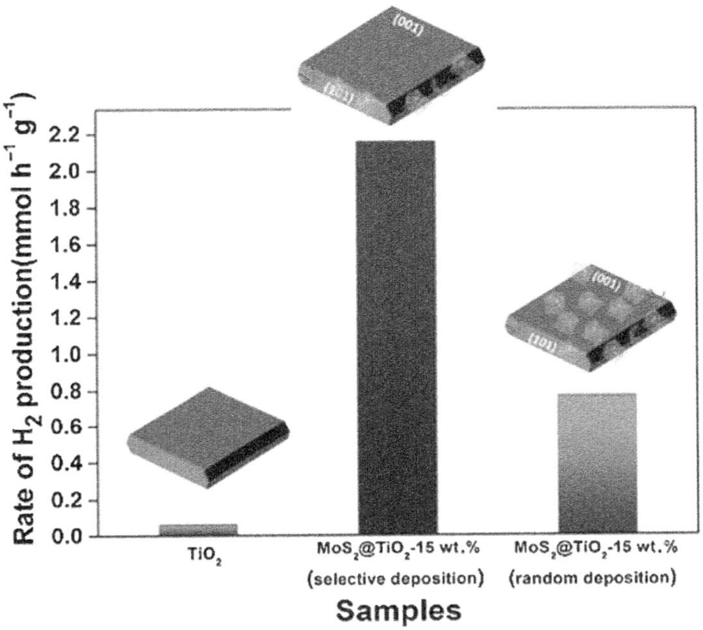

Figure 11.5 Rate of photocatalytic H_2 production from TiO_2 NSs and $MoS_2@TiO_2$ composite (selective and random deposition). *Source:* Reprinted with permission from X. Hu, S. Lu, J. Tian, N. Wei, X. Song, X. Wang, et al., The selective deposition of MoS_2 nanosheets onto (101) facets of TiO_2 nanosheets with exposed (001) facets and their enhanced photocatalytic H_2 production, Appl. Catal. B: Environ. 241 (2019) 329–337.

agents [27]. Photoactivity was drastically affected by the content of MoS_2 and materials containing 50 wt.% MoS_2 showed superior H_2 evolution rates. Similar results in H_2 photocatalytic evolution were observed when MoS_2 were deposited on other SCs, for example, $ZnIn_2S_4$ and carbon nitride [28–30].

Copper-based semiconductors have also received great attention in the past years in H_2 photoproduction mostly because the nature of the meta (3D transition metal presenting various oxidation states) allows the development of different types of nanomaterials (oxides, sulfides, and meta nanoparticles). Furthermore, Cu-based photocatalysts are excited under visible light irradiation. However, the development of 2D Cu-sulfide nanosheets is not an easy task. Zhou et al. developed stable Cu_2S nanosheets of 1.2 nm thickness as cocatalysts for H_2 production [31]. The sulfur precursor was identified as the parameter that controlled and directed the formation of the 2D structure. The improved photoactivity was ascribed to the efficient charge separation via the interface formed between Cu_2S and TiO_2. Porous ZnS nanosheets functionalized with CuS presented impressive H_2 evolution rates under pure visible light irradiation from aqueous solution containing Na_2S/Na_2SO_3 as sacrificial agents with an apparent quantum efficiency of 20% at 420 nm [32]. Once again, the interfacial charge transfer between the two phases together with the partial reduction of CuS to Cu_2S were the reasons for the enhanced photoactivity.

Over the last years graphitic carbon nitride (CN), a 2D semiconductor containing only C, N, and H, has emerged into a promising candidate for solar fuels production. CN presents high chemical and thermal stability, 2D morphology, and visible light absorption. In addition, its polymeric nature allows modification of its textural properties, morphology, and electronic structure in terms of band edge potential and bandgap energy. Based on the band structure, CN can be used for the overall WS process. Most importantly, this can be performed under pure visible irradiation. However, the bulk structure of CN usually presents low photoactivity due to the high charge carrier recombination. However, CN nanosheets have shown considerable activity especially under pure visible light irradiation. Compared with the bulk structure, CN nanosheets show greater surface area, increased E_g which may be also accompanied by shifting the CB to more negative values, improvements in electron transport ability along the 2D morphology (in-plane direction), and prolonged lifetime of the photogenerated charge carriers. These changes result in increased hydrogen production rates. Several approaches have been developed for the exfoliation of bulk CN such as thermal oxidation etching of bulk CN [3,33] and the intercalation of salts in the bulk structure followed by a washing step [34]. Control of the atmosphere during the thermal polycondensation process as well as the exact temperature or the overall thermal process (i.e., two-step thermal treatment) may also affect the final properties [35,36]. For example, Niu et al. using a thermal oxidation etching of bulk CN under air obtained CN nanosheets [33]. The actual conditions applied controlled the thickness of the final material which had critical effect in H_2 production. The formation of a layered nanojunction through the coupling of CN with MoS_2 nanosheets was proven to be more active even from CN deposited with Pt (a highly active cocatalyst in H_2 evolution reaction) as cocatalyst [37]. Recently, we have studied the effect of thermally treated bulk CN on the properties of the final material obtained [36] and we were able to identify the properties linked with H_2 evolution reaction and CO_2 photoreduction (i.e., two competing reactions). By applying a multimethodology characterization approach it was demonstrated that H_2 photocatalytic evolution is controlled by light absorption properties and charge separation efficiency that is allowed by the presence of trap states created by N-vacancies. However, in the case of CO_2

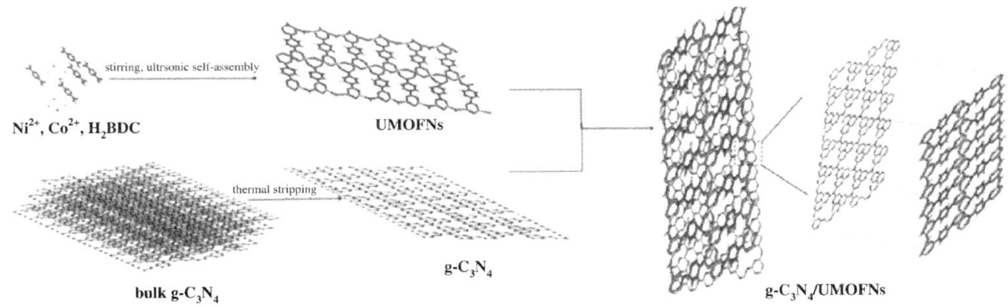

Figure 11.6 Synthesis route of composite CN nanosheets/2D MOF catalyst. *Source*: Reprinted with permission from Y. Liang, R. Shang, J. Lu, W. An, J. Hu, L. Liu, et al., 2D MOFs enriched g-C_3N_4 nanosheets for highly efficient charge separation and photocatalytic hydrogen evolution from water, Int. J. Hydrogen Energy 44 (2019) 2797–2810.

photoreduction, the CO_2 adsorption ability of the catalyst appears as the dominating factor.

Metal-organic frameworks (MOFs) is a relatively new class of materials where metal ions or metal clusters are linked with polydentate organic molecules forming crystalline and highly porous structures. MOFs or MOFs-based materials have been also applied in fuels production through photocatalysis [38,39]. Since the highly porous structure of MOFs is one of the main properties that directed their development, 3D structures have been mostly studied. However, there are few reports where 2D MOFs structures have been applied in H_2 production. Recently, Liang et al. have efficiently coupled an ultrathin 2D MOF structure with CN nanosheets for the development of a 2D–2D heterostructure (Fig. 11.6) [40]. Photocatalytic activity in H_2 evolution under simulated solar light irradiation was significantly improved in the heterostructure compared with the bare CN nanosheets. The optimum content of MOF was determined. The MOF structure played the role of the electron transfer mediator from the CN nanosheets to the Pt cocatalyst due to the highly dispersed MOF moieties (Ni–O and Co–O) on CN and its $\pi-\pi$ conjugated structure that allowed fast electron transfer.

11.5 Concluding remarks

Based on the documented advantages of 2D materials and the specific examples given above, there is no doubt that 2D photocatalysts can play a leading role in the development of efficient photoactive materials for converting solar energy into fuels. Recent results in activity are encouraging and definitely more efforts are required to get a deeper understanding of several aspects. Catalytic studies should be always accompanied by adequate characterization techniques that allow full characterization of the catalyst and can identify parameters that govern activity. This aspect will act as a guiding agent for improving efficiency. Selectivity is another major issue. Selective photocatalysis regarding product formation is currently rapidly developing but requires more efforts. The development of smart materials, devices, and set-ups through for example the rational design of multicomponent materials in order to perform the two half-reactions of WS in different parts may contribute to bypass this issue. Despite the fact that improvements are indispensable for meeting the needs of realistic applications, recent advantages leave no doubt that the suitable photocatalyst will be found. However, our ultimate goal must be the development of photocatalysts that imitate fully natural photosynthesis. This includes the atmosphere where the reaction takes place. In this direction the major challenge lies in the development of materials that can work in the presence of oxygen.

References

[1] K.C. Christoforidis, P. Fornasiero, Photocatalysis for hydrogen production and CO_2 reduction: the case of copper-catalysts, ChemCatChem 11 (2018) 368–382.
[2] A.J. Esswein, D.G. Nocera, Hydrogen production by molecular photocatalysis, Chem. Rev. 107 (2007) 4022–4047.
[3] K.C. Christoforidis, P. Fornasiero, Photocatalytic hydrogen production: a rift into the future energy supply, ChemCatChem 9 (2017) 1523–1544.
[4] J. Schneider, M. Matsuoka, M. Takeuchi, J. Zhang, Y. Horiuchi, M. Anpo, et al., Understanding TiO_2 photocatalysis: mechanisms and materials, Chem. Rev. 114 (2014) 9919–9986.
[5] C. Tan, X. Cao, X.J. Wu, Q. He, J. Yang, X. Zhang, et al., Recent advances in ultrathin two-dimensional nanomaterials, Chem. Rev. 117 (2017) 6225–6331.
[6] T. Su, Q. Shao, Z. Qin, Z. Guo, Z. Wu, Role of interfaces in two-dimensional photocatalyst for water splitting, ACS Catal. 8 (2018) 2253–2276.
[7] A. Crake, K.C. Christoforidis, R. Godin, B. Moss, A. Kafizas, S. Zafeiratos, et al., Titanium dioxide/carbon nitride nanosheet nanocomposites for gas phase CO_2 photoreduction under UV-visible irradiation, Appl. Catal. B: Environ. 242 (2019) 369–378.
[8] Y.-J. Yuan, Z.-J. Ye, H.-W. Lu, B. Hu, Y.-H. Li, D.-Q. Chen, et al., Constructing anatase TiO_2 nanosheets with exposed (001) facets/layered MoS_2 two-dimensional nanojunctions for enhanced solar hydrogen generation, ACS Catal. 6 (2016) 532–541.
[9] H.G. Yang, C.H. Sun, S.Z. Qiao, J. Zou, G. Liu, S.C. Smith, et al., Anatase TiO_2 single crystals with a large percentage of reactive facets, Nature 453 (2008) 638–641.

[10] J. Pan, G. Liu, G.Q. Lu, H.-M. Cheng, On the true photoreactivity order of {001}, {010}, and {101} facets of anatase TiO_2 crystals, Angew. Chem. Int. Ed. 50 (2011) 2133–2137.
[11] G. Liu, C. Sun, H.G. Yang, S.C. Smith, L. Wang, G.Q. Lu, et al., Nanosized anatase TiO_2 single crystals for enhanced photocatalytic activity, Chem. Commun. 46 (2010) 755–757.
[12] G. Liu, H.G. Yang, X. Wang, L. Cheng, J. Pan, G.Q. Lu, et al., Visible light responsive nitrogen doped anatase TiO_2 sheets with dominant {001} facets derived from TiN, J. Am. Chem. Soc. 131 (2009) 12868–12869.
[13] S. Chen, D. Li, Y. Liu, W. Huang, Morphology-dependent defect structures and photocatalytic performance of hydrogenated anatase TiO_2 nanocrystals, J. Catal. 341 (2016) 126–135.
[14] H.N. Kim, T.W. Kim, I.Y. Kim, S.J. Hwang, Cocatalyst-free photocatalysts for efficient visible-light-induced H_2 production: porous assemblies of CdS quantum dots and layered titanate nanosheets, Adv. Funct. Mater. 21 (2011) 3111–3118.
[15] M.R. Allen, A. Thibert, E.M. Sabio, N.D. Browning, D.S. Larsen, F.E. Osterloh, Evolution of physical and photocatalytic properties in the layered titanates $A_2Ti_4O_9$ ($A = K, H$) and in nanosheets derived by chemical exfoliation, Chem. Mater. 22 (2009) 1220–1228.
[16] O.C. Comptons, C.H. Mullet, S. Chiang, F.E. Osterloh, A building block approach to photochemical water-splitting catalysts based on layered niobate nanosheets, J. Phys. Chem. C 112 (2008) 6202–6208.
[17] H. Hata, Y. Kobayashi, V. Bojan, W.J. Youngblood, T.E. Mallouk, Direct deposition of trivalent rhodium hydroxide nanoparticles onto a semiconducting layered calcium niobate for photocatalytic hydrogen evolution, Nano Lett. 8 (2008) 794–799.
[18] Y. Ebina, T. Sasaki, M. Harada, M. Watanabe, Restacked perovskite nanosheets and their Pt-loaded materials as photocatalysts, Chem. Mater 14 (2002) 4390–4395.
[19] Y. Okamoto, S. Ida, J. Hyodo, H. Hagiwara, T. Ishihara, Synthesis and photocatalytic activity of rhodium-doped calcium niobate nanosheets for hydrogen production from a water/methanol system without cocatalyst loading, J. Am. Chem. Soc. 133 (2011) 18034–18037.
[20] K. Maeda, M. Eguchi, T. Oshima, Perovskite oxide nanosheets with tunable band-edge potentials and high photocatalytic hydrogen-evolution activity, Angew. Chem. 53 (2014) 13164–13168.
[21] Y. Zhao, P. Chen, B. Zhang, D.S. Su, S. Zhang, L. Tian, et al., Highly dispersed TiO_6 units in a layered double hydroxide for water splitting, Chemistry 18 (2012) 11949–11958.
[22] G. Zhang, B. Lin, Y. Qiu, L. He, Y. Chen, B. Gao, Highly efficient visible-light-driven photocatalytic hydrogen generation by immobilizing CdSe nanocrystals on ZnCr-layered double hydroxide nanosheets, Int. J. Hydrogen Energy 40 (2015) 4758–4765.
[23] K. Chang, M. Li, T. Wang, S. Ouyang, P. Li, L. Liu, et al., Drastic layer-number-dependent activity enhancement in photocatalytic H_2 evolution over $nMoS_2$/CdS ($n \geq 1$) under visible light, Adv. Energy Mater. 5 (2015) 1402279.
[24] L. Yin, X. Hai, K. Chang, F. Ichihara, J. Ye, Synergetic exfoliation and lateral size engineering of MoS_2 for enhanced photocatalytic hydrogen generation, Small 14 (2018) e1704153.
[25] X. Zhang, Y. Guo, J. Tian, B. Sun, Z. Liang, X. Xu, et al., Controllable growth of MoS_2 nanosheets on novel Cu_2S snowflakes with high photocatalytic activity, Appl. Catal. B: Environ. 232 (2018) 355–364.

[26] X. Hu, S. Lu, J. Tian, N. Wei, X. Song, X. Wang, et al., The selective deposition of MoS_2 nanosheets onto (101) facets of TiO_2 nanosheets with exposed (001) facets and their enhanced photocatalytic H_2 production, Appl. Catal.B: Environ. 241 (2019) 329–337.

[27] W. Zhou, Z. Yin, Y. Du, X. Huang, Z. Zeng, Z. Fan, et al., Synthesis of few-layer MoS_2 nanosheet-coated TiO_2 nanobelt heterostructures for enhanced photocatalytic activities, Small 9 (2013) 140–147.

[28] Z. Zhang, L. Huang, J. Zhang, F. Wang, Y. Xie, X. Shang, et al., In situ constructing interfacial contact $MoS_2/ZnIn_2S_4$ heterostructure for enhancing solar photocatalytic hydrogen evolution, Appl. Catal. B: Environ. 233 (2018) 112–119.

[29] Y. Liu, X. Xu, J. Zhang, H. Zhang, W. Tian, X. Li, et al., Flower-like MoS_2 on graphitic carbon nitride for enhanced photocatalytic and electrochemical hydrogen evolutions, Appl. Catal. B: Environ. 239 (2018) 334–344.

[30] L. Kong, J. Yan, S. Liu, Carbonyl linked carbon nitride loading few layered MoS_2 for boosting photocatalytic hydrogen generation, ACS Sustain. Chem. Eng. 7 (2019) 1389–1398.

[31] Y. Zhou, Y. Lei, D. Wang, C. Chen, Q. Peng, Y. Li, Ultra-thin Cu_2S nanosheets: effective cocatalysts for photocatalytic hydrogen production, Chem. Commun. 51 (2015) 13305–13308.

[32] J. Zhang, J. Yu, Y. Zhang, Q. Li, J.R. Gong, Visible light photocatalytic H_2-production activity of CuS/ZnS porous nanosheets based on photoinduced interfacial charge transfer, Nano Lett. 11 (2011) 4774–4779.

[33] P. Niu, L. Zhang, G. Liu, H.-M. Cheng, Graphene-like carbon nitride nanosheets for improved photocatalytic activities, Adv. Funct. Mater. 22 (2012) 4763–4770.

[34] L. Ma, H. Fan, J. Wang, Y. Zhao, H. Tian, G. Dong, Water-assisted ions in situ intercalation for porous polymeric graphitic carbon nitride nanosheets with superior photocatalytic hydrogen evolution performance, Appl. Catal. B 190 (2016) 93–102.

[35] Y. Chen, B. Wang, S. Lin, Y. Zhang, X. Wang, Activation of $n \rightarrow \pi^*$ transitions in two-dimensional conjugated polymers for visible light photocatalysis, J. Phys. Chem. C 118 (2014) 29981–29989.

[36] E.M. Dias, K.C. Christoforidis, L. Francàs, C. Petit, Tuning thermally treated graphitic carbon nitride for H_2 evolution and CO_2 photoreduction: the effects of material properties and mid-gap states, ACS Appl. Energy Mater. 1 (2018) 6524–6534.

[37] Y. Hou, A.B. Laursen, J. Zhang, G. Zhang, Y. Zhu, X. Wang, et al., Layered nanojunctions for hydrogen-evolution catalysis, Angew. Chem. 52 (2013) 3621–3625.

[38] A. Crake, K.C. Christoforidis, A. Kafizas, S. Zafeiratos, C. Petit, CO_2 capture and photocatalytic reduction using bifunctional TiO_2/MOF nanocomposites under UV–vis irradiation, Appl. Catal. B: Environ. 210 (2017) 131–140.

[39] A. Crake, K.C. Christoforidis, A. Gregg, B. Moss, A. Kafizas, C. Petit, The effect of materials architecture in TiO_2/MOF composites on CO_2 photoreduction and charge transfer, Small 15 (2019) 1805473.

[40] Y. Liang, R. Shang, J. Lu, W. An, J. Hu, L. Liu, et al., 2D MOFs enriched g-C_3N_4 nanosheets for highly efficient charge separation and photocatalytic hydrogen evolution from water, Int. J. Hydrogen Energy 44 (2019) 2797–2810.

12

Application of two-dimensional materials for electrochemical carbon dioxide reduction

Xin Li and Zhenyu Sun

State Key Laboratory of Organic–Inorganic Composites, College of Chemical Engineering, Beijing University of Chemical Technology, Beijing, P.R. China

12.1 Introduction

Since the industrial revolution, the large-scale application of fossil fuels has brought many conveniences to people; meanwhile, it has also caused excessive carbon dioxide (CO_2) emissions. The resulting changes in the global ecological environment, such as the greenhouse effect and ocean acidification, will greatly affect human survival and development. Therefore, reducing the adverse effects of excessive CO_2 emissions is a common problem facing the world today. CO_2, which was the cheapest and abundant carbon-based resource, had particularly interesting to convert it into other value-added products. Generally, the feasible approaches of CO_2 conversion mainly include CO_2 methanation [1–6], reverse water gas shift [7–11], CO_2 hydrogenation to hydrocarbons or alcohols [12–18], electrochemical reduction [19], photochemical reduction [20], biological reduction [21], and radioactive reduction [22]. Among those techniques, ECR powered by renewable energy generation or nuclear energy is considered as a promising method (Fig. 12.1). There are several advantages to using ECR: (1) the selectivity of the product can be controlled by changing the electrolysis conditions (such as electrode, electrolyte, electrode potential, etc.); (2) the power can be provided by energy sources such as solar energy, wind energy, nuclear energy, or tidal energy [23]; (3) the degree of conversion of the reaction can be controlled by controlling the consumption of electricity; (4) the electrochemical reaction system is compact, modular, and on-demand, easy to expand the scale of

2D Nanomaterials for Energy Applications. DOI: https://doi.org/10.1016/B978-0-12-816723-6.00012-5

Figure 12.1 Schematic illustration of the carbon cycle via CO_2 electrolysis powered by renewable energy sources such as wind, solar, and hydro [23].

application; and (5) the reaction can be carried out under mild conditions (normal temperature and pressure). Converting CO_2 into useful chemicals by electrochemical reduction not only helps to reduce CO_2 emissions but also alleviate the sortage of resourses. It has important practical significance for alleviating the dual pressures of energy and environment.

ECR can be carried out by homogeneous or heterogeneous reactions. Accordingly, the electrocatalyst for CO_2 reduction can be broadly classified into homogeneous [24] and heterogeneous [25] catalysts. In a typical CO_2 homogeneous phase catalytic reaction, the catalyst is generally an organic molecule or a metal-organic complex molecule dissolved in an electrolyte solution, and a specific active center (such as a metal cation) acts with a CO_2 molecule [26,27]. Due to their unique molecular structure, homogeneous catalysts generally exhibit very good selectivity for CO_2. Although homogeneous catalysts can achieve high selectivity, they have some deficiencies such as high cost, easy to deactivate, high toxicity, and complex postseparation processes hinder the practical application of homogeneous catalysts [28]. In contrast, heterogeneous catalysts have attracted the attention of researchers due to its unique advantages such as recyclability, mild synthesis conditions, particularly remarkable catalytic efficiency, and large-scale industrial production. During the past few decades, a variety of heterogeneous catalysts have been developed for ECR including metal foils [29–33], metal nanostructures [34–42], oxide-derived metals [43–45], carbon nanomaterials [46–48], as well as bioinspired catalysts [49–52], which are capable of reducing CO_2 into a variety of carbon compounds. However, there are still many problems in the current ECR such as large overpotential (the difference between the equilibrium and applied potential), slow electron transfer kinetics, low efficiency, and unsatisfactory selectivity.

Therefore, the requirement for high-performing electrocatalyst of ECR is significant. In general, good electrocatalyst features high surface area, high activity, good electrical conductivity, and long-term stability. The activity of an electrode strongly depends on the physicochemical properties of the electrocatalyst surface and the electrode–electrolyte interaction. To achieve high catalytic activity, electrocatalysts need to be effective at lowering the energy barriers of electrocatalytic reactions and promoting the rate of surface electron charge transfer. Accordingly, the development of well-defined surfaces and structures is a valued approach to promote electrochemical processes. Moreover, a high electrode surface area and electrical conductivity can be achieved by structuring conductive supports/substrates at the nanoscale. The durability of a catalyst determines the operating life of the electrocatalytic cell which is very important for practical applications. Developing inexpensive electrocatalysts for CO_2 reduction with high activity, efficiency, stability, and selectivity is an important part of CO_2 conversion.

Two-dimensional (2D) nanomaterials, which have a more uniformly exposed lattice and achieving higher concentrations of exposed electrocatalytic active sites under certain electrocatalyst mass loading, have received wide attention in the field of electrocatalysis. Also, the simple and ordered molecular framework of 2D nanomaterials makes the identification of active sites easier from both theoretical and experimental perspectives. The research on 2D nanomaterials was initially focused on graphene owing to its terrific electronic, optical, and mechanical properties. Graphene is one of the extensively investigated 2D nanomaterials as the catalyst or support to various metal nanocatalysts exhibiting excellent catalytic activity for almost all types of chemical reactions such as electrochemical, photochemical as well as traditional heterogeneous reactions. Beyond graphene, there is a wide spectrum of 2D nanomaterials having similar or diverse structural, electronic, and chemical properties to graphene and these have become established as excellent catalysts or catalyst supports. Graphitic carbon nitride (g-C_3N_4), metals, oxides, inorganic hexagonal boron nitride (h-BN), black phosphorus, MXene, metal-organic frameworks compound (MOFs), covalent-organic frameworks

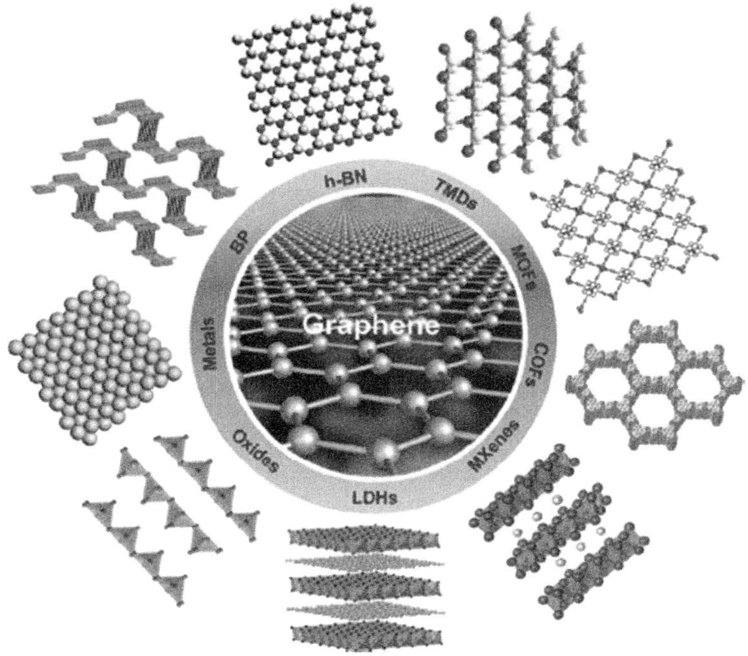

Figure 12.2 Different kinds of typical 2D nanomaterials, such as graphene, h-BN, TMDs, MOFs, COFs, MXenes, LDHs, oxides, metals, and BP [53].

compound, layered double hydroxides (LDHs), and transition-metal dichalcogenides (TMDs), that is, molybdenum sulfide (MoS_2) and tungsten sulfide (WS_2), are the most prominent examples of 2D nanomaterials and these have gained a renewed interest as supports for various metal nanocatalysts. The different kinds of typical ultrathin 2D nanomaterials were shown in Fig. 12.2 [53].

12.2 Background on electrochemical carbon dioxide reduction

As a heterogonous catalytic process in aqueous solution, ECR involves two-electron or multielectron and proton processes occurred on complex electrode-electrolyte three-phase gas/solid/liquid interfaces, which are also accompanied by hydrogen evolution reaction (HER). Three steps are mainly describing the heterogeneously catalyzed reaction process:

(1) The CO_2 molecules dissolved in the electrolyte are combined with the catalyst on the surface of the electrode by chemisorption.

(2) The adsorbed CO_2 molecules react with electrons or protons in the solution to break carbon–oxygen bonds or form carbon–hydrogen bonds.
(3) The reaction intermediate continues to accept electron–proton pairs with possible molecular rearrangement to give the final product, after which the product molecules desorb on the catalyst surface and diffuse to the electrolyte solution.

12.2.1 Thermodynamic process of electrochemical carbon dioxide reduction

General speaking, ECR is subject to the initial step of CO_2 producing an electron to generate $CO_2^{\bullet-}$. However, this step is difficult to perform and the reduction potential required in an aprotic solvent such as dimethylformamide reaches −1.97 V (relative to a standard hydrogen electrode, SHE); the reduction potential in a neutral aqueous solution is also as high as −1.90 V (vs SHE). However, the actual electrochemical reduction process of CO_2 is proton-dependent. In this case, the pure $CO_2^{\bullet-}$ generation process through high energy barrier can be avoided, so that the applied potential of CO_2 electrochemical reduction can be more positive, and thus the theory reduces a large amount of power consumption. According to the proton/electron (H^+/e^-) logarithm of the reaction involved in CO_2 reduction, the reactant CO_2 can be reduced to C1 compound [formic acid (HCOOH, $HCOO^-$ in alkaline electrolyte), carbon monoxide (CO), formaldehyde (HCHO), methanol (CH_3OH), methane (CH_4)], C2 compound [oxalic acid ($H_2C_2O_4$), ethylene (C_2H_4), ethanol (C_2H_5OH) and ethane (C_2H_6)], C3 compound [propanol (C_3H_7OH)], etc. The specific electrochemical half-reactions involved are shown in Eqs. (12.i)–(12.xi) in Table 12.1 (all potentials are relative to SHE at pH 7). Competitive HER is the simplest two-electron reaction process, namely the following Eq. (12.xii) [19,54–56].

According to the basic formula of chemical thermodynamics $\Delta G = -nFE^0$, it is known that the reduction reaction requires a positive potential E^0 to make the cathode reaction easier (n is the electron transfer number of the reduction reaction, and F is the Faraday constant, the value was 96,485 C mol^{-1}). Based on the individual E^0 values given by the above equations, the reduction of CO_2 to hydrocarbons or alcohols should be more thermodynamically easier than reduction to CO, HCOOH, and H_2. However, the actual electrochemical reduction process is

Table 12.1 Cathodic half-reaction of CO_2 electroreduction and HER [63].

Cathodic half-reaction	Standard reduction potential	Number
$CO_2 + 2H^+ + 2e^- \rightarrow HCOOH$	$E^0_{redox} = -0.610$ V	(12.i)
$CO_2 + 2H^+ + 2e^- \rightarrow CO + H_2O$	$E^0_{redox} = -0.530$ V	(12.ii)
$2\ CO_2 + 2H^+ + 2e^- \rightarrow H_2C_2O_4$	$E^0_{redox} = -0.913$ V	(12.iii)
$2\ CO_2 + 2e^- \rightarrow C_2O_4^{2-}$	$E^0_{redox} = -1.003$ V	(12.iv)
$CO_2 + 4H^+ + 4e^- \rightarrow HCHO + H_2O$	$E^0_{redox} = -0.480$ V	(12.v)
$CO_2 + 6H^+ + 6e^- \rightarrow CH_3OH + H_2O$	$E^0_{redox} = -0.380$ V	(12.vi)
$CO_2 + 8H^+ + 8e^- \rightarrow CH_4 + 2\ H_2O$	$E^0_{redox} = -0.240$ V	(12.vii)
$2\ CO_2 + 12H^+ + 12e^- \rightarrow C_2H_4 + 4\ H_2O$	$E^0_{redox} = -0.349$ V	(12.viii)
$2\ CO_2 + 12H^+ + 12e^- \rightarrow C_2H_5OH + 3\ H_2O$	$E^0_{redox} = -0.329$ V	(12.ix)
$2\ CO_2 + 14H^+ + 14e^- \rightarrow C_2H_6 + 4\ H_2O$	$E^0_{redox} = -0.270$ V	(12.x)
$3\ CO_2 + 18H^+ + 18e^- \rightarrow C_3H_7OH + 5\ H_2O$	$E^0_{redox} = -0.310$ V	(12.xi)
$2H^+ + 2e^- \rightarrow H_2$	$E^0_{redox} = -0.42$ V	(12.xii)

not the case, because the ECR depends on the reaction kinetics involving proton concentration and electron transport rate in addition to the thermodynamic barrier. This indicates that an effective electrocatalyst of CO_2 reduction should have an active site for adsorbing protons while providing a fast electron transport channel.

12.2.2 Performance parameters of electrochemical carbon dioxide reduction catalyst

In order to evaluate and compare the performance of different ECR catalysts, early researchers in the field defined some basic performance parameters. Specifically, the following six metrics are included [57].

1. Starting potential (overpotential)

   Since the ECR needs to overcome the kinetic energy barrier, the initial potential of CO_2 reduction is always lower than the thermodynamic standard reduction potential of CO_2 reduction, that is, more negative. The difference between the initial potential and the standard reduction potential is the overpotential (η). The initial potential of the catalyst for the electrolysis reaction can directly reflect the overpotential of CO_2 reduction to a specific product. The magnitude of the onset potential depends on the type of catalytic material on the working electrode and on the intrinsic properties of electrocatalyst.

2. Current density

The total current density is defined as the value of the current divided by the geometric area of the working electrode or the electrochemically active surface area (ECSA) of the catalyst, generally expressed in J (mA cm^{-2}). In addition, especially for catalysts containing a noble metal and a noble metal as an active component, a mass activity is also commonly used to indicate the catalytic activity of the catalyst, which is defined as the magnitude of the current divided by the mass of the precious metal component used in the working electrode, in units of A g^{-1}. At a specific electrolytic potential, the current density is a measure of the reaction rate in the electrocatalytic reaction, directly related to the amount of catalyst supported on a conductive substrate (such as carbon paper), also related to the mass transfer rate of the reactant CO_2 and the product desorption-diffusion rate. Different working electrode conductive substrates are gas diffusion electrode (GDE), fluorine-doped tin oxide conductive film (FTO film), metal foil, carbon paper, glassy carbon electrode (GCE), etc. The reaction exhibits geometric current densities of different orders of magnitude. Current density is an important consideration for the practical application of electrocatalytic reactions because it determines the rate of product formation and the required working electrode and cell size.

3. Faradaic efficiency (FE)

The Faraday efficiency (FE) in ECR refers to the selectivity of the electrolytic reduction product. FE can be calculated by the following formula (12.1):

$$FE = \frac{\alpha nF}{Q} \tag{12.1}$$

where α is the number of electron transfer in the electrochemical reaction (e.g., $\alpha = 2$ when CO_2 is reduced to CO or HCOOH); n (mol) is the amount of a substance that produces a specific product within a certain electrolysis time; F refers to Faraday constant; and Q (C) represents the total amount of charge accumulated during electrolysis. Furthermore, the bias current density can be calculated by the product of the total current density J and the Faraday efficiency of a particular product (e.g., product i), that is, $J_i = FE_i * J$, where J_i is the partial current density of product i.

4. Energy efficiency

In order to measure the degree to which the applied electrical energy is converted into chemical energy, the energetic efficiency parameter is introduced and calculated as follows (12.2):

$$\varepsilon_{\text{energetic}} = \frac{E_{\text{eq}}}{E_{\text{eq}} + \eta} \times \text{FE} \tag{12.2}$$

$\varepsilon_{\text{energetic}}$ is the energy efficiency, E_{eq} is the thermodynamic standard reduction potential (V), η is the overpotential (V) at a specific applied potential, and FE is the Faraday efficiency of the product. Based on this, higher energy efficiency requires less overpotential and higher Faraday efficiency.

5. Tafel slope

The logarithm of the partial current density of the specific reduction product (i.e., $\log[J_i]$) is plotted on the abscissa, and the overpotential at the selected different applied potential is plotted on the ordinate. The slope is the so-called Tafel slope in mV dec^{-1}. The calculated slope value determines the rate-determining step of the CO_2 reduction reaction and the reaction path. For example, when the Tafel slope is ~ 59 mV dec^{-1}, it is indicated that there is a rapid chemical reaction step after CO_2 obtains a fast equilibrium of electrons. When the Tafel slope is ~ 118 mV dec^{-1}, it indicates that the initial CO_2 has an electron change to the $CO_2^{\bullet-}$ intermediate. The intermediate is the final step of the reduction process. Such a discussion is generally applicable to metal electrodes [42,45]. In summary, high-efficiency ECR catalysts must meet the requirements of low overpotential, high current density and selectivity, and excellent stability that are not easily deactivated [19].

12.2.3 Product generation path of electrochemical carbon dioxide reduction

From a theoretical and experimental point of view, researchers have done a lot of research work on the activation and electroreduction mechanism of CO_2 [54,58]. However, regarding the determination of the formation route of CO_2 reduction products, the results of most existing literature are based on the Tafel slope obtained by electrochemical tests and the results of density functional theory (DFT) calculations. For a comprehensive and in-depth explanation of the mechanism of ECR, it is necessary to further combine in situ spectroscopic

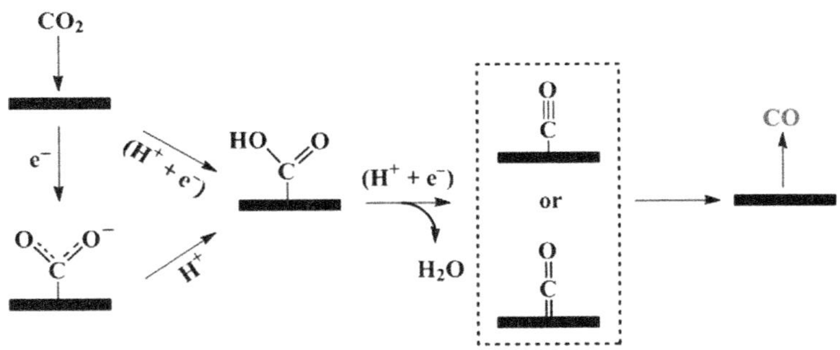

Figure 12.3 The possible reaction path for CO_2 electrochemical reduction to CO [54].

characterization techniques with experimental results and DFT calculations. Aiming at the target product of electrochemical reduction of CO_2 by catalyst, the formation path of CO and C_2H_4 was briefly introduced.

1. Possible path of CO generation

The reduction of CO_2 to CO is a simple two-electron transfer reaction process (see Fig. 12.3). A proton and an electron are respectively transferred from the electrolyte solution and the external circuit to the CO_2 molecule adsorbed on the surface of the catalyst, and the CO_2 is first reduced to a carboxyl radical (*COOH). Then, the second pair of H^+/e^- can attack the oxygen atoms (hydroxyl oxygen) in the *COOH to form liquid H_2O and absorbed-state of CO. Finally, CO is desorbed as a product from the catalyst and diffused to the electrolyte solution [58]. In general, the conversion of CO_2 to *COOH in the first step is easily limited by the weaker binding energy of the catalyst for the middle *COOH, while the energy barrier of COOH* to *CO is not high, this step is easy to carry out. In addition, the last-step CO desorption process is also hindered by the strong binding force between the catalyst and CO, so the total process of CO_2 reduction to CO is higher and the selectivity is not high. The first step or the last step is also considered as a speed-determining step from the perspective of theoretical calculations [59]. Another possible path to produce *COOH is through stepwise electron and proton transfer involving the formation of $CO_2^{\bullet-}$ and adsorption on the catalyst surface. There are two steps of generating *COOH show different pH dependence. Increasing the pH or CO_2 pressure promotes *COOH formation [60]. Metal catalysts such as Gold(Au), silver(Ag), palladium(Pd), and zinc (Zn) are sufficiently strong to bind to *COOH such that the intermediate can continue to be reduced to *CO, and then *CO is desorbed as a product CO on the electrode surface due to weak adsorption on the surface of these metal electrodes.

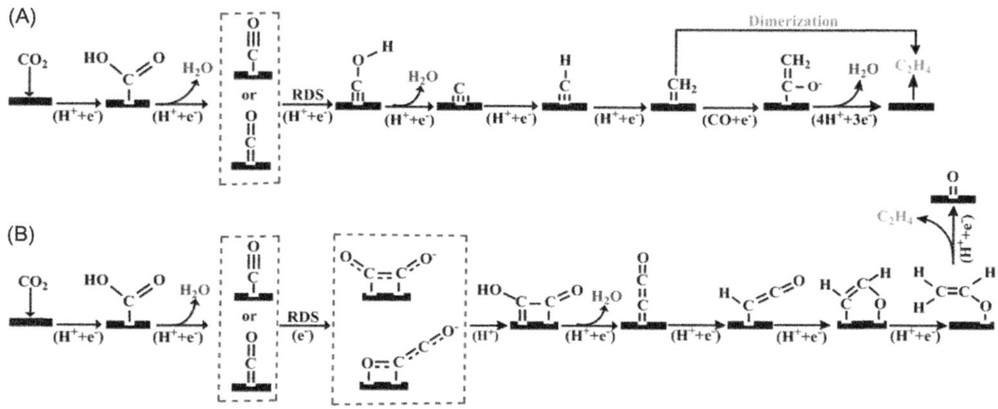

Figure 12.4 The possible reaction pathways for CO_2 electrochemical reduction to C_2H_4: (A) via coupling of two $^*CH_2$ species or the CO insertion reaction in a Fischer–Tropsch like step [61] and (B) via the *CO dimerization step to form active $^*C_2O_2^-$ [62].

2. Possible path of C_2H_4 production

At present, there is no consensus on the specific reaction route and mechanism to formation C_2H_4. It is generally believed that the ease of formation and stability of the intermediates of the adsorbed state such as *CO, *CHO (or *COH), $^*CH_2$, etc. play a key role in further generating the simplest C2 product C_2H_4. The formation of C_2H_4 requires generating carbon–carbon bonds (single or double bonds). There are two main types of carbon–carbon bonds generation paths (see Fig. 12.4): (1) *CO was hydrogenated to generate *COH (this step is a rate control step to generate C2 product including C_2H_4). $^*CH_2$ was obtained by multistep protonated and reduced, and further dimerization or CO insertion and multistep proton and electron transfer reactions. Finally, the C_2H_4 was produced [63]. (2) Adjacent adsorption state *CO was dimerized and $^*C_2O_2^-$ was generate by obtaining an electron (this step is the rate control step for generating C2 product). Then, the $^*C_2O_2^-$ was protonated to obtain *CO-COH, and C_2H_4 was produced by further multistep protons and electrons ($5\,H^+ + 5e^-$) transfer reaction. In the reaction pathway involving "carbene ($\bullet CH_2$)", the two adsorbed states of $^*CH_2$ were combining to form C_2H_4 [58]. In the path involving the $^*C_2O_2^-$dimerization intermediate, the reactive intermediates *CO-COH and vinyl alcohol were sequentially reduced by the dimer plus proton, and the unstable vinyl alcohol was finally hydrodeoxygenated to C_2H_4 [62]. On the surface of the metallic copper catalysts, C_2H_4 was the most easily formed C2 product [64].

12.3 Two-dimensional materials for electrochemical carbon dioxide reduction

Searching for catalysts with low cost but high activity and selectivity remains a challenge to promote the kinetically sluggish CO_2 reduction process for large-scale production. Recently, 2D nanomaterials featuring a large fraction of low-coordinated surface atoms have gained much attention for CO_2 reduction (Table 12.2).

12.3.1 Elements

12.3.1.1 Heteroatom-doped graphene

When graphene was first prepared via the "Scotch tape" method in 2004, it has emerged as the subject of intense research, especially as the key component of electrode materials [90]. However, pristine graphene has low activity requiring high CO_2 activation free energy (over 1.3 eV). Deliberately doping heteroatoms (e.g., nitrogen, boron, and oxygen) into the graphitic network can effectively alter their chemical as well as electronic properties, therefore significantly enhances their catalytic activity toward specific electrochemical reactions.

The first example of a low-cost and metal-free nitrogen-doped graphene-based (N-graphenes) material for electrochemical reduction of CO_2 to HCOOH was reported by Tianwei Tan group [65]. The catalyst was synthesized by graphene oxide as a carbon template and melamine as a precursor containing nitrogen using pyrolysis. HCOOH with Faradaic efficiencies reached a maximum (73%) at the moderately negative potential (−0.84 V) versus reversible hydrogen electrode (RHE) in saturated 0.5 M $KHCO_3$. The current density maintained steady at around 7.5 mA cm^{-2} after the 12 h test period. In addition, they further compared the overpotentials and FEs of HCOOH formation for a series of metal electrodes, including Pb, Hg, In, Sn, Cd, Tl, and nanostructured metal catalysts (nano-SnO_2 [37]; Sn/SnO_2 [44]; Cu_2O-derived Cu [43]) in literatures with that of N-graphenes, as shown in Fig. 12.5. In order to reach 73% FEs of formate, all the metal electrodes [91] required a minimum of 1.1 V of overpotential to accomplish the transformation. However, N-graphenes catalyst only needed −0.84 V (versus RHE), which exhibits a much lower overpotential. The only exception is the Cu_2O-derived Cu and Sn/SnO_2, which can drive the reduction reaction with less overpotentials. But, the FEs of

Table 12.2 Summary of 2D-based materials reported for electrocatalytic CO_2 reduction.

Catalysts	Electrolytes	J^a ($mA \cdot cm^{-2}$)	Onset potential (V) or η^b (V)	Main products (FE)[c]	References
N-doped graphene	0.5 M $KHCO_3$	7.5 mA cm^{-2} @ −0.84 V (vs RHE)	Onset potential: −0.3 V (vs RHE)	$HCOO^-$: 73% @ −0.84 V (vs RHE)	[65]
B-doped graphene	0.1 M $KHCO_3$	~2.0 mA cm^{-2} @ −1.4 V (vs SCE)	Onset potential: −1.1 V (vs SCE)	$HCOO^-$: 66% @ −1.4 V (vs SCE)	[66]
A single site zinc coordinated nitrogen co-dope graphene	0.5 M $KHCO_3$	11.2 mA cm^{-2} @ −0.8 V (vs RHE)	η: 0.39 V @ 11.2 mA cm^{-2}	CO: 91% @ −0.5 V(vs RHE)	[67]
Ni^{2+}@NG	0.5 M $KHCO_3$	10.2 mA cm^{-2}@ −0.68 V (vs RHE)	Onset potential: − 0.38 V (vs RHE)	CO: 92% @ −0.68 V (vs RHE)	[68]
Sn sheets confined in graphene	0.1 M $NaHCO_3$	21.1 mA cm^{-2} @ −1.8 V (vs SCE)	Onset potential: −0.85 V (vs SCE)	$HCOO^-$: 89% @ −1.8 V (vs SCE)	[69]
Ag nanosheets	0.5 M $NaHCO_3$	10 mA cm^{-2} @ −0.8 V (vs RHE)	η: 0.29 V @ 5 mA cm^{-2}	CO: 95% @ −0.7 V (vs RHE)	[70]
Triangular silver nanoplates	0.1 M $KHCO_3$	over 5.5 mA cm^{-2} (vs RHE)	η: 0.746 V @ 5.5 mA cm^{-2}	CO:96.8% @ −0.855 V (vs RHE)	[71]
Ultrathin Pd nanosheets	0.1 M $KHCO_3$	400 mA cm^{-2} @ −0.9 V (vs RHE)	Onset potential: −0.20 V (vs RHE)	CO: 94% @ −0.5 V (vs RHE)	[72]
Partially oxidized Co 4-atomic-layers	0.1 M Na_2SO_4	10.59 mA cm^{-2} @ −0.85 V (vs SCE)	η: 0.24 V @ 10.59 mA cm^{-2}	$HCOO^-$: 90.1% @ −0.85 V (vs SCE)	[73]
Ultrathin Bi nanosheets	0.5 M $NaHCO_3$	24 mA cm^{-2} @ −1.7 V (vs SCE)	Onset potential: −1.3 V (vs SCE)	$HCOO^-$: about 100% @ −1.7 V (vs SCE)	[74]
BiOCl derived Bi nanosheets	0.5 M $KHCO_3$	3.7 mA cm^{-2} @ −1.8 V (vs SCE)	Onset potential: −1.3 V (vs SCE)	$HCOO^-$: 92% @ −1.5 V (vs SCE)	[75]
N-based Cu(I)/C-doped boron nitride	[Emim]BF_4/LiI/H_2O	13.9 mA cm^{-2} @ −2.2 V (vs Ag/AgCl)	η: 0.52 V @ 13.9 mA cm^{-2}	CH_3COOH: 80.3% @ −.2.2 V (vs Ag/AgCl)	[76]
g-C_3N_4/MWCNTs	0.1 M $KHCO_3$	0.55 mA cm^{-2} (J_{CO}) @ −0.75 V (vs RHE)	η: 0.64 V @ 0.92 mA cm^{-2}	CO: 60% @ −0.75 V (vs RHE)	[77]
A polarized ultrathin g-C_3N_4	2 M $KHCO_3$	3.05 mA cm^{-2} @ −1.2 V (vs Ag/AgCl)	Onset potential: −0.9 V (vs Ag/AgCl)	CO: ~ 80% @ −1.2 V (vs Ag/AgCl)	[78]
WSe_2 nanoflakes	50 vol% EMIM-BF_4 aqueous solution	330 mA cm^{-2} @ −0.764 V (vs RHE)	η: 54 mV @ 18.95 mA/cm^2	CO: 90% @ −0.764 V (vs RHE)	[79]

(*Continued*)

Table 12.2 (Continued)

Catalysts	Electrolytes	J^a (mA · cm^{-2})	Onset potential (V) or η^b (V)	Main products (FE)^c	References
MoSeS alloy monolayers	4 mol% EMIM-BF_4 aqueous solution	43 mA cm^{-2} @ −1.15 V (vs RHE)	Onset potential: −0.55 V (vs RHE)	CO: 45.2% @ −1.15 V (vs RHE)	[80]
Nb-doped vertically aligned MoS_2	50 vol% EMIM-BF_4 aqueous solution	237 mA cm^{-2} @ −0.8 V (vs RHE)	Onset potential: −0.142 V (vs RHE)	CO: 82% @ −0.8 V (vs RHE)	[81]
Oxygen-deficient Co_3O_4 atomic layers	0.1 M $KHCO_3$	2.7 mA cm^{-2} @ −0.87 V (vs SCE)	Onset potential: −0.78 V (vs SCE)	$HCOO^-$: 87.6% @ −0.87 V (vs SCE)	[82]
Mesoporous SnO_2 nanosheets	0.5 M $NaHCO_3$	50 mA cm^{-2} @ −1.6 V (vs Ag/AgCl)	η: 0.88 V @ 45 mA cm^{-2}	$HCOO^-$: 89% @ −1.6 V (vs Ag/AgCl)	[83]
Re-SURMOF thin films	0.1 M TBAH in ACN with 5 vol% trifluoroethanol	~ 2.5 mA cm^{-2} @ −1.6 V (vs normal hydrogen electrode, NHE)	Onset potential: −1.3 V (vs NHE)	CO: ~ 93% @ −1.6 V (vs NHE)	[84]
SnS_2/rGO nanosheets	0.5 M $NaHCO_3$	~ 13.7 mA cm^{-2} @ −0.75 V (vs RHE)	η: 0.68 V @ 13.9 mA cm^{-2}	$HCOO^-$: 84.5% @ −0.76 V (vs RHE)	[85]
N-doped graphene foam	0.1 M $KHCO_3$	~ 1.8 mA cm^{-2} @ −0.58 V (vs RHE)	Onset potential: −0.3 V (vs RHE)	CO: ~ 85% @ −0.58 V (vs RHE)	[86]
Ni-N modified graphene	0.1 M $KHCO_3$	~ 0.2 mA cm^{-2} @ −0.65 V (vs RHE)	η: 0.54 V @ 0.2 mA cm^{-2}	CO: > 90% @ −0.7 ~ −0.9 V (vs RHE)	[87]
Mo-Bi bimetallic chalcogenide nanosheets	[BMIM]BF_4/MeCN	12.1 mA cm^{-2} @ −0.7 V (vs SHE)	η: 0.16 V @ 1.0 mA cm^{-2}	CH_3OH: 71.2% @ −0.7 V (vs SHE)	[88]
N-doped graphene-like materials	[BMIM]BF_4 with 3 wt% H_2O	3.26 mA cm^{-2} @ −1.4 V (vs SHE)	η: 1.16 V @ 3.26 mA cm^{-2}	CH_4: 93.5% @ −1.4 V (vs SHE)	[89]

[a] J refers to geometric current density or partial current density with specific subscript.
[b] η refers to overpotential to generate a specific product.
[c] FE refers to maximum faradic efficiency of the main products.

HCOOH production for both catalysts only has 40%. In contrast, the N-graphenes has well-balanced two aspects by achieving both the good selectivity of HCOOH and relatively low overpotential simultaneously. The results show that metal-free nitrogen-doped graphene is a promising alternative to expensive precious metal electrocatalysts, achieving higher energy efficiency and better selectivity for the desired product by commercial reduction of CO_2.

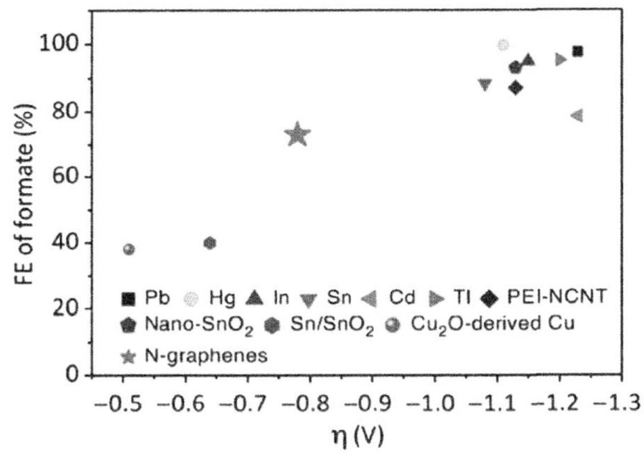

Figure 12.5 Comparison of overpotentials and FEs of formate formation of N graphenes with other electrocatalysts in literatures: Pb, Hg, In, Sn, Cd, and Tl [91]; PEI-NCNT [47]; nano-SnO_2 [37]; Sn/SnO_2 [44]; and Cu_2O-derived Cu [43].

Narayanaru Sreekanth [66] reported the electrocatalytic activity of boron-doped graphene (BG) which was prepared by calcination of graphene oxide and boric acid at 900°C for the reduction of CO_2 to $HCOOH/HCOO^-$. By examining the electrocatalytic activity of BG, $HCOO^-$ with a faradic efficiency of 66% was attained at −1.4 V versus SCE in CO_2-saturated 0.1 M $KHCO_3$ aqueous solution. In order to realize the mechanism of BG for ECR, they used $C_{42}H_{16}$ and $C_{41}BH_{16}$ to simulate the structure of graphene and BG for DFT calculation (Fig. 12.6). The calculations suggested that the presence of boron introduces asymmetric charge and spin density distribution throughout the ground state geometry, resulting in a high spin density. Those B and C atoms with positive spin densities were speculated to be catalytically active sites, favoring the chemisorption of protonated CO_2 (COOH*).

Although the nitrogen-doped graphene-based catalysts had high catalytic activity [46,92], it is always a great challenge to prove the catalytic activity center accurately, even though the mechanism for electrochemical reduction of CO_2 on nitrogen-doped carbon catalysts has been corroborated by computational methods [86,93–95]. Therefore, Chen et al. [67] first reported a single site zinc coordinated nitrogen co-dope graphene (Zn-N-G) catalyst for electrochemical reduction of CO_2 to CO. They developed four kinds of Zn-N-G electrocatalysts under different high-temperature pyrolysis (700°C, 800°C, 900°C, and 1000°C) and acid leaching in order to understanding of the active sites for electrochemical reduction of CO_2 to CO

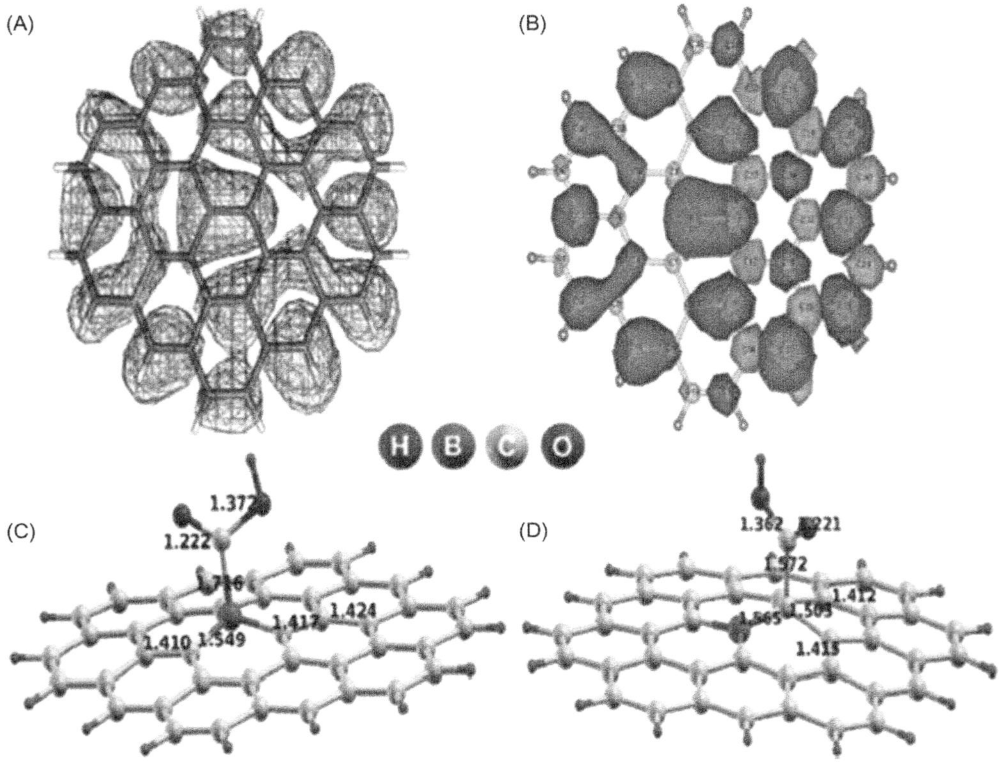

Figure 12.6 (A) HOMO of $C_{41}BH_{16}$, (B) spin density mapping, (C) chemisorption of *COOH on $C_{41}BH_{16}$ via B atoms, and (D) chemisorption of *COOH on $C_{41}BH_{16}$ via C atoms [66].

on the Zn-N-G catalysts. The results have showed that the Zn-N-G-800 catalyst exhibits an excellent intrinsic activity toward CO_2 reduction, reaching a Faradaic efficiency (FE) of 91% for CO production at a low overpotential of 0.39 V in CO_2-saturated 0.5 M $KHCO_3$ electrolyte. DFT calculation results reveal the Zn-Nx sites can facilitate the formation of COOH* intermediates and desorption of CO*, which is the origin of enhanced activity for electrochemical reduction of CO_2 to CO (Fig. 12.7).

Zhang et al. [96] synthesized that atomic iron dispersed on nitrogen-doped graphene (Fe/NG) catalyst for CO_2 reduction to CO. They used graphene oxide (GO) as the precursor and immobilized the Fe atoms on graphene through N bonding by annealing a mixture of GO and ferric chloride ($FeCl_3$) in an Ar/ NH_3 atmosphere at 700°C–800°C (Fig. 12.8) The Fe/NG catalyst

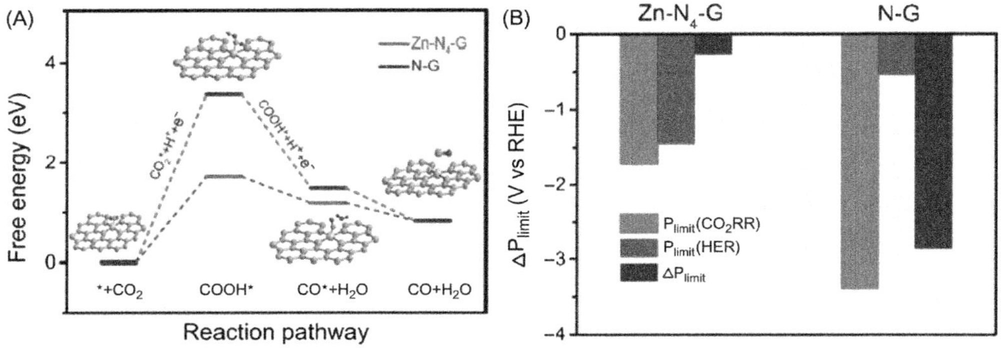

Figure 12.7 DFT calculations for the electrochemical reduction CO_2 to CO process on Zn-N_4-G and N-G. Calculated free energy diagram (A). Calculated limiting potentials for CO_2RR and HER (B) [67].

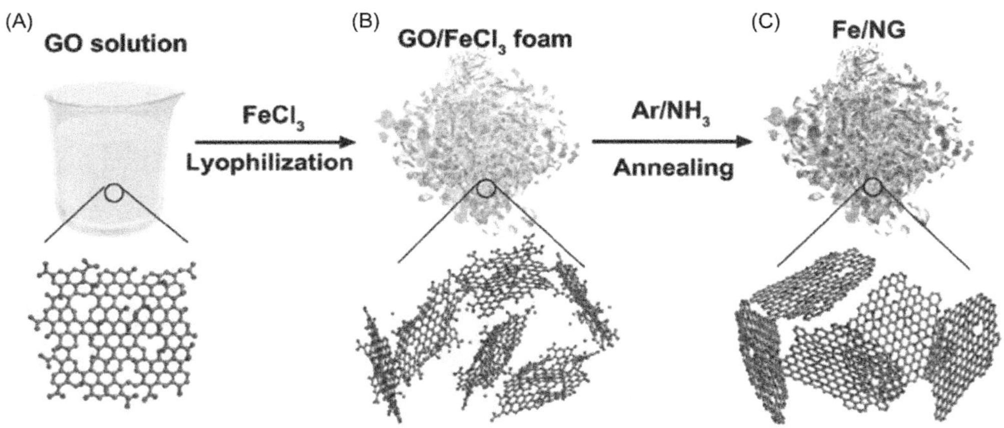

Figure 12.8 Schematic of the synthesis process of the Fe/NG catalyst. (A) GO suspension in deionized water, (B) Fe/NG precursor foam obtained from GO through a lyophilization process, and (C) the final Fe/NG product obtained through the Ar/NH_3 annealing process at 650°C–800°C [96].

has high Faradic efficiency up to 80% at −0.60 V (vs RHE) in CO_2-saturated 0.1 M $KHCO_3$ electrolyte. They found that the pyridinic Fe-N_4 structure and nitrogen-doping enhanced CO_2 adsorption and improved CO_2 activation by DFT calculations.

However, in the current literature reports, the preparation of all metal-N-doped graphene catalysts require high-temperature (800°C–1000°C) pyrolysis of the raw materials. High-temperature pyrolysis procedure and additional acid leaching, which are either highly energy-intensive or environmentally unfriendly, greatly hinder the practical implementation. Hence,

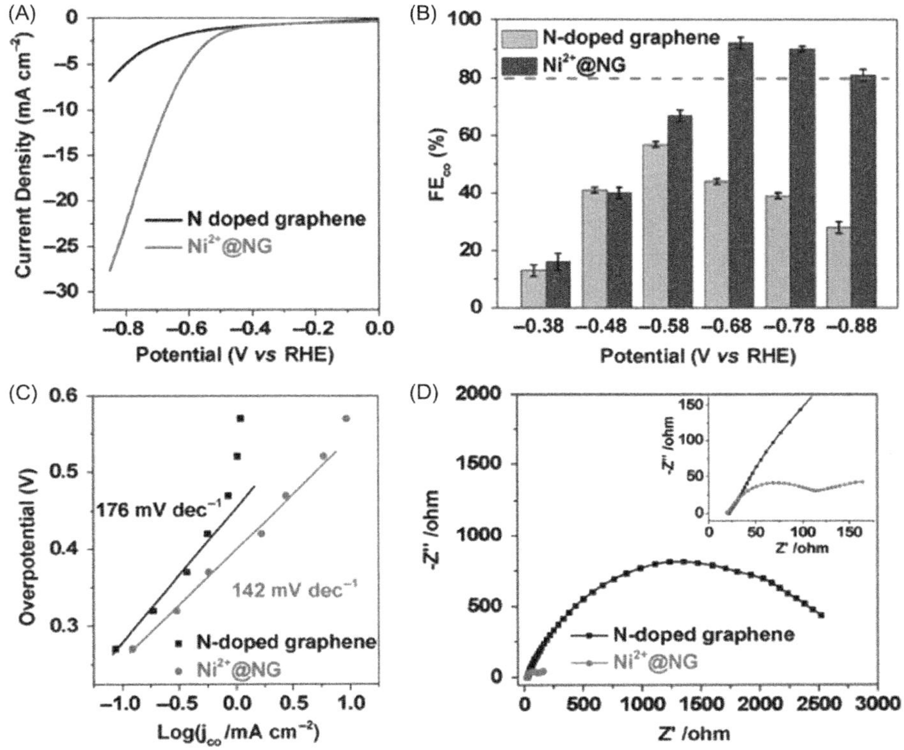

Figure 12.9 Catalytic performance of Ni^{2+}@NG and N-doped graphene. (A) Linear sweep voltammetric curves carried out in the CO_2-saturated 0.5 M $KHCO_3$ aqueous solution. (B) Faradaic efficiencies for CO production at different applied potentials. (C) Tafel plots for CO production. (D) Nyquist plots. The inset shows a local amplified view [68].

a facile ion-adsorption strategy was reported to construct a highly active catalyst in which Ni^{2+} was immobilized on N-doped graphene (Ni^{2+}@NG) for CO_2 reduction to CO by Wentuan Bi [68]. The synthesis Ni^{2+}@NG achieves the optimal faradaic efficiency of 92% for CO production at -0.68 V versus RHE with a current density of 10.2 mA cm^{-2} in CO_2-saturated 0.5 M $KHCO_3$ solution. The electrochemical results verified the high-efficient and selective CO_2 reduction over N-doped graphene after the immobilization of Ni^{2+}compared to pristine N-doped graphene (Fig. 12.9). Free from the conventional harsh pyrolysis and acid-leaching procedures, this solution-chemistry strategy is easy to scale up and of general applicability, thus paving a rational avenue for the design of high-efficiency catalysts for CO_2 reduction and beyond.

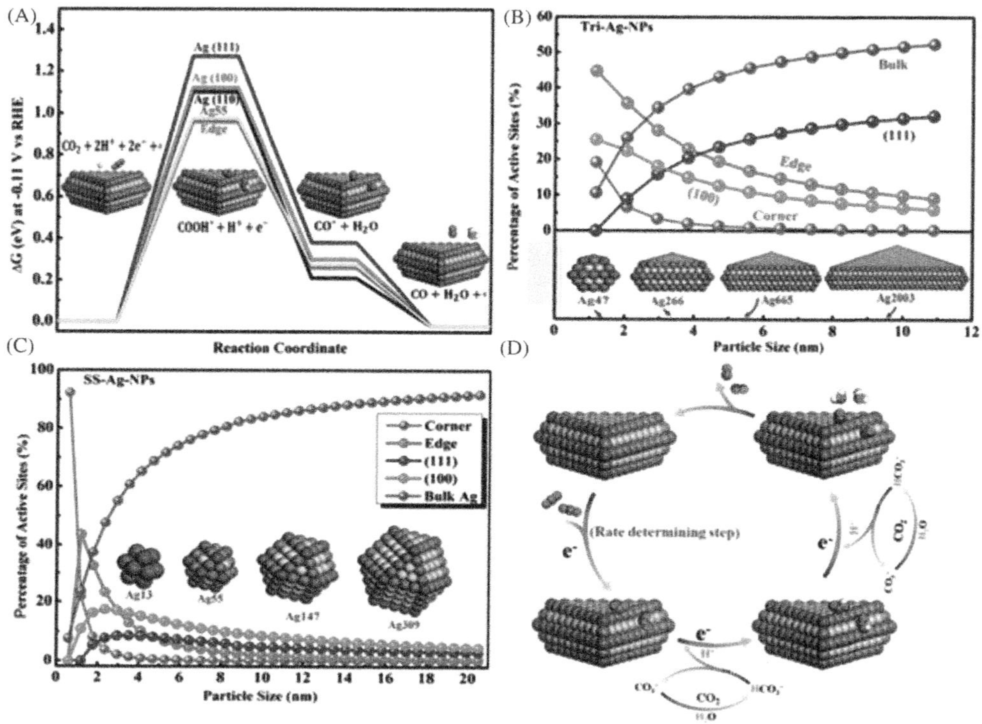

Figure 12.10 (A) Free energy diagrams for CO_2RR to CO on different facets and Ag 55 cluster at −0.11 V; active adsorption site density on (B) Tri-Ag-NPs and (C) SS-Ag-NPs as a function of particle size; and (D) proposed mechanism for CO_2RR to CO on Tri-Ag-NPs [71].

12.3.1.2 Metals

There are few research works about noble metal nanosheets used for selective ECR. Recently, Subiao Liu [71] synthesized triangular silver nanoplates (Tri-Ag-NPs) for electrochemical reduction of CO_2 to CO in 0.1 M $KHCO_3$ by using a direct chemical reduction method. Tri-Ag-NPs showed the best ECR performance and reached its maximum value of 96.8% with a much lower overpotential of 0.746 V as compared to similar-sized Ag nanoparticles SS-Ag-NPs (65.4%, 0.846 V) and bulk Ag (57.2%, 0.946 V). Additionally, CO starts to be observed at an ultralow overpotential of 96 mV. DFT calculations reveal that the significantly enhanced electrocatalytic activity and selectivity at lowered overpotential originate from the shape-controlled structure. This not only provides the optimum edge-to-corner ratio but also dominates at the facet of Ag (100) where it requires lower energy to initiate the rate-determining step. This study demonstrates a promising approach to tune (Fig. 12.10).

Wenjin Zhu [72] synthesized ultrathin palladium nanosheets (PdNs) with different edge lengths and explored their structure-sensitive performance of converting CO_2 into CO in 0.1 M $KHCO_3$ solution. The hexagonal PdNs with 5 atomic thickness and 5.1 nm edge length reached CO faradaic efficiency of 94% at −0.5 V, without any decay after a stability test of 8 h. Compared to similar-sized Pd nanoparticles, the PdNs with thickness of 5.1 nm showed an over fivefold greater mass activity under −0.9 V, of 140 Ag^{-1}. DFT calculations attributed the enhanced performance to more exposed atoms with a generalized coordination number of around 5. It appears to be the most efficient among all of Pd-based catalysts toward CO_2ER.

Xie and colleagues [73] recently prepared freestanding five-atom-thick Co sheets with and without surface Co oxide by using a ligand-confined growth strategy. The authors showed an onset potential of −0.68 V (vs SCE) for HCOOH production in a CO_2-saturated 0.1 M Na_2SO_4 solution on partially oxidized Co nanosheets (about 0.84 nm), with an overpotential of only 0.07 V.

Fengjiao Quan [97] synthesized Zn nanoplates by a facile electrochemical strategy. They found that the popular electrolyte $NaHCO_3$ was not optimum for CO_2 reduction here, and salt anions were found to greatly influence CO_2 conversion efficiency on Zn catalysts for the first time. The results have showed that the activity of a Zn nanoplate could be dramatically enhanced toward the reduction of CO_2 to CO by the optimum combination of catalyst size and electrolyte anions. A faradic efficiency of 93% at −1.6 V (vs SCE) was achieved for CO in CO_2-saturated aqueous NaCl solution whereas the CO is only 18% on the bulk Zn electrode in $NaHCO_3$ solution.

Bi catalyst possessing the features of low toxicity, cost-effectiveness, as well as excellent stability, has attracted much attention for ECR. An indirect approach to prepare ultrathin Bi nanosheets via in situ topotactic transformation of bismuth oxyiodide (BiOI) naonsheet template was reported by Na Han [98]. The BiOI nanosheet (the thickness was measured to be ~8.7 nm) was comprised of hierarchical microflowers that were assembled from very thin nanosheets oriented to different angles (Fig. 12.11). These ultrathin Bi nanosheets can act as an efficient ECR catalyst for the conversion of CO_2 to HCOOH with impressive activity and selectivity close to 100% over a broad potential window, as well as satisfactory durability. DFT calculations suggested that the high reaction selectivity toward HCOOH was due to its stabilized intermediate ($OCHO^*$) on Bi (001) surface relative to $COOH^*$ or H^*.

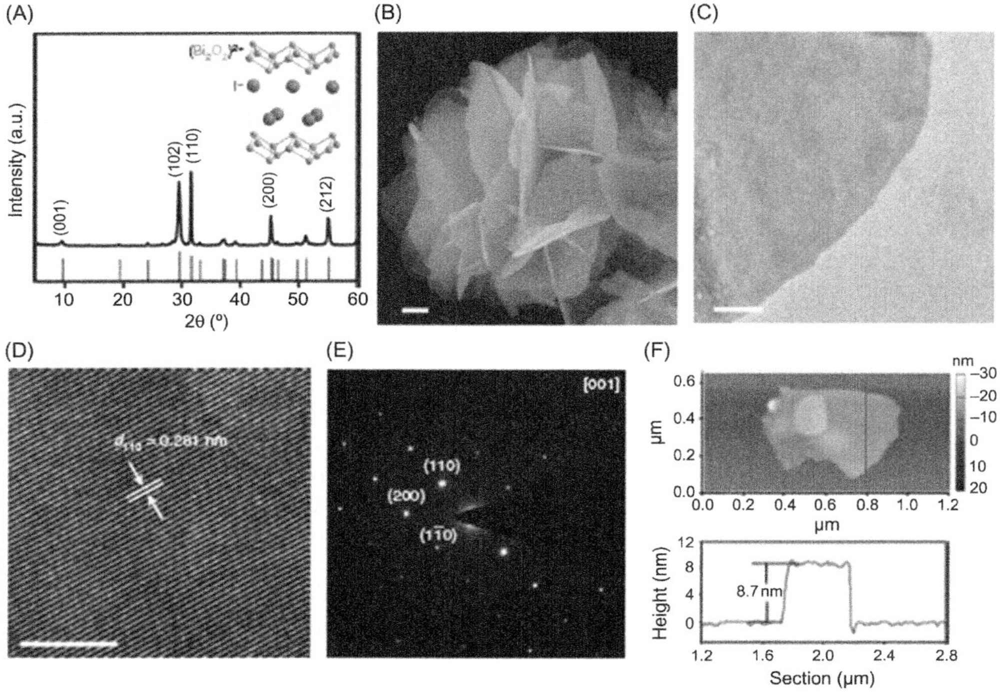

Figure 12.11 Structural characterizations of BiOI nanosheets. (A) XRD pattern and (inserted) schematic crystal structure; (B) SEM image; (C, D) TEM images at different magnifications; (E) SAED pattern; and (F) AFM image and corresponding height profile. Scale bar: 200 nm (B); 50 nm (C); and 5 nm (D) [98].

Antimony (Sb) has received much less attention as an electrocatalyst for the ECR, despite it being adjacent to tin and bismuth in the periodic table. This lack of attention is probably due to the fact that the catalytically active sites are not highly exposed in the bulk material. For the first time, Fengwang Li [99] demonstrated the preparation of few-layer Sb nanosheets (SbNSs) from bulk crystal by a cathodic exfoliation method. This method has been proven to be a simple, economical, and environmentally friendly, yet efficient approach for the mass production of classic 2D nanomaterials. In contrast to the bulk counterpart, SbNSs were found to selectively and efficiently catalyze the reduction of CO_2 to HCOOH. This catalytic activity of Sb is known to increased number of active edge sites in the SbNSs. Furthermore, when coupled with the anodic exfoliation of graphite, the few-layer Sb nanosheet-graphene (SbNS-G) composite can be readily formed in an in situ manner. This composite showed an enhanced performance toward the ECR, compared to the SbNSs, because of a synergistic effect resulting from the strong electronic interaction between graphene and the SbNSs. The FE of SbNSs for HCOOH reaches a maximum of about 84% at −1.06 V.

12.3.2 Nonmetallic compounds

12.3.2.1 Boron nitride-based hybrids

Boron nitride (BN) has been found to be composed of alternating B and N atoms in a honeycomb arrangement consisting of sp^2-bonded 2D layers, which has a similar structural lattice to that of the C atoms of graphene. Xiaofu Sun [76] prepared N-based Cu(I)/C-doped boron nitride (BN-C) composites (Cu(I)/BN-C) for ECR to CH_3COOH. They found when the electrolyte was 1-ethyl-3-methylimidazolium tetrafluoroborate ([Emim]BF_4)-LiI-water solution, the Faradaic efficiency of CH_3COOH could reach as high as 80.3% with a current density of 13.9 mA cm^{-2}. The synergistic effect between BN-C, the Cu metal center, the N-based ligand, and ionic liquid electrolyte ([Emim]BF_4/LiI/H_2O solution) leads to excellent performance of the Cu(I)/BN-C catalyst for producing CH_3COOH.

Transition metals Ni, Co, and Cu are all earth-abundant, and numerous researchers have studied the electrochemical CO_2 reduction at these metal electrodes. Guoxiang Hu [100] investigate the electrocatalytic activities of h-BN/Ni, h-BN/Co, and h-BN/Cu interfaces for CO_2 reduction by DFT calculations. They examine the geometry and electronic structures of h-BN/Ni, h-BN/Co, and h-BN/Cu. And then compute the binding energies of key intermediates during HER and CO_2 reduction on the unsupported h-BN, three metal surfaces, and the h-BN/metal composite (Fig. 12.12). H adsorption is greatly reduced on the h-BN/metal surface, while HCOO adsorption is only slightly affected, in comparison with adsorption on the pure metal surface. This differential adsorption led to selectivity control for ECR. The calculated free-energy profiles show that h-BN/Ni and h-BN/Co will have very good electrocatalytic activities for CO_2 reduction to HCOOH by avoiding HER.

12.3.2.2 Graphitic carbon nitride (g-C_3N_4)

Materials based on g-C_3N_4, the most stable phase among all the allotropes of carbon nitride, have attracted much attention owing to their wide applications as sustainable heterogeneous catalysts for artificial photocatalysis [101–104] and electrocatalysis [105–107]. Theoretical studies and experimental results

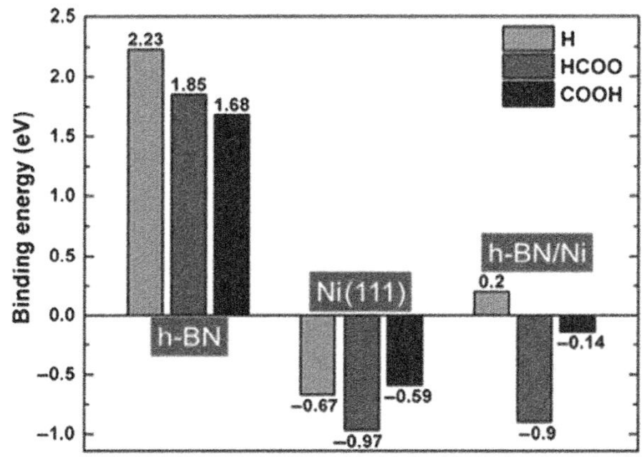

Figure 12.12 Binding energies (eV) of H, HCOO, and COOH on the h-BN monolayer, Ni(111), and h-BN/Ni(111) [100].

also suggested that modified g-C_3N_4 has an appropriate energy potential for CO_2 reduction.

Recently, Xunyu Lu [77] first reported the potential of utilizing g-C_3N_4/multi-walled carbon nanotube (MWNT) composites as a stable and selective electrocatalyst for CO_2 reduction in 0.1 M $KHCO_3$. The as-prepared composites exhibited a CO faradic efficiency of 60% at −0.75 V (vs RHE) in 0.1 M $KHCO_3$, and no decay in catalytic activity was observed after 50 h. The covalent C−N bonds formed between g-C_3N_4 and MWNTs were proposed as active sites for ECR. In addition, high specific surface area and improved material conductivity of the composite led to enhancement in CO_2 reduction.

Bing Zhang [78] prepared a polarized two-dimensional ultrathin g-C_3N_4 (2D-pg-C_3N_4) layer (thickness: ~ 1 nm) by a hydrothermal-thermal exfoliation method. The ultrathin g-C_3N_4 can function as an efficient metal-free electrocatalyst for selective reduction of CO_2 to CO with high Faradaic efficiency of ~ 80% in CO_2-saturated 2 M $KHCO_3$ electrolyte. The current density of the 2D-pg-C_3N_4 achieved 3.05 mA cm^{-2} at −1.2 V (vs Ag/AgCl), which was nearly 30 times larger than that over the bulk g-C_3N_4 catalyst. The polarized surface of ultrathin g-C_3N_4 layers with a more reductive conduction band yields excellent electrochemical activity for CO_2 reduction.

12.3.3 Metallic compounds

12.3.3.1 Transition-metal dichalcogenides

Transition-metal dichalcogenides (TMDs) are compounds composed of transition metal elements and chalcogen elements (sulfur, selenium, or tellurium). The layers are connected by van der Waals force. The TMD single layer contains three atomic layers and the transition metal elements are sandwiched chalcogens element layers. Theoretical and empirical studies have suggested that TMDs may constitute promising catalysts for ECR. A series of TMDs including Tungsten diselenide (WSe_2), molybdenum diselenide ($MoSe_2$), tungsten disulfide (WS_2), and molybdenum disulfide (MoS_2) nanoflakes (NFs) were prepared for electrocatalytic CO_2 reduction in the ionic liquids 1-ethyl-3-methylimidazolium tetrafluoroborate (EMIM-BF_4) [79]. The current density of bulk Ag was 3.3 mA cm^{-2} with 0% CO selectivity and the current density of Ag NPs was 10 mA cm^{-2} with 65% CO selectivity. By contrast with bulk Ag and Ag NPs, they found the four TMDs had very high current densities above 130 mA cm^{-2} (up to 330 mA cm^{-2} for WSe_2 nanosheets) with 90% CO selectivity (Fig. 12.13). Moreover, WSe_2 NFs gave the best performance and exhibited a current density of 18.95 mA cm^{-2}, CO faradic efficiency of 24%, and CO TOF of 0.28 s^{-1} at an overpotential of 54 mV. The current density decreased in the order of $WSe_2 > MoSe_2 > WS_2 > MoS_2$, which might correlate with their respective electron transfer properties. WSe_2 has the lowest work function and thus performs the best for CO_2 activation in these TMDs. It was further inferred that the metal edges of these TMDs had strong binding of COOH* and CO* intermediates during the catalytic reaction to maintain a high turnover rate. These TMD nanosheet/ionic liquids systems exhibit a synergy between metallic edge CO binding and bulk IL CO_2 solubility, thus providing promising TMD-based electrocatalysts for CO production.

Pedram Abbasi [108] reported on the activity of single crystals and thin films of MoS_2 for the reduction of CO_2 dissolved in an aqueous electrolyte. Reduction of CO_2 at the surface of MoS_2 electrodes held potentiostatically at −0.59 V (vs RHE) while in contact with aqueous electrolytes at room temperature yielded 1-propanol as the major CO_2 reduction product with Faradaic efficiencies of 2%–5%. Formate, *t*-butanol, and ethylene glycol were also produced at efficiencies (<1%) lower than that for 1-propanol. The yields of 1-propanol were higher for single-crystal-terrace electrodes and for thin films with low edge-site densities than for thin films with high densities of edge sites, suggesting that the active sites for reduction of CO_2 to 1-propanol are terraces rather than edges.

In addition, increasing the number of exposed edges by doping with metal atoms to modify the binding energies of intermediates can effectively enhance the CO_2 conversion efficiency

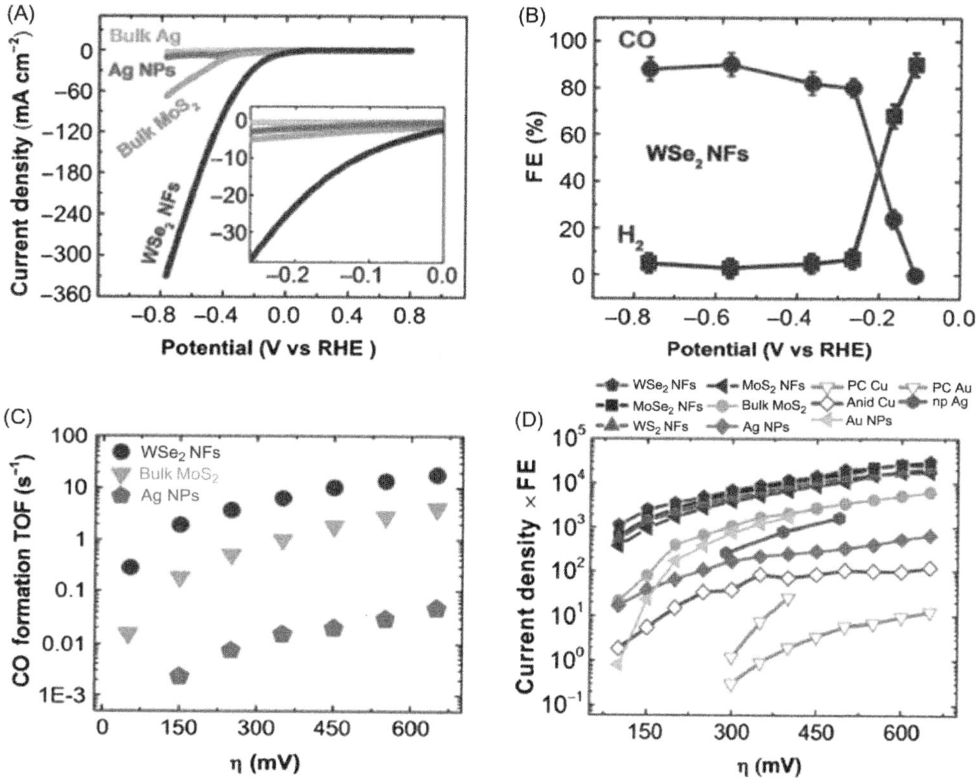

Figure 12.13 CO_2 reduction performance of the TMDC catalysts, Ag NPs, and bulk Ag in the EMIM-BF_4 solution. (A) Cyclic voltammetry (CV) curves for WSe_2 NFs, bulk MoS_2, Ag nanoparticles (Ag NPs), and bulk Ag in CO_2 environment. Inset shows the current densities in low overpotentials. (B) CO and H_2 overall faradaic efficiency (FE) at different applied potentials for WSe_2 NFs. The error bars represent SD of four measurements. (C) CO formation TOF of WSe_2 NFs, bulk MoS_2, and Ag NPs in IL electrolyte at overpotentials of 54 to 650 mV. At 54 mV overpotential, Ag NPs' result is zero. (D) Overview of different catalysts' performance at different overpotentials (h) [79].

of TMDs. Pedram Abbasi [81] also prepared Niobium (Nb)-doped MoS_2 nanosheets. The electrochemical results indicate that 5% Nb-doped MoS_2 nanosheets were demonstrated to exhibit an order of magnitude higher CO formation TOF than pristine MoS_2 in ionic liquids at an overpotential range of 50–150 mV. This catalyst yield a very low onset overpotential of 31 mV for ECR.

12.3.3.2 Oxides

Recently, the transition-metal oxides with naturally abundant and chemically stable have been explored as potential ECR catalysts. Because of the environmentally friendly, low cost, and an abundance of reserves, the spinel-type oxide of Co_3O_4 with

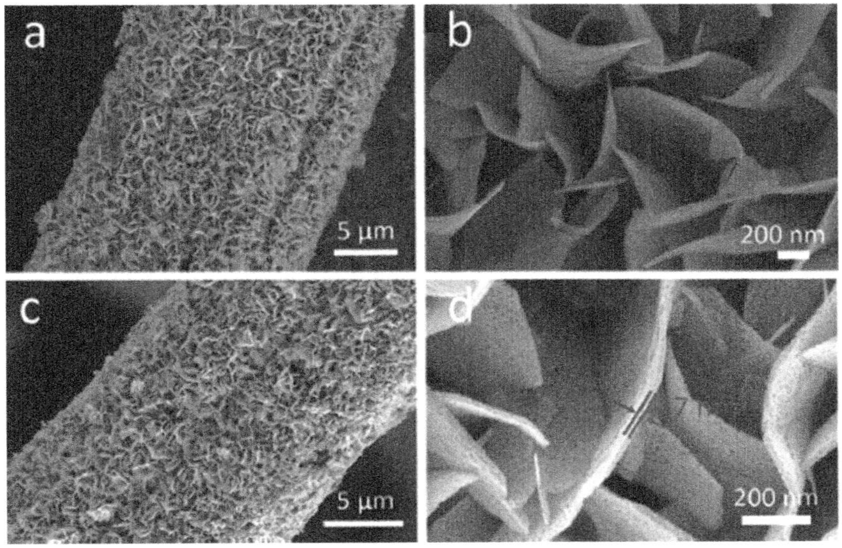

Figure 12.14 SEM images of SnS_2 nanosheets on (A and B) CC and porous SnO_2 nanosheets on (C and D) CC [83].

Co^{2+} ions in tetrahedral interstices and Co^{3+} ions in octahedral interstices attracted the attention of researchers. Shan Gao [82] successfully synthesized Co_3O_4 atomic layers with a thickness of 1.72 and 3.51 nm through fast-heating the respective precursors of ultrathin $Co(CO_3)_{0.5}(OH) \cdot 0.11H_2O$ layers. 1.72-nm-thick Co_3O_4 layers exhibited a CO_2 reduction current density of 0.68 mA cm^{-2} at −0.88 V (vs SCE), 1.5 and 20 times higher than those of 3.51-nm-thick Co_3O_4 layers and bulk Co_3O_4, respectively. Moreover, 1.72 nm thick Co_3O_4 layers showed HCOOH Faradaic efficiency of over 60% in 20 h.

Fengwang Li [83] fabricated mesoporous SnO_2 nanosheets with hierarchical structure by a facile combination of hydrothermal reaction and calcination (Fig. 12.14). A high partial current density for HCOO- (45 mA cm^{-2}) at a moderate overpotential (0.88 V) with large faradic efficiency (89% ± 2%) in CO_2-saturated 0.5 M $NaHCO_3$ electrolyte was achieved. This superior performance was correlated to the porous hierarchical structure, which provided a large surface area and facilitated charge and mass transfer.

12.3.4 Organics

12.3.4.1 Metal-organic frameworks

Metal-organic framework compounds (MOFs) are a class of porous crystal compounds formed by the coordination of metal

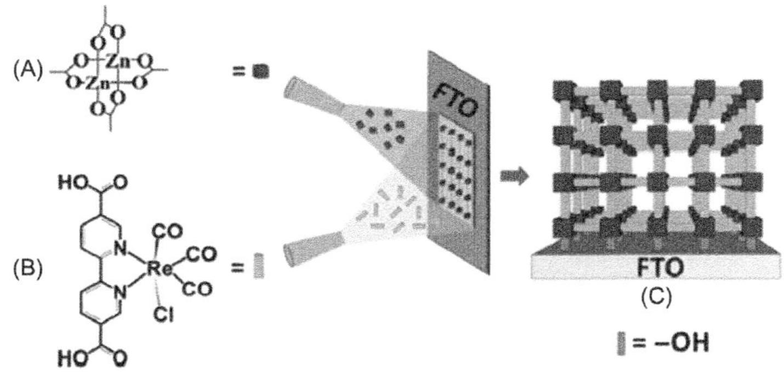

Figure 12.15 Schematic representation of the fabrication process of Re-SURMOF on functionalized FTO substrate in a layer-by-layer fashion. (A) Zn acetate; (B) Re-linker, and (C) idealized structure of Re-SURMOF on FTO [84].

ions or clusters with organic ligand molecules. The first MOF catalyst for the electrocatalytic conversion of aqueous CO_2 to CO was reported by Nikolay Kornienko [109]. They have chosen a stable thin cobalt porphyrin MOF film where these porphyrin units are linked with aluminum oxide rods to form a 3D porous structure with pores of 6×11 Å^2. ECR studies showed that this MOF catalyst for electroreduction of CO_2 to CO with CO selectivity in excess of 76% at -0.7 V versus RHE in aqueous electrolytes and with high turnover number (TON = 1400).

Lu Ye [84] reported the first work on the incorporation of a molecular catalyst for CO_2 reduction, $ReL(CO)_3Cl$ (L = 2,20-bipyridine-5,50-dicarboxylic acid), into highly oriented surface-grafted metal-organic framework (SURMOF) thin films grown on fluorine-doped tin oxide (FTO) glass electrodes using a liquid phase epitaxy method (Fig. 12.15). This catalyst displayed an FE of up to $93\% \pm 5\%$ toward CO production at -1.6 V (vs NHE), exceeding those reported for most electrocatalytically active MOF thin films. However, the stability of such Re-based MOF catalysts needs to be further improved for future applications.

Riming Wang [110] chosen a prototypical metal-organic framework ZIF-8 with a nitrogen-rich organic linker as a sacrificial template for the synthesis of N-doped carbon catalysts through the pyrolysis and acid treatment for ECR. The result shows Faradaic efficiencies to CO as high as ~78% with H_2 being the only byproduct. The catalytic performance of the catalyst was studied by changing the pyrolysis temperature and the results showed that the higher the pyrolysis temperature, the better the catalytic performance. They thought the high activity of catalyst can be attributed to the presence of a large amount of pyridinic-N and quaternary-N species in the carbon structure, which are known to lower the energy barrier for the formation of COOH*, an intermediate to produce CO.

Dae-Hyun Nam [111] reported a strategy to optimize the ECR activity of Cu clusters by using MOFs to promoting under-coordinated Cu sites toward multiple-carbon product generation. The electrochemical test results showed that the MOF-derived Cu cluster catalysts achieved an FE of 45% C_2H_4 with the current density of 262 mA cm^{-2} at −1.07 V versus RHE in 1 M KOH. This value is the highest C_2H_4 FE reported among Cu organometallic and MOF based Cu catalysts to date [112–116].

12.3.4.2 Covalent-organic frameworks

Song Lin [25] reported modular optimization of covalent organic frameworks (COFs), in which the building units are cobalt porphyrin catalysts linked by organic struts through imine bonds, to prepare a catalytic material for aqueous electrochemical reduction of CO_2 to CO. The catalysts exhibit high Faradaic efficiency (90%) and turnover numbers (up to 290,000, with initial turnover frequency of 9400 h^{-1}) at pH 7 with an overpotential of −0.55 volts, equivalent to a 26-fold improvement in activity compared with the molecular cobalt complex, with no degradation over 24 h.

12.4 Strategies for improving electrochemical carbon dioxide reduction activity of two-dimension materials

2D nanomaterials electrocatalytic activity can be improved by taking into account the following: (1) enhancing current density at low overpotential toward a single energy-rich product especially toward C2-C4 products, (2) slowing down the competitor reactions such as HER, and (3) improving long-term stability.

12.4.1 Surface modification

The surface modification of 2D nanomaterials catalysts with heteroatoms is successful in tuning their adsorption energy, thus boosting their intrinsic properties for CO_2 reduction.

Modified atomically thin SnS_2 nanosheets with Ni doping for enhanced performance toward ECR was reported by An Zhang [117]. The introduction of Ni into SnS_2 nanosheets significantly enhanced the current density and Faradaic efficiency for carbonaceous product relative to pristine SnS_2 nanosheets. When the Ni content was 5 atm%, the Ni-doped SnS_2 nanosheets achieved a remarkable Faradaic efficiency of 93% for carbonaceous product with a current density of 19.6 mA cm^{-2} at −0.9 V (vs RHE). A mechanistic study revealed that the Ni doping gave rise to a defect level and lowered the work function of SnS_2 nanosheets, resulting in the promoted CO_2 activation and thus improved performance in ECR.

Guodong Shi [118] was successfully synthesized flower-like MoS_2 modified by ultrasmall Cu nanoparticles for efficient ECR. This is the first time MoS_2 was modified with Cu nanoparticles for the improvement of the faradaic efficiency and selectivity of ECR. Cu/MoS_2-2, where the weight percentages of element Cu was 12.76%, exhibited an extremely large current density of 17 mA cm^{-2} at −1.7 V (vs SCE) in CO_2-saturated 0.1 M aqueous $KHCO_3$. Cu/MoS_2-2 with the overall C-containing compound faradaic efficiency of 55.2% (CO, CH_4, and C_2H_4 were 35.19%, 17.08% and 2.93%, respectively), over 1.9 times higher than the bare flower-like MoS_2. The designed optimal hierarchical Cu/MoS_2 composite not only exhibited remarkably enhanced electronic conductivity and specific surface area but also possessed improved CO_2 adsorption capacity, resulting in a significant increase in overall faradaic efficiency and a sevenfold augmentation of the faradaic efficiency of CH_4 in comparison with bare MoS_2. In addition, the Cu/MoS_2 composite had superior stability with high efficiency retained for 48 h in the electrochemical process.

Chenglong Li [119] synthesized a novel ZnS/Zn/ZnS (S-Zn-S) sandwich nanostructure consisting of inner Zn NSs (4 nm thickness) and a ZnS porous subnanometer layer (PSL, 0.68 nm thickness) via chemical vapor deposition. The experimental and theoretical investigations demonstrate that the ZnS PSL on the Zn NSs can activate the adsorbed CO_2 molecules and enhance the adsorption of intermediates, and as a result, the S-Zn-S catalyst achieves 94.2% faradaic efficiency of CO formation at −0.8 V (vs RHE). Moreover, the S-Zn-S catalyst exhibits high long-term durability, the faradaic efficiency remains above 90% after 15 h testing while the current density remains unchanged.

So far the majority of Pd-based nanomaterials just produce CO or $HCOO^-$ through $2e^-$ transfer in ECR; therefore, 2D hierarchical Pd/SnO_2 structure, ultrathin Pd nanosheets partially capped by SnO_2 nanoparticles, was designed to enable multi-electron transfer for selective electroreduction of CO_2 into CH_3OH [120]. Such a structure design not only enhances the adsorption of CO_2 on SnO_2 but also weakens the binding strength of CO on Pd due to the as-built Pd–O–Sn interfaces, which is demonstrated to be critical to improve the electrocatalytic selectivity and stability of Pd catalysts. A maximum faradaic efficiency (FE) of 54.8% ± 2% for CH_3OH was achieved at −0.24 V versus RHE in CO_2-saturated 0.1 M $NaHCO_3$ solution by electrochemical test.

12.4.2 Surface-structure tuning

Four effective strategies have been proposed to boost the electrocatalytic performance of structural tuning of 2D nanomaterials: (1) surface porosity, (2) surface defects, (3) exposed crystal facets, and (4) surface phase.

Introduction of nanosized pores to carbons can provide developed surface area as well as decrease the diffusion distance of the electrolyte ion. The utilization of 2D materials in electrochemical reduction of CO_2 is in its infancy. Hongqiang Lia [121] developed a facile yet efficient NH_3 etching strategy to synthesize pitch derived nitrogen-doped porous carbon nanosheets with highly wrinkled surface for CO_2 reduction by using NaCl as templates. NH_3 etching not only increases specific surface area by creating nanopores but also introduces nitrogen atoms into the carbon framework. The as-obtained wrinkled N-doped carbon nanosheets (WNCNs) feature 2D structures with highly exposed active sites, abundant nanopores, and highly conductive carbon frameworks for the transport of electrolyte ion and electron, thus further resulting in the fast reaction kinetics. Benefiting from the well-balanced micro/mesostructure and surface chemical composition, the catalyst exhibits negligible onset overpotential (−0.19 V) for CO production and a maximum Faradaic efficiency (FE) of 84% at an overpotential of −0.49 V. The relationship between N types, surface morphology, porous structure, and CO_2 reduction activity was investigated. DFT calculations confirm that the pyridinic-N can offer favorable binding sites to COOH* intermediates and facilitate their subsequent catalytic reduction.

Taotao Zhang [122] also reported a loose and porous structure Zn electrode consisting of multilayered Zn nanosheets (MZnNSs) with a high density of edge sites. The prepared MZnNSs achieved a superior CO Faradaic efficiency about 86% with a CO partial current density above 5.2 mA cm^{-2} at moderately negative potential 1.13 V versus RHE, which is almost 9 times and 14 times higher than that of bulk Zn foil, respectively. Furthermore, the MZnNSs electrode showed good stability with negligible deterioration during 7 h continuous operation. DFT calculations suggest that the high density of edge sites is a benefit for the improvement of the activity and selectivity of MZnNSs for CO_2 reduction.

The introduction of oxygen vacancies has provided a powerful method to enhance the performance of CO_2 conversion in heterogeneous catalysis by promoting CO_2 adsorption and activation. Zhigang Geng [123] developed an efficient strategy to facilitate CO_2 activation by introducing oxygen vacancies into electrocatalysts with electron rich surface. In ECR, ZnO nanosheets rich in oxygen vacancies exhibited a current density of -16.1 mA cm^{-2} with the Faradaic efficiency of 83% for CO production at -1.1 V versus RHE. The introduction of oxygen vacancies into ZnO can increase the charge density around the valence band maximum, resulting in the enhancement of CO_2 activation by DFT calculations. Mechanistic studies further revealed that the enhanced CO production by introducing oxygen vacancies into ZnO nanosheets originated from the assisted CO_2 activation.

By taking advantage of 2D layered precursor materials bismuth oxyhalides that can lead to selective reconstruction of in situ electroreduction, the high-surface-area petal-structured bismuth electrode was formed by F. Pelayo García de Arquer [124]. This synthesis method results in the preferential exposure of more active Bi (110) facets. Thereby, the CO_2 reduction reaction selectivity is increased to over 90% Faradaic efficiency and simultaneously stable current densities of up to 200 mA cm^{-2} were achieved.

R.A. Geioushy [125] synthesized the graphene (GN)/ZnO/Cu2O electrodes by using a simple wet-chemical approach. The C_3H_7OH as the dominant liquid product fraction was produced on the GN/ZnO/Cu2O catalyst. The electrochemical test showed that the maximum Faradaic efficiency for CO_2 conversion to C_3H_7OH was 30% for GN/ZnO/Cu_2O (ZnO/Cu_2O weight ratio of 2:1) at lower overpotential of -0.9 V (vs Ag/AgCl) in 0.5 M NaHCO3 electrolyte.

12.5 Conclusion

Increased utilization of fossil fuels has brought about record-breaking levels of atmospheric CO_2, which causes climate change and environmental issues. ECR points to a promising direction not only in decreasing CO_2 accumulation but also in converting intermittent renewable electricity into energy-dense fuels. The electrochemical reduction of CO_2 to C_2^+ products has been extensively studied over various electrode surfaces. 2D nanomaterials have attracted more and more attention due to their unique structure and electronic properties. For commercial applications of 2D nanomaterials, a great challenge remains in developing highly active and cost-effective electrocatalysts toward ECR. This review is to provide a summary of the recent progress on 2D nanomaterials for ECR. The electronic properties of particular 2D materials can be tuned to improve the efficiency of electron transfer and of CO_2 reduction. The combination of theory and experiment will further aid catalyst design to increase the number of active sites and intrinsic activity of each active site as well as the development of novel robust 2D catalysts for this clean energy reaction. Exploration of low-cost, non-noble metal 2D nanosheets that can convert CO_2 to target products at a sufficiently high reaction rate and efficiency continues to be a focus of the research.

Acknowledgment

This work was supported by National Natural Science Foundation of China (NSFC, No. 21972010); Beijing Natural Science Foundation (No. 2192039); the State Key Laboratory of Organic–Inorganic Composites (No. oic-201901001); and Beijing University of Chemical Technology (XK180301).

References

[1] G. Du, S. Lim, Y. Yang, C. Wang, L. Pfefferle, G.L. Haller, Methanation of carbon dioxide on Ni-incorporated MCM-41 catalysts: the influence of catalyst pretreatment and study of steady-state reaction, J. Catal. 249 (2007) 370–379.

[2] J.-N. Park, E.W. McFarland, A highly dispersed Pd-Mg/SiO_2 catalyst active for methanation of CO_2, J. Catal. 266 (2009) 92–97.

[3] S. Sharma, Z. Hu, P. Zhang, E.W. McFarland, H. Metiu, CO_2 methanation on Ru-doped ceria, J. Catal. 278 (2011) 297–309.

[4] H.H. Shin, L. Lu, Z. Yang, C.J. Kiely, S. McIntosh, Cobalt catalysts decorated with platinum atoms supported on barium zirconate provide enhanced activity and selectivity for CO_2 methanation, ACS Catal. 6 (2016) 2811–2818.

[5] F. Wang, S. He, H. Chen, B. Wang, L. Zheng, M. Wei, et al., Active site dependent reaction mechanism over Ru/CeO_2 catalyst toward CO_2 methanation, J. Am. Chem. Soc. 138 (2016) 6298–6305.

[6] W. Wang, S. Wang, X. Ma, J. Gong, Recent advances in catalytic hydrogenation of carbon dioxide, Chem. Soc. Rev. 40 (2011) 3703–3727.

[7] F. Bustamante, R.M. Enick, A.V. Cugini, R.P. Killmeyer, B.H. Howard, K.S. Rothenberger, et al., High-temperature kinetics of the homogeneous reverse water-gas shift reaction, AIChE J. 50 (2004) 1028–1041.

[8] Y.A. Daza, R.A. Kent, M.M. Yung, J.N. Kuhn, Carbon dioxide conversion by reverse water-gas shift chemical looping on perovskite-type oxides, Ind. Eng. Chem. Res 53 (2014) 5828–5837.
[9] L. Dietz, S. Piccinin, M. Maestri, Mechanistic insights into CO_2 activation via reverse water-gas shift on metal surfaces, J. Phys. Chem. C 119 (2015) 4959–4966.
[10] A.A. Upadhye, I. Ro, X. Zeng, H.J. Kim, I. Tejedor, M.A. Anderson, et al., Plasmon-enhanced reverse water gas shift reaction over oxide supported Au catalysts, Catal. Scl. Technol. 5 (2015) 2590–2601.
[11] W. Luhui, L. Hui, L. Yuan, C. Ying, Y. Shuqing, Effect of precipitants on Ni-CeO_2 catalysts prepared by a co-precipitation method for the reverse water-gas shift reaction, J. Rare Earth. 31 (2013) 969–974.
[12] G. Centi, S. Perathoner, Opportunities and prospects in the chemical recycling of carbon dioxide to fuels, Cata. Today 148 (2009) 191–205.
[13] V. Iablokov, S.K. Beaumont, S. Alayoglu, V.V. Pushkarev, C. Specht, J. Gao, et al., Size-controlled model Co nanoparticle catalysts for CO_2 hydrogenation: synthesis, characterization, and catalytic reactions, Nano lett. 12 (2012) 3091–3096.
[14] R.E. Owen, D. Mattia, P. Plucinski, M.D. Jones, Kinetics of CO_2 hydrogenation to hydrocarbons over iron-silica catalysts, ChemPhysChem 18 (2017) 3211–3218.
[15] B. Rungtaweevoranit, J. Baek, J.R. Araujo, B.S. Archanjo, K.M. Choi, O.M. Yaghi, et al., Copper nanocrystals encapsulated in Zr-based metal-organic frameworks for highly selective CO_2 hydrogenation to methanol, Nano lett. 16 (2016) 7645–7649.
[16] S. Saeidi, N.A.S. Amin, M.R. Rahimpour, Hydrogenation of CO_2 to value-added products: a review and potential future developments, J. CO_2 Util. 5 (2014) 66–81.
[17] R. Satthawong, N. Koizumi, C. Song, P. Prasassarakich, Comparative study on CO_2 hydrogenation to higher hydrocarbons over Fe-based bimetallic catalysts, Top. Catal. 57 (2014) 588–594.
[18] C. Xie, C. Chen, Y. Yu, J. Su, Y. Li, G.A. Somorjai, et al., Tandem catalysis for CO_2 hydrogenation to C2-C4 hydrocarbons, Nano lett. 17 (2017) 3798–3802.
[19] D.D. Zhu, J.L. Liu, S.Z. Qiao, Recent advances in inorganic heterogeneous electrocatalysts for reduction of carbon dioxide, Adv. Mater. 28 (2016) 3423–3452.
[20] S.N. Habisreutinger, L. Schmidt-Mende, J.K. Stolarczyk, Photocatalytic reduction of CO_2 on TiO_2 and other semiconductors, Angew. Chem. Int. Edi. 52 (2013) 7372–7408.
[21] J. Shi, Y. Jiang, Z. Jiang, X. Wang, X. Wang, S. Zhang, et al., Enzymatic conversion of carbon dioxide, Chem. Soc. Rev 44 (2015) 5981–6000.
[22] J. Grodkowski, P. Neta, Copper-catalyzed radiolytic reduction of CO_2 to CO in aqueous solutions, J. Phys. Chem. B 105 (2001) 4967–4972.
[23] K.P. Kuhl, T. Hatsukade, E.R. Cave, D.N. Abram, J. Kibsgaard, T.F. Jaramillo, Electrocatalytic conversion of carbon dioxide to methane and methanol on transition metal surfaces, J. Am. Chem. Soc. 136 (2014) 14107–14113.
[24] C.W. Machan, S.A. Chabolla, J. Yin, M.K. Gilson, F.A. Tezcan, C.P. Kubiak, Supramolecular assembly promotes the electrocatalytic reduction of carbon dioxide by Re (I) bipyridine catalysts at a lower overpotential, J. Am. Chem. Soc. 136 (2014) 14598–14607.
[25] S. Lin, C.S. Diercks, Y.-B. Zhang, N. Kornienko, E.M. Nichols, Y. Zhao, et al., Covalent organic frameworks comprising cobalt porphyrins for catalytic CO_2 reduction in water, Science 349 (2015) 1208–1213.

[26] E.E. Benson, C.P. Kubiak, A.J. Sathrum, J.M. Smieja, Electrocatalytic and homogeneous approaches to conversion of CO_2 to liquid fuels, Chem. Soc. Rev 38 (2009) 89–99.
[27] J. Schneider, H. Jia, J.T. Muckerman, E. Fujita, Thermodynamics and kinetics of CO_2, CO, and H^+ binding to the metal centre of CO_2 reduction catalysts, Chem. Soc. Rev 41 (2012) 2036–2051.
[28] X. Feng, K. Jiang, S. Fan, M.W. Kanan, Grain-boundary-dependent CO_2 electroreduction activity, J. Am. Chem. Soc. 137 (2015) 4606–4609.
[29] Y. Hori, A. Murata, R. Takahashi, Formation of hydrocarbons in the electrochemical reduction of carbon dioxide at a copper electrode in aqueous solution, J. Chem. Soc., Faraday Trans. 1 (85) (1989) 2309–2326.
[30] D.W. DeWulf, T. Jin, A.J. Bard, Electrochemical and surface studies of carbon dioxide reduction to methane and ethylene at copper electrodes in aqueous solutions, J. Electrochem. Soc. 136 (1989) 1686–1691.
[31] B.A. Rosen, A. Salehi-Khojin, M.R. Thorson, W. Zhu, D.T. Whipple, P.J.A. Kenis, et al., Ionic liquid-mediated selective conversion of CO_2 to CO at low overpotentials, Science 334 (2011) 643–644.
[32] K.P. Kuhl, E.R. Cave, D.N. Abram, T.F. Jaramillo, New insights into the electrochemical reduction of carbon dioxide on metallic copper surfaces, Energ. Envirom. Sci. 5 (2012) 7050–7059.
[33] N. Hoshi, M. Kato, Y. Hori, Electrochemical reduction of CO2 on single crystal electrodes of silver Ag (111), Ag (100) and Ag (110), J. Electroanal. Che. 440 (1997) 283–286.
[34] D. Kim, J. Resasco, Y. Yu, A.M. Asiri, P. Yang, Synergistic geometric and electronic effects for electrochemical reduction of carbon dioxide using gold-copper bimetallic nanoparticles, Nat. Commun. 5 (2014) 4948–4956.
[35] W. Zhu, R. Michalsky, On Metin, H. Lv, S. Guo, C.J. Wright, et al., Monodisperse Au nanoparticles for selective electrocatalytic reduction of CO_2 to CO, J. Am. Chem. Soc. 135 (2013) 16833–16836.
[36] W. Zhu, Y.-J. Zhang, H. Zhang, H. Lv, Q. Li, R. Michalsky, et al., Active and selective conversion of CO_2 to CO on ultrathin Au nanowires, J. Am. Chem. Soc. 136 (2014) 16132–16135.
[37] S. Zhang, P. Kang, T.J. Meyer, Nanostructured tin catalysts for selective electrochemical reduction of carbon dioxide to formate, J. Am. Chem. Soc. 136 (2014) 1734–1737.
[38] K. Manthiram, B.J. Beberwyck, A.P. Alivisatos, Enhanced electrochemical methanation of carbon dioxide with a dispersible nanoscale copper catalyst, J. Am. Chem. Soc. 136 (2014) 13319–13325.
[39] D. Gao, H. Zhou, J. Wang, S. Miao, F. Yang, G. Wang, et al., Size-dependent electrocatalytic reduction of CO_2 over Pd nanoparticles, J. Am. Chem. Soc. 137 (2015) 4288–4291.
[40] J. Medina-Ramos, R.C. Pupillo, T.P. Keane, J.L. DiMeglio, J. Rosenthal, Efficient conversion of CO_2 to CO using tin and other inexpensive and easily prepared post-transition metal catalysts, J. Am. Chem. Soc. 137 (2015) 5021–5027.
[41] J. Medina-Ramos, J.L. DiMeglio, J. Rosenthal, Efficient reduction of CO_2 to CO with high current density using in situ or ex situ prepared Bi-based materials, J. Am. Chem. Soc. 136 (2014) 8361–8367.
[42] Q. Lu, J. Rosen, Y. Zhou, G.S. Hutchings, Y.C. Kimmel, J.G. Chen, et al., A selective and efficient electrocatalyst for carbon dioxide reduction, Nat. Commun. 5 (2014) 3242–3247.

[43] C.W. Li, M.W. Kanan, CO_2 reduction at low overpotential on Cu electrodes resulting from the reduction of thick Cu_2O films, J. Am. Chem. Soc. 134 (2012) 7231–7234.
[44] Y. Chen, M.W. Kanan, Tin oxide dependence of the CO_2 reduction efficiency on tin electrodes and enhanced activity for tin/tin oxide thin-film catalysts, J. Am. Chem. Soc. 134 (2012) 1986–1989.
[45] Y. Chen, C.W. Li, M.W. Kanan, Aqueous CO_2 reduction at very low overpotential on oxide-derived Au nanoparticles, J. Am. Chem. Soc. 134 (2012) 19969–19972.
[46] J. Wu, R.M. Yadav, M. Liu, P.P. Sharma, C.S. Tiwary, L. Ma, et al., Achieving highly efficient, selective, and stable CO_2 reduction on nitrogen-doped carbon nanotubes, ACS nano 9 (2015) 5364–5371.
[47] S. Zhang, P. Kang, S. Ubnoske, M.K. Brennaman, N. Song, R.L. House, et al., Polyethylenimine-enhanced electrocatalytic reduction of CO_2 to formate at nitrogen-doped carbon nanomaterials, J. Am. Chem. Soc. 136 (2014) 7845–7848.
[48] B. Kumar, M. Asadi, D. Pisasale, S. Sinha-Ray, B.A. Rosen, R. Haasch, et al., Renewable and metal-free carbon nanofibre catalysts for carbon dioxide reduction, Nat. Commun. 4 (2013) 2819–2827.
[49] C. Liu, J.J. Gallagher, K.K. Sakimoto, E.M. Nichols, C.J. Chang, M.C.Y. Chang, et al., Nanowire-bacteria hybrids for unassisted solar carbon dioxide fixation to value-added chemicals, Nano lett. 15 (2015) 3634–3639.
[50] J.P. Torella, C.J. Gagliardi, J.S. Chen, D.K. Bediako, B. Colón, J.C. Way, et al., Efficient solar-to-fuels production from a hybrid microbial-water-splitting catalyst system, Proc. Natl. Acad. Sci. U. S. A. 112 (2015) 2337–2342.
[51] T. Reda, C.M. Plugge, N.J. Abram, J. Hirst, Reversible interconversion of carbon dioxide and formate by an electroactive enzyme, Proc. Natl. Acad. Sci. U. S. A. 105 (2008) 10654–10658.
[52] M. Beller, U.T. Bornscheuer, CO_2 fixation through hydrogenation by chemical or enzymatic methods, Angew. Chem. Int. Edi. 53 (2014) 4527–4528.
[53] H. Zhang, Ultrathin two-dimensional nanomaterials, ACS nano 9 (2015) 9451–9469.
[54] R. Kortlever, J. Shen, K.J.P. Schouten, F. Calle-Vallejo, M.T.M. Koper, Catalysts and reaction pathways for the electrochemical reduction of carbon dioxide, J. Phys. Chem. Lett. 6 (2015) 4073–4082.
[55] Y. Oh, X. Hu, Organic molecules as mediators and catalysts for photocatalytic and electrocatalytic CO_2 reduction, Chem. Soc. Rev 42 (2013) 2253–2261.
[56] J. Qiao, Y. Liu, F. Hong, J. Zhang, A review of catalysts for the electroreduction of carbon dioxide to produce low-carbon fuels, Chem. Soc. Rev 43 (2014) 631–675.
[57] S. Ma, P.J.A. Kenis, Electrochemical conversion of CO_2 to useful chemicals: current status, remaining challenges, and future opportunities, Curr. Opin. Chem. Eng. 2 (2013) 191–199.
[58] X. Nie, M.R. Esopi, M.J. Janik, A. Asthagiri, Selectivity of CO_2 reduction on copper electrodes: the role of the kinetics of elementary steps, Angew. Chem. Int. Ed. 125 (2013) 2519–2522.
[59] H.A. Hansen, J.B. Varley, A.A. Peterson, J.K. Nørskov, Understanding trends in the electrocatalytic activity of metals and enzymes for CO_2 reduction to CO, J. Phys. Chem. Lett. 4 (2013) 388–392.
[60] R.P.S. Chaplin, A.A. Wragg, Effects of process conditions and electrode material on reaction pathways for carbon dioxide electroreduction with particular reference to formate formation, J. Appl. Electrochem. 33 (2003) 1107–1123.

[61] Y. Hori, R. Takahashi, A. Yuzuru Yoshinami, A. Murata, Electrochemical reduction of CO at a copper electrode, J. Phys. Chem. B 101 (1997) 7075–7081.
[62] K.D. Yang, C.W. Lee, K. Jin, S.W. Im, K.T. Nam, Current status and bioinspired perspective of electrochemical conversion of CO_2 to a long-chain hydrocarbon, J. Phys. Chem. Lett. 8 (2017) 538–545.
[63] Z. Sun, T. Ma, H. Tao, Q. Fan, B. Han, Fundamentals and challenges of electrochemical CO_2 reduction using two-dimensional materials, Chem 3 (2017) 560–587.
[64] E. Bertheussen, A. Verdaguer-Casadevall, D. Ravasio, J.H. Montoya, D.B. Trimarco, C. Roy, et al., Acetaldehyde as an intermediate in the electroreduction of carbon monoxide to ethanol on oxide-derived copper, Angew. Chem. Int. Edi. 55 (2016) 1450–1454.
[65] H. Wang, Y. Chen, X. Hou, C. Ma, T. Tan, Nitrogen-doped graphenes as efficient electrocatalysts for the selective reduction of carbon dioxide to formate in aqueous solution, Green Chem. 18 (2016) 3250–3256.
[66] N. Sreekanth, M.A. Nazrulla, T.V. Vineesh, K. Sailaja, K.L. Phani, Metal-free boron-doped graphene for selective electroreduction of carbon dioxide to formic acid/formate, Chem. Commun. 51 (2015) 16061–16064.
[67] Z. Chen, K. Mou, S. Yao, L. Liu, Zinc-coordinated nitrogen-codoped graphene as an efficient catalyst for selective electrochemical reduction of CO_2 to CO, ChemSusChem 11 (2018) 2944–2952.
[68] W. Bi, X. Li, R. You, M. Chen, R. Yuan, W. Huang, et al., Surface immobilization of transition metal ions on nitrogen-doped graphene realizing high-efficient and selective CO_2 reduction, Adv. Mater. 30 (2018) 1706617–1706623.
[69] F. Lei, W. Liu, Y. Sun, J. Xu, K. Liu, L. Liang, et al., Metallic tin quantum sheets confined in graphene toward high-efficiency carbon dioxide electroreduction, Nat. Commun. 7 (2016) 12697–12705.
[70] C.Y. Lee, Z. Yong, C. Wang, D.R.G. Mitchell, G.G. Wallace, Rapid formation of self-organised Ag nanosheets with high efficiency and selectivity in CO_2 electroreduction to CO, Sustain. Energ. Fuels 1 (2017) 1023–1027.
[71] S. Liu, H. Tao, L. Zeng, Q. Liu, Z. Xu, Q. Liu, et al., Shape-dependent electrocatalytic reduction of CO_2 to CO on triangular silver nanoplates, J. Am. Chem. Soc. 139 (2017) 2160–2163.
[72] W. Zhu, L. Zhang, P. Yang, C. Hu, Z. Luo, X. Chang, et al., Low-coordinated edge sites on ultrathin palladium nanosheets boost carbon dioxide electroreduction performance, Angew. Chem. Int. Ed. 57 (2018) 11544–11548.
[73] S. Gao, Y. Lin, X. Jiao, Y. Sun, Q. Luo, W. Zhang, et al., Partially oxidized atomic cobalt layers for carbon dioxide electroreduction to liquid fuel, Nature 529 (2016) 68–72.
[74] N. Han, Y. Wang, H. Yang, J. Deng, J. Wu, Y. Li, et al., Ultrathin bismuth nanosheets from in situ topotactic transformation for selective electrocatalytic CO_2 reduction to formate, Nat Commun 9 (2018) 1320–1328.
[75] H. Zhang, Y. Ma, F. Quan, J. Huang, F. Jia, L. Zhang, Selective electro-reduction of CO2 to formate on nanostructured Bi from reduction of BiOCl nanosheets, Electrochem. Comm. 46 (2014) 63–66.
[76] X. Sun, Q. Zhu, X. Kang, H. Liu, Q. Qian, J. Ma, et al., Design of a Cu(i)/C-doped boron nitride electrocatalyst for efficient conversion of CO_2 into acetic acid, Green Chem. 19 (2017) 2086–2091.
[77] X. Lu, T.H. Tan, Y.H. Ng, R. Amal, Highly selective and stable reduction of CO_2 to CO by a graphitic carbon nitride/carbon nanotube composite electrocatalyst, Chem.- Eur. J. 22 (2016) 11991–11996.

[78] B. Zhang, T.-J. Zhao, W.-J. Feng, Y.-X. Liu, H.-H. Wang, H. Su, et al., Polarized few-layer g-C_3N_4 as metal-free electrocatalyst for highly efficient reduction of CO_2, Nano Res. 11 (2018) 2450–2459.
[79] M. Asadi, K. Kim, C. Liu, A.V. Addepalli, P. Abbasi, P. Yasaei, et al., Nanostructured transition metal dichalcogenide electrocatalysts for CO_2 reduction in ionic liquid, Science 353 (2016) 467–470.
[80] J. Xu, X. Li, L. Wei, Y. Sun, Z. Ju, Y. Tao, et al., Carbon dioxide electroreduction into syngas boosted by a partially delocalized charge in molybdenum sulfide selenide alloy monolayers, Angew. Chem. Int. Edi. 129 (2017) 9249–9253.
[81] P. Abbasi, M. Asadi, C. Liu, S. Sharifi-Asl, B. Sayahpour, A. Behranginia, et al., Tailoring the edge structure of molybdenum disulfide toward electrocatalytic reduction of carbon dioxide, ACS nano 11 (2016) 453–460.
[82] S. Gao, X. Jiao, Z. Sun, W. Zhang, Y. Sun, C. Wang, et al., Ultrathin Co_3O_4 layers realizing optimized CO_2 electroreduction to formate, Angew. Chem. Int. Ed. 55 (2016) 698–702.
[83] F. Li, L. Chen, G.P. Knowles, D.R. MacFarlane, J. Zhang, Hierarchical mesoporous SnO_2 nanosheets on carbon cloth: a robust and flexible electrocatalyst for CO_2 reduction with high efficiency and selectivity, Angew. Chem. Int. Ed. 56 (2017) 505–509.
[84] L. Ye, J. Liu, Y. Gao, C. Gong, M. Addicoat, T. Heine, et al., Highly oriented MOF thin film-based electrocatalytic device for the reduction of CO_2 to CO exhibiting high faradaic efficiency, J. Mater. Chem. A 4 (2016) 15320–15326.
[85] F. Li, C. Lu, M. Xue, T. Williams, Z. Ying, D.R. Macfarlane, et al., Towards a better Sn: efficient electrocatalytic reduction of CO_2 to formate by Sn/SnS_2 derived from SnS_2 nanosheets, Nano Energy 31 (2017) 270–277.
[86] J. Wu, M. Liu, P.P. Sharma, R.M. Yadav, L. Ma, Y. Yang, et al., Incorporation of nitrogen defects for efficient reduction of CO_2 via two-electron pathway on three-dimensional graphene foam, Nano lett. 16 (2015) 466–470.
[87] P. Su, K. Iwase, S. Nakanishi, K. Hashimoto, K. Kamiya, Nickel-nitrogen-modified graphene: an efficient electrocatalyst for the reduction of carbon dioxide to carbon monoxide, Small 12 (2016) 6083–6089.
[88] X. Sun, Q. Zhu, X. Kang, H. Liu, Q. Qian, Z. Zhang, et al., Molybdenum-bismuth bimetallic chalcogenide nanosheets for highly efficient electrocatalytic reduction of carbon dioxide to methanol, Angew. Chem. Int. Edi. 55 (2016) 6771–6775.
[89] X. Sun, X. Kang, Q. Zhu, J. Ma, G. Yang, Z. Liu, et al., Very highly efficient reduction of CO_2 to CH_4 using metal-free N-doped carbon electrodes, Chem. Sci. 7 (2016) 2883–2887.
[90] K.S. Novoselov, A.K. Geim, S.V. Morozov, D.A. Jiang, Y. Zhang, S.V. Dubonos, et al., Electric field effect in atomically thin carbon films, Science 306 (2004) 666–669.
[91] Y. Hori, H. Wakebe, T. Tsukamoto, O. Koga, Electrocatalytic process of CO selectivity in electrochemical reduction of CO_2 at metal electrodes in aqueous media, Electrochim. Acta 39 (1994) 1833–1839.
[92] J. Xu, Y. Kan, R. Huang, B. Zhang, B. Wang, K.H. Wu, et al., Revealing the origin of activity in nitrogen-doped nanocarbons towards electrocatalytic reduction of carbon dioxide, ChemSusChem 9 (2016) 1085–1089.
[93] P.P. Sharma, J. Wu, R.M. Yadav, M. Liu, C.J. Wright, C.S. Tiwary, et al., Nitrogen-doped carbon nanotube arrays for high-efficiency electrochemical reduction of CO_2: on the understanding of defects, defect density, and selectivity, Angew. Chem. Int. Edi 54 (2015) 13701–13705.

[94] M.H. Seo, S.M. Choi, E.J. Lim, I.H. Kwon, J.K. Seo, S.H. Noh, et al., Toward new fuel cell support materials: a theoretical and experimental study of nitrogen-doped graphene, Chemsuschem 7 (2015) 2609–2620.
[95] D. Kwak, A. Khetan, S. Noh, H. Pitsch, B. Han, First principles study of morphology, doping level, and water solvation effects on the catalytic mechanism of nitrogen-doped graphene in the oxygen reduction reaction, ChemCatChem 6 (2014) 2662–2670.
[96] C. Zhang, S. Yang, J. Wu, M. Liu, S. Yazdi, M. Ren, et al., Electrochemical CO_2 reduction with atomic iron-dispersed on nitrogen-doped graphene, Adv. Energy Mater. 8 (2018) 1703487–1703496.
[97] F. Quan, D. Zhong, H. Song, F. Jia, L. Zhang, A highly efficient zinc catalyst for selective electroreduction of carbon dioxide in aqueous NaCl solution, J. Mater. Chem. A 3 (2015) 16409–16413.
[98] N. Han, Y. Wang, H. Yang, J. Deng, J. Wu, Y. Li, et al., Ultrathin bismuth nanosheets from in situ topotactic transformation for selective electrocatalytic CO_2 reduction to formate, Nat. Commun. 9 (2018) 1320–1328.
[99] F. Li, M. Xue, J. Li, X. Ma, L. Chen, X. Zhang, et al., Unlocking the electrocatalytic activity of antimony for CO2 reduction by two-dimensional engineering of the bulk material, Angew. Chem. Int. Edi. 129 (2017) 14910–14914.
[100] G. Hu, Z. Wu, S. Dai, D. Jiang, Interface engineering of earth-abundant transition metals using boron nitride for selective electroreduction of CO_2, ACS Appl. Mater. Inte. 10 (2018) 6694–6700.
[101] X. Wang, K. Maeda, A. Thomas, K. Takanabe, G. Xin, J.M. Carlsson, et al., A metal-free polymeric photocatalyst for hydrogen production from water under visible light, Nature Mater. 8 (2009) 76–80.
[102] Y.Y. Cai, X.H. Li, Y.N. Zhang, X. Wei, K.X. Wang, J.S. Chen, Highly efficient dehydrogenation of formic acid over a palladium-nanoparticle-based Mott–Schottky photocatalyst, Angew. Chem. Int. Edi 125 (2013) 12038–12041.
[103] Q. Liang, Z. Li, Y. Bai, Z.-H. Huang, F. Kang, Q.-H. Yang, Reduced-sized monolayer carbon nitride nanosheets for highly improved photoresponse for cell imaging and photocatalysis, Sci. China Mater. 60 (2017) 109–118.
[104] S. Yang, Y. Gong, J. Zhang, L. Zhan, L. Ma, Z. Fang, et al., Exfoliated graphitic carbon nitride nanosheets as efficient catalysts for hydrogen evolution under visible light, Adv. Mater. 25 (2013) 2452–2456.
[105] Q. Han, B. Wang, Y. Zhao, C. Hu, L. Qu, A graphitic-C3N4 “seaweed” architecture for enhanced hydrogen evolution, Angew. Chem. Int. Edi. 127 (2015) 11595–11599.
[106] X.-H. Li, J.-S. Chen, X. Wang, J. Sun, M. Antonietti, Metal-free activation of dioxygen by graphene/g-C_3N_4 nanocomposites: functional dyads for selective oxidation of saturated hydrocarbons, J. Am. Chem. Soc. 133 (2011) 8074–8077.
[107] T.Y. Ma, S. Dai, M. Jaroniec, S.Z. Qiao, Graphitic carbon nitride nanosheet-carbon nanotube three-dimensional porous composites as high-performance oxygen evolution electrocatalysts, Angew. Chem. Int. Edi. 126 (2014) 7409–7413.
[108] S.A. Francis, J.M. Velazquez, I.M. Ferrer, D.A. Torelli, D. Guevarra, M.T. McDowell, et al., Reduction of aqueous CO_2 to 1-propanol at MoS_2 electrodes, Chem. Mate. 30 (2018) 4902–4908.
[109] N. Kornienko, Y. Zhao, C.S. Kley, C. Zhu, D. Kim, S. Lin, et al., Metal-organic frameworks for electrocatalytic reduction of carbon dioxide, J. Am. Chem. Soc. 137 (2015) 14129–14135.

[110] R. Wang, X. Sun, S. Ould-Chikh, D. Osadchii, F. Bai, F. Kapteijn, et al., Metal-organic-framework-mediated nitrogen-doped carbon for CO_2 electrochemical reduction, ACS Appl. Mater. Inte. 10 (2018) 14751–14758.
[111] D.-H. Nam, O.S. Bushuyev, J. Li, P. De Luna, A. Seifitokaldani, C.-T. Dinh, et al., Metal-organic frameworks mediate Cu coordination for selective CO_2 electroreduction, J. Am. Chem. Soc. 140 (2018) 11378–11386.
[112] Z. Weng, Y. Wu, M. Wang, J. Jiang, K. Yang, S. Huo, et al., Active sites of copper-complex catalytic materials for electrochemical carbon dioxide reduction, Nat. Commun. 9 (2018) 415–424.
[113] J. Albo, D. Vallejo, G. Beobide, O. Castillo, P. Castaño, A. Irabien, Copper-based metal-organic porous materials for CO_2 electrocatalytic reduction to alcohols, ChemSusChem 10 (2017) 1100–1109.
[114] Z. Weng, J. Jiang, Y. Wu, Z. Wu, X. Guo, K.L. Materna, et al., Electrochemical CO_2 reduction to hydrocarbons on a heterogeneous molecular Cu catalyst in aqueous solution, J. Am. Chem. Soc. 138 (2016) 8076–8079.
[115] S. Kusama, T. Saito, H. Hashiba, A. Sakai, S. Yotsuhashi, Crystalline copper (II) phthalocyanine catalysts for electrochemical reduction of carbon dioxide in aqueous media, ACS Catal. 7 (2017) 8382–8385.
[116] C.-W. Kung, C.O. Audu, A.W. Peters, H. Noh, O.K. Farha, J.T. Hupp, Copper nanoparticles installed in metal–organic framework thin films are electrocatalytically competent for CO_2 reduction, ACS Energy Lett. 2 (2017) 2394–2401.
[117] A. Zhang, R. He, H. Li, Y. Chen, T. Kong, K. Li, et al., Nickel doping in atomically thin tin disulfide nanosheets enables highly efficient CO_2 reduction, Angew. Chem. Int. Edi. 57 (2018) 10954–10958.
[118] G. Shi, L. Yu, X. Ba, X. Zhang, J. Zhou, Y. Yu, Copper nanoparticle interspersed MoS 2 nanoflowers with enhanced efficiency for CO_2 electrochemical reduction to fuel, Dalton T. 46 (2017) 10569–10577.
[119] C. Li, G. Shen, R. Zhang, D. Wu, C. Zou, T. Ling, et al., Zn nanosheets coated with a ZnS subnanometer layer for effective and durable CO_2 reduction, J. Mater. Chem. A (2019).
[120] W. Zhang, Q. Qin, L. Dai, R. Qin, X. Zhao, X. Chen, et al., Electrochemical reduction of carbon dioxide to methanol on hierarchical Pd/SnO_2 nanosheets with abundant Pd-O-Sn interfaces, Angew. Chem. Int. Edi. 57 (2018) 9475–9479.
[121] H. Li, N. Xiao, M. Hao, X. Song, Y. Wang, Y. Ji, et al., Efficient CO_2 electroreduction over pyridinic-N active sites highly exposed on wrinkled porous carbon nanosheets, Chem. Eng. J. 351 (2018) 1385–8947.
[122] T. Zhang, X. Li, Y. Qiu, P. Su, W. Xu, H. Zhong, et al., Multilayered Zn nanosheets as an electrocatalyst for efficient electrochemical reduction of CO_2, J. Catal. 357 (2018) 154–162.
[123] Z. Geng, X. Kong, W. Chen, H. Su, Y. Liu, F. Cai, et al., Oxygen vacancies in ZnO nanosheets enhance CO_2 electrochemical reduction to CO, Angew. Chem. Int. Edi. 57 (2018) 6054–6059.
[124] F.P. García de Arquer, O.S. Bushuyev, P. De Luna, C.T. Dinh, A. Seifitokaldani, M.I. Saidaminov, et al., 2D metal oxyhalide-derived catalysts for efficient CO_2 electroreduction, Adv. Mater. 30 (2018) 1802858.
[125] R.A. Geioushy, M.M. Khaled, K. Alhooshani, A.S. Hakeem, A. Rinaldi, $Graphene/ZnO/Cu_2O$ electrocatalyst for selective conversion of CO_2 into n-propanol, Electrochim. Acta 245 (2017) 456–462.

Index

Note: Page numbers followed by "*f*" and "*t*" refer to figures and tables, respectively.

C

D

E

F

N

T

U

V

CPI Antony Rowe
Eastbourne, UK
December 21, 2019